Materials Science and Engineering
It's Nucleation and Growth

Materials Science and Engineering
It's Nucleation and Growth

Proceedings of a conference held at
Imperial College of Science, Technology
and Medicine, London, UK
14–15 May 2001

Edited by
Malcolm McLean

FOR THE INSTITUTE OF MATERIALS

Book 774
First Published in 2002 by
Maney Publishing
1 Carlton House Terrace
London SW1Y 5DB

for the Institute of Materials

© The Institute of Materials 2002
All rights reserved

ISBN 1-902653-59-9

Typeset in the UK by
Keyset Composition, Colchester
Printed and bound in the UK at
The University Printing House, Cambridge

Contents

Introduction and Background   ix
MALCOLM McLEAN

List of Contributors   xvii

Nucleation of the Subject: Key Contributions

The Origins of Metallurgy and the RSM   1
J. A. CHARLES

The Contributions of William Chandler Roberts-Austen to
Metallurgy and to the Arts   13
MICHAEL R. NOTIS

The Nuffield Research Group in Extraction Metallurgy (NRGEM):
Founding and Evolution   41
C. B. ALCOCK

The Early Years of Transmission Electron Microscopy at RSM   53
PETER R. SWANN

Transition from Metallurgy to Materials Science   73
ROBERT W. CAHN

Driving Force for Development of Materials Science & Engineering

Structural Metals: Is there a Future?   81
JEFF W. EDINGTON

Role of Advanced Materials in Modern Aero Engines (abstract only)   103
PETER PRICE

The Past, Present and Future of Magnetic Data Storage   105
WILLIAM O'KANE

On the Horizon – Some Material Advances in the Nuclear Energy   113
SUE ION

Materials in Medicine   135
R. C. BIELBY, L. D. K. BUTTERY and J. M. POLAK

Mechanism for Development

Research Council Priorities (abstract only)   153
RICHARD BROOK

Formula 1 Materials Engineering   155
G. SAVAGE

Mature and Incremental Developments Diffusional Growth

Some Features of Iron- and Steelmaking in Integrated Steel Plants   185
AMIT CHATTERJEE

Interaction of Materials Science with Extractive Metallurgy   213
DEREK FRAY

Single Crystal Superalloys (abstract only)   225
KEN HARRIS

Materials Processing: The Enabling Discipline (abstract only)   227
L. CHRISTODOULOU and J. A. CHRISTODOULOU

The Materials Science of Cement and Concrete:
An Industrial Perspective   229
KAREN L. SCRIVENER

Materials for Ceramic Ion Conducting Membrane Devices   243
J. A. KILNER

Novel Revolutionary Developments Displacive Transformations

Grain Boundaries in High T_C Superconductors:
The key to Applications (abstract only)   257
DAVID LARBALESTIER

Lightweight Materials and Structures   259
A. G. EVANS

The Engineering of Cells and Tissues (abstract only)   281
WILLIAM BONFIELD

A Genetic Basis for Biomedical Materials   283
LARRY L. HENCH, IOANNIS D. XYNOS, ALASDAIR J. EDGAR,
LEE D. K. BUTTERY and JULIA M. POLAK

Modelling

Controversial Concepts in Alloy Theory Revisited 297
DAVID PETTIFOR

Virtual Materials (atstract only) 337
ALAN WINDLE

Atomistic Simulation of Materials 339
ROBIN W. GRIMES, K. J. W. ATKINSON, MATTHEW O. ZACATE
and MOSHIN PIRZADA

Author and Subject Index 355

Introduction and Background

MALCOLM McLEAN
Department of Materials, Imperial College of Science, Technology and Medicine, Prince Consort Road, London SW7 2BP, UK

This book is a record of the conference, entitled 'Materials Science and Engineering: Its Nucleation and Growth' held at Imperial College on 14 and 15 May 2001 as one of the events held to mark the 150th Anniversary of the founding of the Royal School of Mines (RSM). The present Department of Materials at Imperial College has been in continuous existence, under various names, since the establishment in 1851 of the Department of Metallurgy in the Government School of Mines and Science Applied to the Arts, the initial title of the RSM. The Department is the oldest of the discipline in the UK, and we believe remains one of the most active in both teaching and research in metallurgy and materials science and engineering.

The impetus for the formation of the RSM and its constituent Department of Metallurgy came from the rapid technological developments that occurred during the late 18th and early 19th centuries. An entrepreneurial spirit had taken advantage of the available scientific knowledge and had exploited it to tremendous industrial advantage. However, in the early 19th century there was growing concern that continuing industrial advances were being limited by a lack of systematic scientific understanding. The development of the steam engine is an excellent example. James Watt, through his contacts with Joseph Black at the University of Glasgow, understood the evolving concepts of latent heat and modified the Newcomen steam engine by the inclusion of a condenser to greatly improve its efficiency. As the Boulton–Watt steam engines were progressively developed commercially, Watt was well aware of the thermodynamic benefits that would result from increasing the steam temperature (or using *strong steam*). However, he also understood that steel-making technology was not then available to allow this development. There were similar barriers to progress in limitations in basic understanding of chemistry.

Intense lobbying of Government led to the establishment of a Museum of Economic Geology in 1841 and to the Royal College of Chemistry in 1845. Further pressure, led by Sir Henry de la Beche, culminated in the establishment of the Government School of Mines and Science Applied to the Arts in new premises in Jermyn Street (close to Piccadilly Circus) in 1851 under his direction. *Crafts* would, more accurately describe in modern English usage the term *Arts* that was used at that time. John Percy, a physician, was appointed

Lecturer (later Professor) of Metallurgy, a position that he held for some 28 years, and he led the Department towards an improved knowledge of mineral resources and the winning of metals from them. During his tenure, the name of the institution changed: to the Metropolitan School of Science Applied to Mining and the Arts (1853) and the Royal School of Mines (1863). The Royal School of Mines was an independent institution until 1907 when it combined with the Royal College of Science and the City and Guilds College to form the Imperial College of Science and Technology. It remains one of the constituent schools of the current expanded Imperial College of Science, Technology and Medicine. The Department, which has been in continuous existence since 1851, has changed its name less frequently than RSM; it became a Department of Metallurgy and Materials Science in 1970 and adopted its present name, Department of Materials, in 1986.

1851 was, coincidentally, the year of The Great Exhibition, which brought to public notice the growing importance of science and technology. This was a great financial, as well as critical, success and Prince Albert was largely responsible for the profits being devoted to educational and cultural development. A trust administered by the Royal Commissioners of the Exhibition of 1851, which still exists, was responsible for establishing on an 86 acre site in South Kensington the educational establishments (including Imperial College), museums and concert hall that continue to dominate this area.

Of course nothing goes smoothly, particularly when academics are involved. John Percy was vehemently opposed to widening the interests of RSM to more general scientific subjects; others, such as T. H. Huxley, advocated its development into a broad-based technical college incorporating chemistry and biology. It was this proposal, coupled with a planned move from Jermyn Street to South Kensington in 1879 that caused John Percy to resign. His successor as Head of Department was Sir William Chandler Roberts-Austen, whose scientific contributions are the subjects of one of the articles in this volume. He, and his successor, William Gowland in their 23 and 11 year tenures of leadership of the department contributed very significantly to the development of concepts of physical metallurgy and to establishing an international recognition of the Department's activities; both were also Masters of the Mint. With the exception of W. A. Carlyle who died in post after two years, subsequent Heads of Department also had long tenures: H. C. H. Carpenter (28 years), C. W. Dannatt (16 years), J. G. Ball (22 years) and D. W. Pashley (11 years). The writer having only completed 10 years feels he has not displayed the stamina of his predecessors; perhaps the incumbent, John Kilner will be able to re-establish this longevity.

Others have played key roles in the evolution of scientific and technological concepts underlying the subject of Materials Science and Engineering. Among them Hume-Rothery was a PhD student under the (nominal?) supervision of

Carpenter, developing his early concepts of electron alloy theory before returning to Oxford; Andrade on retiring from the Royal Institution continued his seminal contributions on the creep of metals; Richardson founded and nurtured the Nuffield group on metallurgical thermodynamics that had a lasting and world-wide impact on extraction metallurgy; Constance Tipper quietly worked on the metallographic aspects of fracture that led to her war-time contributions to understanding the causes of catastrophic failure of welds in the liberty ships. An important factor in the Department's continuing success has been its willingness to evolve the subject matter covered in both its teaching and research. This forward-looking characteristic of the Department still exists where our current priorities embrace materials for environmentally friendly energy production, modelling of materials structure/processing/performance and materials for medical applications with a particular emphasis on tissue engineering.

Conferences such as that reported in this volume inevitably focus on current research and, indeed, the Department of Materials has a strong research activity. However, it places at least equal emphasis on the content and quality of its undergraduate teaching. It was the only UK department awarded the maximum possible grading in the UK Quality Assurance Agency assessment of materials teaching. The recognition of changing student demands and industrial requirements in the late 1960s led to the introduction of a Materials Science syllabus, in parallel with the traditional Metallurgy course. Professors Brian Steele and Peter Pratt pioneered this initiative, one of the first of its type, while building strong research activities in structural and electrical ceramics, which remain active and are represented in this volume. Metallurgy and Materials Science have long been incorporated in a single syllabus, which is complemented by courses with Management Studies and Foreign Language. There are also speciality courses in Aerospace Materials and (from 2002) Tissue Engineering. The quality and flexibility of these courses are probably responsible for the growth in student numbers and quality (as indicated by A-level scores) that has occurred in recent years, at a time when student recruitment into the discipline has been declining.

The Department is proud of its history and of its past contributions to the subject. Most of the contributors to this volume were either students or staff in the department who have proceeded to make important contributions elsewhere. This provides a measure of the influence it has had on an international and national dimension. However, we believe that the meeting acknowledges only the first 150 years of the Department's contributions to Materials Science and Engineering. There is a buoyant undergraduate course, a forward-looking research programme carried out by enthusiastic, and predominantly young, members of academic staff. History can have its drawbacks and conducting state-of-the art research in a century-old building has been one of them.

However, the Department is currently in the midst of a major refurbishment that will leave us with the historic exterior of the RSM Building, but modern laboratories and teaching facilities on the inside. The prospects for the future are excellent.

The Department has benefited through the years from many interactions with industry, government bodies and agencies. The meeting that has led to publication of this volume was made possible by generous support from a number of bodies, which we wish to thank. These include:

- US Office of Naval Research International Field Office and European Office of Aerospace Research and Development, Air Force Office of Scientific Research, United States Air Force Research Laboratory.
- Corus PLC
- British Nuclear Fuels
- The National Physical Laboratory
- Kobe Steel Limited
- JEOL Ltd
- The Institute of Materials

Introduction and Background xiii

John Percy
1851–1879

Sir William Chandler Roberts-Austen
1880–1902

William Gowland
1902–1911

W. A. Carlyle
1911–1913

xiv Introduction and Background

Sir Henry Cort Harold Carpenter
1913–1940

C. W. Dannatt
1940–1956

J. G. Ball
1957–1979

D. W. Pashley
1979–1990

Introduction and Background xv

M. McLean
1990–2000

J. A. Kilner
2000–present

List of Contributors

Ben Alcock was a member of the Nuffield Group of the Department of Metallurgy between 1950 and 1969, having been appointed to a Chair of Metallurgical Chemistry in 1964. He was at the University of Toronto between 1969 and 1987, serving as Chairman of the Department of Metallurgy and Materials Science from 1969 to 1977. He was Freimann Professor at Notre Dame between 1987 and 1994. His distinctions include Fellowships of the Royal Society of Canada, the Royal Society of Chemistry, the Royal Society of Arts and the American Institute of Metallurgical Engineers.

Bill Bonfield gained a BSc(Eng) and PhD in Metallurgy from Imperial College in 1958 and 1961 respectively. After a period in industry in the USA he returned to England to a Readership at Queen Mary College. He remained there in various capacities until 2000, notably developing a world-renowned group on biomaterials. Latterly he was Founder Director of the IRC in Biomaterials at QMW. He now holds a Chair in the Department of Materials Science and Metallurgy at the University of Cambridge. Among his many honours are a Fellowship of the Royal Academy of Engineering and the *Acta Materialia et Metallurgia*.

Richard Brook obtained a BSc in Ceramics from the University of Leeds and a DSc from MIT. He has spent periods at the University of Southern California (1966–1970), AERE, Harwell (1970–1974) and as Head of Department of Ceramics at the University of Leeds (1974–1988). In 1988 he became Director of the Max-Planck Institut für Metallforschung in Stuttgart and then Cookson Professor and Head of Department of Materials at Oxford in 1992. He has been Chief Executive of the Engineering and Physical Science Research Council since 1994. Professor Brook is a Fellow of the Royal Academy of Engineering, a Distinguished Life Member of the American Ceramic Society, Membre d'Honneur de la Société Française de Métallurgie et Matériaux.

Robert Cahn holds three degrees from Cambridge: BA (1945), PhD (1950), DSc (1963). After a brief period at AERE, Harwell he moved to academia holding various positions in the Department of Physical Metallurgy at Birmingham. He was Professor of Materials Science at the University of Wales at Bangor, Professor and Head of Department of Materials Science at the University of Sussex and Professor of Physical Metallurgy at Université de Paris-Sud. His many awards include Fellowship of the Royal Society, the Griffiths Medal and the Luigi Losana Gold Medal. His recent book traces the evolution of the discipline of Materials Science.

Jim Charles obtained his bachelor's degree in Metallurgy from the Royal School of Mines in 1947. After a career in industry he returned to academia to the Department of Metallurgy at the University of Cambridge in 1960 where he remained until he retired in 1990. His honours include Fellowship of the Royal Academy of Engineering, the Beilby Medal, the Hadfield Medal and the Kroll Medal. He has taken a particular interest in the history of metallurgy and his recent autobiography *Out of the Fiery Furnace* recalls his experiences in the Royal School of Mines.

Amit Chatterjee was awarded a BSc in Metallurgical Engineering from Banaras Hindu University in 1966, a PhD in Process Metallurgy from Imperial College in 1970 and a DSc in Engineering from the University of London in 1988. He has worked for Tata Steel in various capacities since 1977, apart from a 5 year period as Managing Director Ipitata Sponge Iron Company. He is currently General Manager (Technology) for Tata Steel. His technical achievements have been recognised by a number of prestigious awards, including the National Metallurgist Award of the Government of India, the Application to Practice Award of TMS and the Thomas Medal of the Institute of Materials.

Julie Christodoulou was awarded a PhD in Materials from Imperial College in 1999. She worked for the Office of Naval Research International Field Office while in London. She currently has a research position with the US Navy in Washington.

Leo Christodoulou obtained both his BSc and PhD in Metallurgy from Imperial College. After a post-doctoral research appointment at Carnegie-Mellon University, he joined the Martin Marrieta Laboratories where he was responsible for the development of XD composites. He is currently on leave-of-absence at DARPA from his current position as Reader of Materials Processing in the Department of Materials at Imperial College. He was a recipient of the Grunfield Prize of the Institute of Materials.

Jeff Edington gained his Bachelor and Doctoral degrees in Metallurgy from the University of Birmingham. His career has included periods in universities (Cambridge, Delaware), research laboratories (Battelle) and industry (Alcan, British Steel/Corus). Until his recent retirement he was Executive Director for Technology for the Corus Group. He is a Fellow of the Royal Academy of Engineering and was President of the Institute of Materials.

Tony Evans received his BSc and PhD degrees from the Department of Metallurgy at Imperial College in 1964 and 1967 respectively. He then spent 10 years as a research scientist at AERE Harwell, NBS and Rockwell International Science Center. Since 1978 he has held senior academic positions at the University of California Berkeley and Santa Barbara, Harvard University and (currently) Princeton University. His many awards include the Matthew Prize of Imperial College, the Griffith Medal of The Institute of Materials, Membership of the National Academy of Engineering and Distinguished Life Member (American Ceramic Society). He was elected to Fellowship of the Royal Society in 2001.

Derek Fray entered Imperial College in 1958 with a Royal Scholarship, graduating in 1961 with a BSc(Eng), ARSM and in 1965 with a PhD, DIC and the Matthey Prize. He is currently Professor of Materials Chemistry and Head of Department, Department of Materials Science and Metallurgy, University of Cambridge. He is a Fellow of the Royal Academy of Engineering.

Dr Robin Grimes is Reader in Atomistic Simulation in the Department of Materials at Imperial College, which he joined in January 1995 as a Governor's lecturer after 5 years at the Royal Institution of Great Britain. He obtained his PhD in Chemistry from Keele University, his MS in Materials Science from Case Western Reserve University and BSc in Mathematical Physics from Nottingham University. At the present time he is on leave from Imperial College as the Matthias Scholar at Los Alamos National Laboratory. His research group is concerned primarily with predicting the behaviour of ceramic materials at an atomistic level using computer simulation techniques.

Ken Harris received his BSc in Metallurgy from Imperial College in 1965. He has spent all of his subsequent career in the metallurgical industry. He has been responsible for the development of the most commercially widely-used single-crystal superalloys which power modern aero-engines. He is a Fellow of ASM International.

Larry Hench graduated with bachelor and doctoral degrees from the Ohio State University. He was appointed to the Faculty of the University of Florida in 1965 where he remained until accepting a Chair in the Department of Materials in 1994. He is widely recognised as one of the pioneers of biomaterials; his development of Bioglass and championing it to medical applications has been particularly important. His many honours include election to the US National Academy of Engineering and the Von Hippel award of the Materials Research Laboratory.

Sue Ion gained both her BSc degree in Materials Science and her PhD in Dynamic Recrystallisation of a Magnesium Alloy from Imperial College, London. She joined British Nuclear Fuels in 1979 working in various roles before becoming Director of Technology & Operations in 1992. Dr Ion was awarded the Hinton Medal by the Institution of Nuclear Engineers in 1993 for an outstanding contribution to nuclear engineering and was elected a Fellow of the Royal Academy of Engineering in 1996. Dr Ion is a member of the European Union Science and Technology Committee and she also represents the United Kingdom on the IAEA's Standing Advisory Group on Nuclear Energy.

John Kilner obtained a first degree in Physics from the University of Birmingham and a PhD in Physical Metallurgy in 1975. He then took up a post-doctorial position at the University of Leeds working on ion conducting ceramics with Richard Brook. He moved to Imperial College in 1979 as Wolfson Fellow and was then awarded an EPSRC Advanced Research Fellowship to work on ion beam synthesised materials. In 1987 he joined the academic staff of the then Department of Metallurgy, becoming Dean of the Royal School of Mines in 1998. Professor Kilner is currently the Head of Department of the Materials Department of Imperial College and a Fellow of the Institute of Materials.

David Larbalestier graduated BSc (1965) and PhD (1970) from Imperial College/Royal School of Mines. He has worked in Switzerland, the UK and the USA since leaving RSM. He has been on the faculty of the Department of Materials Science and Engineering, University of Wisconsin Madison since 1976 where he holds the I. V. Shubnikov and David Grainger chairs. His career in superconductivity started during his PhD studies and has continued to this day. He is also Director of the Applied Superconductivity Center which brings together about 5 groups and some 50 students and staff to research the science and applications of low and high temperature superconductors.

Michael R. Notis is a Professor of Materials Science and Engineering at Lehigh University in Bethlehem, Pennsylvania, USA, and his PhD is from this same institution. His main research work concerns mass transport and phase equilibria in multicomponent systems, and he is currently involved in studies on microstructure development at reaction interfaces between metal substrates and lead-free solders. He has a keen interest in the history of materials technology.

William O'Kane gained his BSc, MSc and PhD in Materials Science from UMIST in 1990, 1991 and 1994 respectively. He joined Seagate Technologies in 1994 as a Senior Research and Development Engineer and has held various positions in the company, both in the USA and Ireland, since that time. He was appointed to his current post as Director for Research and Development at the Seagate Technologies wafer fabrication facility in Derry, Northern Ireland in August 2000.

David Pettifor received his BSc from the University of Witswatersrand in 1967 and PhD from the University of Cambridge in 1970. After brief periods at the University of Dar es Salaam, the Cavendish Laboratory and Bell Laboratories he joined the Department of Mathematics at Imperial College where he remained between 1978 and 1992. He was appointed Wolfson Professor of Metallurgy in 1992. Professor Pettifor is a Fellow of the Royal Society.

Julia Polak obtained her Medical Degree in Buenos Aires, Argentina in 1966. She came to the UK in 1968 and worked at the Hammersmith Hospital, first as a Junior Doctor and then as a Head of the Department of Histochemistry (after the retirement of Professor Prearce), from 1980 onwards. She is currently the Director of the Imperial College Tissue Engineering Centre, with dedicated laboratories and offices at Chelsea and Westminster Hospital. She works closely with Professor Larry Hench, the co-Director of the Centre. She has published over 1000 original papers, 24 books and is the recipient of numerous national and international honours. She is the Founder President of the Tissue Engineering Society in Great Britain and the European Editor of *Tissue Engineering*.

Peter Price gained a BSc in Aeronautics and Astronautics from Southampton University in 1980. He is a Chartered Engineer and Fellow of the Royal Aeronautical Society. He joined Rolls-Royce in 1980 as a Graduate Trainee and has held various positions in the company since then, being appointed to his current post as Director of Engineering for the Defence (Europe) business of Rolls-Royce, based in Bristol, in January 1999.

Gary Savage graduated with both BSc and PhD from the Department of Metallurgy at Imperial College. He was a Research Physicist with ICI Advanced Materials from 1985 to 1990 working on the processing of composite materials. He became involved with the application of composites to Formula 1 cars in 1990 and has worked with McLaren International, Arrows, Prost Grand Prix and (currently) BAR Grand Prix. He is a Fellow of the Institute of Materials and the Institution of Mechanical Engineers. He is Visiting Professor at the University of Modena.

Karen Scrivener gained her BSc from the University of Cambridge and her PhD from the Department of Materials, Imperial College. She was awarded a Royal Society Research Fellowship, tenable in the Department of Materials at Imperial College where she was appointed to a Lectureship. She accepted a senior research position in Lafarge in 1994 and has recently been appointed to a professorship at EPFL in Lausanne.

Peter Swann obtained his PhD at Cambridge in 1960 and, after 6 years at US Steel's Fundamental Research Center, became a Reader and then a Professor in the Department of Metallurgy at RSM. In 1978 he left the Department to found Gatan Inc., a scientific instrument company based in Pennsylvania, USA. He has now retired and is living on Jumby Bay Island in the West Indies

Alan Windle graduated with BSc in Metallurgy from Imperial College, in 1963, and with a PhD from Cambridge University. He returned to Imperial, first as an ICI fellow under Peter Pratt and then as a Lecturer, and arising from a period in Andrew Keller's laboratory in Bristol developed a new interest in polymers. Back in Cambridge in 1975, as University Lecturer and Fellow of Trinity, he built up a group concentrating on structural studies of non-crystalline polymers. Interest developed in polymer diffusion and in liquid crystalline polymers which has led to a development of computer molecular modelling. Professor Windle has been awarded the Bessemer Medal and the Royal Society of Arts Silver Medal (1963), the Rosenhain Medal (1987) and the Swinburne medal and prize (1992) and he was elected Fellow of the Royal Society in 1997. He is currently Executive Director of the Cambridge-MIT Institute and a Commissioner of the Royal Commission for the 1851 Exhibition.

The Origins of Metallurgy and the RSM

J. A. CHARLES
Department of Materials Science and Metallurgy, University of Cambridge, Pembroke St, Cambridge, UK

ABSTRACT
The origin of man's use of metals through mineralogical associations, and the subsequent development of metallurgical technology through history, are briefly reviewed. The rate of change in the understanding and application of metals greatly increased from straightforward observation and deduction, through alchemical experimentation and then particularly rapidly after the new atomistic philosophy and the pursuit of science for its own sake in the seventeenth century, as reflected in the foundation of the Royal Society. The development of a scientific basis for understanding during the eighteenth and particularly the nineteenth century was to result in an explosion of metallurgical technology, as for example in steelmaking, and as evidenced by the Great Exhibition and the Government's creation of the School of Mines, both in 1851, the latter to become the RSM in 1863. The initial emphasis in the School was mineralogical, chemical and analytical, an understandable bias at the time that was, however, to persist for many years. The early history of the School and the role of John Percy, the 'father' of English metallurgy as an applied science, are considered.

Metallurgy as a technology has been around for ~6000 years. Neolithic man moved into the recovery of metals and their use through observation and deduction and the Metal Age began. In the beginning the earth's surface must have been an Aladdin's cave of brightly coloured minerals released at each outcrop by geological processes. As well as looking for suitable stones for weapons and tools we know that attractively coloured minerals such as malachite and azurite or interestingly structured minerals such as galena (Plates I, II, III) were collected and used as pigments and stored in bone tubes or pottery vases. In the Neolithic period the use of moulded clay, dried in the sun and hardened by fire, and then the development of kilns for firing pottery, reflected an increasing level of combustion control, essential for what was to follow. Some earthy materials would melt and glazes were developed.

When collecting stones and minerals, Neolithic man would have found intriguing 'stones' that had totally different properties, deforming plastically when hammered. These native metals, gold from stream beds and native copper associated with copper mineralisation, represent the first metal usage, gradually broadening into the Chalcolithic period during 5000–4000 BC as native copper was purposely sought and the transition to smelting for copper began. Perhaps 'fire setting' to separate associated rock and copper minerals from the native copper resulted in a realisation that more copper was being generated, with the oxide copper minerals easily reduced in charcoal fires at above 700°C, when CO rather than CO_2 is available from combustion. In laboratory tests with CO supplied directly, malachite is completely reduced to copper before the charge is even red hot. It would also be recognised that all these copper sources produced a green flame. The first smelting sites at Timna in the southern Arabah desert (Israel/Jordan border) are dated 4240 BC. With furnace control (draughting etc.) temperatures in excess of 1080°C would be generated giving molten copper. When cooled down it would have been seen that it had taken the shape of the furnace bottom, and the concept of casting to a required shape followed. The incorporation of predominantly brown gossan (limonite) from the top of a weathered pyritic copper deposit, often still showing some green copper mineralisation (Plates IV, V) and a green tinted flame, would have given improved liquid/liquid separation of the copper from 'earths' and the concept of fluxing had arrived, with the subsequent use of ochres generally. If copper could be melted, so could gold, although it was recognised that cold welding could be used to build up bulk in gold, unlike copper.

Such development did not occur everywhere at the same time, although the order of development subsequently was generally similar (Fig. 1) although a Cu–As phase is not always evident. It seems that there were independent centres of origin, notably the Mesopotamian region and the Balkans, with initially limited diffusion of ideas. The Balkan activity is exemplified by cast copper axes from the Vinca culture, ~4000 BC (Plate VI). The influence of heating on worked copper to give resoftening was known even in the Chalcolithic period as evidenced by general and localised crystal twinning associated with forging artefacts to shape. Deformed annealing twins indicate further work after annealing. We should never underestimate the human powers of observation and deduction demonstrated by ancient peoples, although the speed of change thus generated may have been slow as compared to later millennia, when the ability to record and communicate rapidly existed, with very different social conditions.

After copper, copper arsenic alloys giving superior properties were produced for about 600 years based probably on the determinative mineralogy of basic copper arsenates ($Cu_3 As_2 O_8 nCu (OH)_2$: a green mineral colour such as chalcophyllite (Plate VII), still a green flame, but now with an associated, easily recognised, garlic smell on heating – just the tests we employed as students at the RSM in the 1940s when the identification of minerals was

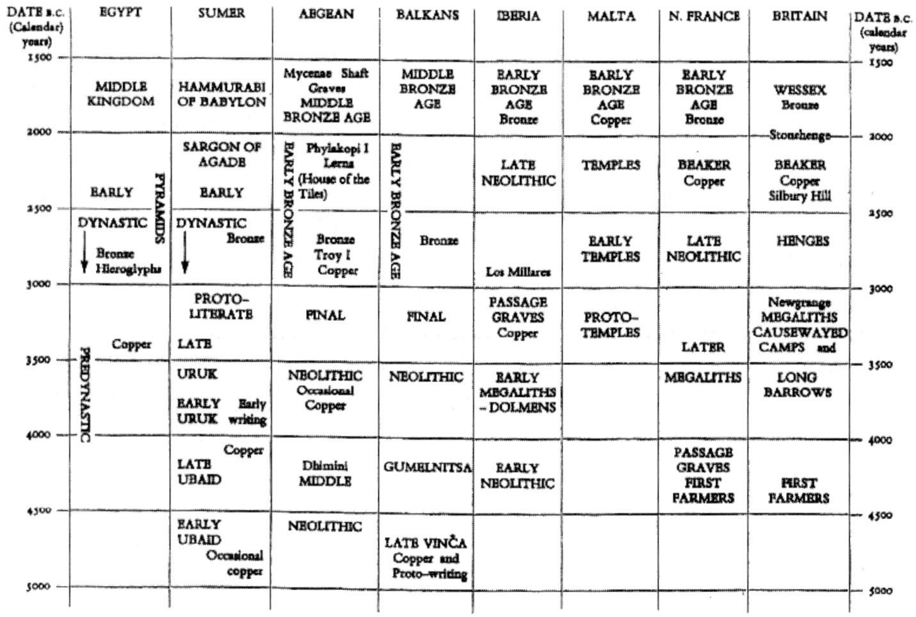

The Origins of Metallurgy and the RSM

DATE B.C. (Calendar years)	EGYPT	SUMER	AEGEAN	BALKANS	IBERIA	MALTA	N. FRANCE	BRITAIN	DATE B.C. (calendar years)
1500			Mycenae Shaft Graves	MIDDLE BRONZE AGE	EARLY BRONZE AGE Bronze	EARLY BRONZE AGE	EARLY BRONZE AGE		1500
	MIDDLE KINGDOM	HAMMURABI OF BABYLON	MIDDLE BRONZE AGE					WESSEX Bronze	
2000								Stonehenge	2000
		SARGON OF AGADE	Phylakopi I Lerna (House of the Tiles)		LATE NEOLITHIC	TEMPLES	BEAKER Copper	BEAKER Copper Silbury Hill	
	EARLY	EARLY							
2500	DYNASTIC Bronze Hieroglyphs	DYNASTIC Bronze	Bronze Troy I Copper	Bronze	Bronze Los Millares	EARLY TEMPLES	LATE NEOLITHIC	HENGES	2500
3000		PROTO-LITERATE	FINAL	FINAL	PASSAGE GRAVES Copper	PROTO-TEMPLES		Newgrange MEGALITHS CAUSEWAYED CAMPS and	3000
3500	Copper	LATE URUK EARLY URUK Early writing	NEOLITHIC Occasional Copper	NEOLITHIC	EARLY MEGALITHS – DOLMENS		LATER MEGALITHS	LONG BARROWS	3500
4000	PREDYNASTIC	Copper LATE UBAID	Dhimini MIDDLE	GUMELNITSA	EARLY NEOLITHIC		PASSAGE GRAVES FIRST FARMERS	FIRST FARMERS	4000
4500		EARLY UBAID Occasional copper	NEOLITHIC	LATE VINČA Copper and Proto-writing					4500
5000									5000

Fig. 1 Chronology of development in prehistory, Renfrew.

still largely concerned with heating on a charcoal block with a blowpipe flame. Probably at some point, as well as using green copper-arsenates to give the alloys, the green iron arsenates derived from arsenopyrite and then arsenopyrite itself, both still giving the garlic odour, would have been employed to provide the arsenic addition to the copper. Eventually the copper-arsenic alloys gave way to tin bronze of similar properties, probably in an evolutionary sense in relation to the well-being of the smiths. The link from copper-arsenic to the use of cassiterite (SnO_2) added to copper to give bronze may have been that stannite, the sulph-arsenide of tin, could also have given the garlic odour on heating, but it could also have been the frequent juxtaposition of the heavy minerals arsenopyrite (metallic grey) and cassiterite (brown/black) in stream beds (Plate VIII), or perhaps through the presence of immobile cassiterite in gossans, a mineral readily recognised by the weight in the hand. Another possible link is that cassiterite in the powder form, as on a streak plate or a vanning shovel, is very much the same red/brown colour as cuprite (Cu_2O), already identified as a source of copper.

As furnaces improved and gave more strongly reducing conditions in copper smelting furnaces, the adventitious occurrence of iron would have increased and it would be a short step to operate the furnace solely as a bloomery, producing solid iron from an iron-rich charge. On Cyprus, for example, the late Bronze Age advanced almost seamlessly into a period of producing and working iron, and iron nail or small tool has even been found in the protective, reducing, environment of a Bronze Age trackway in Holland. The control of the carbon content of iron, forging and the heat treatment of steel was to develop steadily over subsequent centuries.

This whole period of metal development to Roman times was some four thousand years and the rate of change was obviously slow, but steady none-the-less. The Egyptians of the Late Kingdom (∼500 BC), for example, had mastered the technique of lost wax investment casting, even to the point of awareness of the value of high lead additions to bronze to improve fluidity and castability for thin-walled castings (Fig. 2).

In the Greek and Roman periods, as by then recorded by various writers, techniques of manufacture and the application of known metals improved steadily, but without any major change of alloy development, except for the purposeful production of brass by the calamine process in Roman times. Post Roman, for several hundred years metallurgy moved into the secretive alchemical experimentation period, with emphasis on the precious metals, amalgams and chemical properties. No doubt some advances were made and it could be said that the emergence of chemistry as an independent discipline could be traced to the observed need for assaying in relation to standards for trade, as instanced in the work of Agricola and Erkar, writing in the sixteenth century.

Fig. 2 Investment castings, Egypt, Late Kingdom.

In the seventeenth century the alchemical approach was still extant and even Newton continued to have some faith in transmutation, not a good idea for the Master of the Mint! A major acceleration came, however, with the pursuit of science for its own sake in this century of enlightenment, with improved methods of communication through printed books and scholarship, where 'intelligencers' collected and distributed knowledge and met frequently. In Britain this culminated in the foundation of the Royal Society in 1660. Just as Archimedes had foreseen in Alexandria, it was the creation of a truth-seeking philosophy, not the results of immediate practical value, which were seen to be most important in the long term. Men that we now class as scientists began to have a major influence on the way metallurgy developed, for example, in Great Britain Francis Bacon (1561–1626), Robert Boyle (1627–1691), Isaac Newton (1642–1727) and Joseph Priestley (1733–1804) and then Michael Faraday (1791–1867) to name a few. Robert Hooke (1635–1703) began appreciation of the inter-relation between metal properties and their structure through microscopy and contributed largely to the early concepts of mechanical properties, memorably Hooke's law of elastic behaviour.

With steam-power and the Industrial Revolution blossoming in the eighteenth and early nineteenth centuries it became clear that further development required greater scientific understanding of the processes involved, in order that they could be made more efficient in a climate of growing commercial competition, particularly on mainland Europe in iron and steel making. Another feature of the times was the great growth in scientific and technical interest and self education in the population, particularly during the second half of Victoria's reign, which had to be recognised by politicians.

In 1841 the Museum of Economic Geology was opened, attached to the Mining Record office in Whitehall. Amongst its contents were exhibits showing how minerals were treated to produce metals. In this Museum there was rather informal instruction in analytical chemistry, mineral analysis and various metallurgical processes. It has to be understood that analytical methods and their development were essential features of the desire to understand and control existing problems. In 1851 a new building for the Museum, the Geological Survey and for a new Government School of Mines was provided between Piccadilly and Jermyn Street, much due to the efforts of Henry Thomas de la Beche, the mining engineer. It was, in fact, the first important structure in Britain designed for a purely technological or scientific institution and was opened by the Prince Consort less than a fortnight after the Queen had opened the Great Exhibition in Hyde Park. Evening lectures for the public were very well attended, an interest reflecting the first publication of *Boys Playbook of Science* and then the *Playbook of*

6   Materials Science and Engineering: Its Nucleation and Growth

Metals by J. H. Pepper in 1861. It is interesting to note that Pepper seemed still to adhere to alchemical connections in both the book cover (Fig. 3) and preface of this latter work. It is none-the-less a very interesting 'popular' text of the time. With Royal patronage the title of the School was changed to Royal School of Mines in 1863.

John Percy (Fig. 4) had been appointed, first as Lecturer and then as the first Professor in metallurgy in the country. He was by repute an excellent lecturer and teacher, innovative and methodical, building up a great collection of mineral and metal specimens to enliven his courses, which collection now resides in the Science Museum. His series of books commencing in 1861, constituting a treatise on metallurgy, sought underlying science in defining practice. He clearly saw the importance of metallurgy to the nation. A famous quote from his first lecture at the RSM: 'In proportion to the success with which the metallurgic art is practised in this country will the interests of the whole population, directly or indirectly, in no inconsiderable degree, be promoted'.

At its establishment the School was under the control of the Department of Science and Art of the Board of Trade, and then under the Education

Fig. 3   J. H. Pepper *The Playbook of Metals*, 1861.

Fig. 4 John Percy.

Department. From 1863 scholarships for up to three years at the RSM were awarded to the best students, leading to the Associateship Examinations in metallurgy. These were controlled by the Science and Art Department, with scripts submitted from a number of places, including Middlesbrough and Sheffield. Numbers for the Associateship were relatively low and most students at the RSM were sent by industrial firms for short courses. Until 1925 separate examinations had to be taken for the University of London B.Sc(Eng.) degree.

An important event soon after the foundation of the RSM was the Paris Exhibition in 1867 which demonstrated a gulf opening between British industrial capacity and that of our French neighbours, which greatly alarmed politicians and industrialists alike. Disraeli therefore set up a Select Committee in 1868 'to enquire into the provisions for giving instruction in theoretical and applied science to the industrial classes', with Bernard Samuelson, a practising ironmaster and engineer as chairman. There seems to have been a realisation that there was a need for a scientific basis of understanding for the development of technology at a competitive rate to our European neighbours. Even in the first third of the twentieth century it was considered that a knowledge of scientific German and French was an almost vital skill, so far ahead had they become at the turn of the century. Percy in his first book, 1861, underlines the importance of the German scientific contribution at that time (Fig. 5). As recently as 1943 I attended a German course at Imperial College as part of the Inter BSc curriculum.

The forward-looking work of this Select Committee was an important turning point in providing the necessary push for the technical education which enabled twentieth century industrial Britain. A conclusion was that the chief obstacles standing in the way of a technologically-informed and competitive nation were the wholly inadequate provisions for both primary and secondary schooling; nothing seems to change! The report was followed by a Royal Commission on Scientific Instruction and the Advancement of Science set up during Gladstone's first term of office (1870–1875), under the chairmanship of William Cavendish, 7th Duke of Devonshire, an important figure in the ferrous industry and the first President of the Iron and Steel Institute.

One result of this was the introduction of the Higher Grade Schools, the first State-aided secondary schools, which were required to teach science to a recognised educational standard in order to qualify for a grant and recognition. One such school in Cambridge was attended by my father at the end of the nineteenth century. Such schools were a direct result of Gladstone's Education Act of 1870 and, as regards the emphasis on science, the report from the Cavendish Committee. A favourite quote from the report of the Committee – 'considering the increasing importance of science to the material

> The chief writers on Metallurgy are the Germans, to whom we owe two of the most remarkable works on the subject, namely, the treatise of Agricola, in Latin, which appeared in 1555; and the System of Metallurgy of Karsten, in German, published in 1831. The monographs, contributions to periodicals, and compendious treatises relating to the science and practice of Metallurgy which have been published in the German language, are very numerous. We are, probably, indebted to the Germans, to a greater extent than is commonly supposed, for the development of our mineral resources, since the introduction of German miners and metallurgists into England, about three centuries ago, through the wisdom of Elizabeth.

Fig. 5 Percy writing on the significance of German metallurgy, 1861.

interests of the country one cannot but regard its almost total exclusion from the training of the upper and middle classes as little short of a national misfortune'. In relation to the RSM and the Royal College of Chemistry, the Royal Commission's first report was concerned with broadening the curriculum to general science with a primary emphasis shifted to training teachers. This seemed bound to reduce the importance of metallurgy in the courses and, as Almond has pointed out in his excellent thesis Metallurgical Education 1851–1950, it was a miracle that it remained a distinct subject, surviving into the twentieth century. W. W. Smyth, in charge of mining, and J. Percy, the metallurgist, strongly opposed this broadening, and the latter resigned rather than submit to the changes, which involved also moving to a new RSM, in the Huxley Building, South Kensington, offering to rebuild it himself in Jermyn Street. In the aftermath of the Great Exhibition, Parliament had voted £150,000 to add to the Exhibition's profits to enable three Kensington centres to be purchased – in due course to be the homes of the RSM, Science Museum, Geological Museum and the City and Guilds.

Although Percy had resigned from the RSM in 1879, rather than move to the Huxley Building, fearing that metallurgy would lose its identity in a general School of Science ruled over by the younger Thomas Huxley, he should have hung on, since friends rallied round the RSM, influencing the Treasury so that there should not be a complete merging of 'the strictly technical and professional school of mining in a more general scientific institution'. The justification given was 'the development of the mineral riches of this country and of its colonies and dependencies was the foremost object to which the Government intended by its measures in 1851–1853 to direct researches of science and apply their results'. Thus in 1881, after a lengthy period of transition, and no doubt strife, the Normal School of Science (later the RCS) and the Royal School of Mines came into existence in South Kensington, in what we now know as the Huxley Building.

Roberts-Austen succeeded his mentor Percy in 1880 whilst retaining his post as chemist and Assayer at the Royal Mint, to be followed by Gowland, Carlyle, and then Carpenter in 1913, the last supervising the fine new building in 1915 in Prince Consort Road, and then Dannatt in 1940. The relocation of the RSM on this present site commenced with the Bessemer Laboratory of fond memory, where we carried out virtually full-scale mineral dressing experiments (Figs 6, 7) in my time, 1944–1947.

The contribution of the RSM to the metallurgy profession and metallurgical industry has been enormous. For 30 years it was the only school of metallurgy in the country. A close association with the Royal Mint was maintained over many years, from Percy onwards. The foundation of many of the other schools of metallurgy which developed later in the UK was through Associates of the RSM (Table 1), and the alumni list is truly impressive. William Chandler,

10   Materials Science and Engineering: Its Nucleation and Growth

Fig. 6   View of Bessemer Laboratory (*Engineering*, 1951).

Fig. 7   View of Bessemer Laboratory (*Engineering*, 1951).

Plate I Malachite $CuCO_3 (CuOH)_2$.

Plate II Azurite $2CuCO_3 (CuOH)_2$.

Plate III Galena PbS.

Plate IV Green flame from copper/copper minerals.

Plate V Gossan or 'iron hat' overlying copper mineralisation.

Plate VI Cast copper axes, ~4000 BC Balkans.

Plate VII   Chalcophyllite, copper arsenate, $Cu_3As_2O_8nCu(OH)_2$.

Plate VIII   Juxtaposition of arsenopyrite FeAsS (grey) and cassiterite SnO_2 (dark brown).

Roberts-Austen, Clement Le Neve Foster, Sidney Gilchrist Thomas, Percy Gilchrist, George James Snelus, William Gowland, Charles Vernon Boys, Harold Carpenter, Miss Constance Elam (Mrs Tipper), F. W. Harboard, Thomas Kirke Rose, William Hume-Rothery, Andrew McCance, Leonard Pfeil, to name but a few of those in earlier years. May its direct influence and that of its alumni continue to be of great importance to the nation.

Table 1 The Influence of RSM-trained men on British Metallurgical Instruction (after J. K. Almond)

Institution Offering Instruction		RSM-trained Staff in Post
Before 1900		
Class of Artillery Officers, Woolwich	c1861–1888	J. Percy
	1888–1906	H. Bauerman, Lecturer on Metallurgy
Science and Art Department, Kensington	1864–1888	J. Percy
(national exams)	1889–1902	W. C. Roberts-Austen
London, King's College	1879–1919	A. K. Huntingdon, Professor of Metallurgy
Birmingham, Mason's College	1883–1886	T. Turner, Demonstrator in Chemistry
	1886–1894	T. Turner, Lecturer in Metallurgy
Sheffield, Technical School	1885–1889	W. H. Greenwood, Professor of Metallurgy and Engineering
Glasgow, West Scotland Technical College	1886–	A. H. Sexton (not associate), Professor
Camborne, Cornwall. School of Mines		J. J. Beringer, in charge of assaying
Staffordshire, County Council	1894–1902	T. Turner, Director of Technical education
	1892–	A. McWilliam, Lecturer on Metallurgy
Newcastle, Armstrong College	1890s	H. Louis, Professor Mining and Lecturer
(University of Durham)		on Metallurgy
Nottingham, University College	1895–1897	G. Melland, Lecturer on Chemistry and Metallurgy
The Exception:		
Cardiff, University College		No apparent RSM influence
(established 1894)		
After 1901		
Birmingham, The University	1902–1926	T. Turner, Professor of Metallurgy
	1906–1912	D. M. Levy, Assistant Lecturer
Sheffield, The Technical School	1902–1911	A. McWilliam, Lecturer/Assistant Professor
(and University)		
Newcastle, Armstrong College	1901–1905	G. H. Stanley, Demonstrator in Metallurgy and Surveying
(University of Durham)		
Woolwich, The Polytechnic	1904–1927	G. Melland, Head of Chemical and Metallurgy Department
London, Sir John Cass Technical Institute	1903–1914	C. O. Bannister, Head of Metallurgy Department
	1926–1945(?)	G. Patchin, Principal and Head of Metallurgy Department
London, Chelsea Polytechnic Institute	1911–1940(?)	W. A. Naish, Metallurgy, Lecturer in Department of Chemistry/Head of Metallurgy Section
Liverpool, The University	1920–1941	C. O. Bannister, First Professor of Metallurgy
	1928–1960s	S. J. Kennett, Lecturer
Swansea, Technical College	1901–1904	G. Melland, Head of Metallurgy Department
Swansea, University College (from 1920)	1921–1930	L. B. Pfeil, Assistant Lecturer/Senior Lecturer
	1927–1938	L. Taverner, Assistant Professor, Metallurgy
	1936–1938	G. L. Jones, Assistant Lecturer, Metallurgy
Llanelly, Technical College	1921–1926	H. Etherington, Lecturer in Metallurgy
Leeds, the University	1919–1932	P. F. Summers, Lecturer in Metallurgy
Abroad:		
Montreal, McGill University	1901–	A. Stansfield, Professor of Metallurgy
Johannesburg, University of Witwatersrand	1905–1939	G. H. Stanley, Professor of Metallurgy
	1939–	L. Tavener, Professor of Metallurgy

Bibliography

J. K. Almond, 'Factors Influencing Education in Metallurgy in England and Wales 1851–1950', M.Ed. Thesis, University of Durham., March 1983.

C. Renfrew, *Before Civilisation*, Johnathan Cape, London, 1973.

J. A. Charles, *Out of the Fiery Furnace*, Institute of Materials – Book 729, 2000.

D. R. F. West and J. E. Harris, *Metals and the Royal Society*, Institute of Materials – Book 568, 1999.

J. H. Pepper, *The Playbook of Metals*, Routledge, Warne & Routledge, London, 1861.

J. Percy, *Metallurgy*, John Murray, London, five volumes from 1861.

Anon, *A Short History of Imperial College 1845–1945 on the Occasion of Its Centenary in 1945*, Booklet.

The Contributions of William Chandler Roberts-Austen to Metallurgy and to the Arts

MICHAEL R. NOTIS
Department of Materials Science and Engineering, Lehigh University,
Bethlehem, Pennsylvania 18015, USA
e-mail: mrn1@lehigh.edu

ABSTRACT

If there is such a persona as 'the metallurgical hero', Professor William Chandler Roberts-Austen is its epitome. As just one example of the many areas of research to which he made major contributions, is his work on the accurate quantitative measurement of solid state diffusion in metals, decades before anyone else. This led to his discovery that diffusion in solids is strongly temperature dependent, and slower than diffusion in liquids by orders of magnitude. His view of molecular motion in solids allowed him to develop an advanced understanding of what we now call diffusion controlled phase transformations.

However, not only was he a man of science but he had a strong interest in the metallurgical arts. At a time when *Japonisme* and its art-styles were in vogue in Europe and America, and new metals such as aluminium were just coming into use, he played a significant role in fostering interaction between the sciences and the arts, and between the West and Asia. Roberts-Austen also had an abiding love for the Royal School of Mines, and he connected all aspects of his life to it as an institution.

Preface

In 1974 I had the pleasure of spending my sabbatical leave from Lehigh University at Imperial College. Just before arriving in London, I had met with Cyril Stanley Smith who gave me a copy of his *History of Metallography*, and that is where I first read about the wide range of the research performed by Sir William Chandler Roberts-Austen.

My office was near that of Professor D. R. F. West, with whom I am glad to say I have kept in contact with over the years. I am especially happy to see his new book (together with J. E. Harris) *Metals and the Royal Society* in print.

Introduction

Sir William Chandler Roberts-Austen (1843–1902), became an Associate of the Royal School of Mines in 1865. He was for many years both Professor of Metallurgy at the Royal School of Mines (1880–1902), and also Chemist and Assayer (Director) of the Royal Mint (1865–1902). He played a dominant role as a leader of metallurgy at the end of the nineteenth century, a time when this field held special importance to the economic development of England. His work on the famous Alloys Research Committee led to the establishment of the Department of Metallurgy at the National Physical Laboratory (1901). The significance of his contributions to the development of physical metallurgy, especially in the areas of our understanding of phase equilibria, solid-state diffusion, and phase transformations, cannot be overestimated. In addition, he was a man of diverse interests, and he took every opportunity to extend his profession into other areas of society. He therefore became a leader in the development of ties between his science, metallurgy, and the arts. Numerous publications have been written concerning the wide range and different aspects of his research, and of his interactions with others.[1-11]

Very little in science is achieved out of context, and so it was with Roberts-Austen. The two people who most influenced him, John Percy (whom he succeeded at the School of Mines) and Thomas Graham (whom he succeeded at the Royal Mint) both played significant roles in the directions that he was to follow in metallurgy and in the arts.

Percy was very interested in the historic development of smelting technology, as is easily seen from his book, *Metallurgy: The Art of Extracting Metals From Their Ores*. Percy and Roberts-Austen both had significant interactions with Heinrich Schliemann, the excavator of Troy and surrounding regions. In 1880, Schliemann[12] writes (p. 251), 'My friend Mr. W. Chandler Roberts, F.R.S., assayer at the Royal Mint, and Professor of Metallurgy in the Royal School of Mines, kindly analysed the metals of this first city, and wrote for me the following valuable report on the subject: [continues].' The report concerns a copper knife gilded with gold, and two copper nails or pins, and points out that the copper is much harder than modern commercial copper, and gives opinion as to the historic period in which the objects were made. On p. 409 Schliemann includes a description of clay smelting crucibles and boats for refining gold or silver, which were examined both by Roberts-Austen and Percy. In 1884, Schliemann[13] describes (pp. 101–104) an interesting response to his questions on hardening of copper and steel, and the definition of the term *tempering* in the context of steel metallurgy.

Roberts-Austen worked together with Thomas Graham for many years and was directly exposed to Graham's work on diffusion in gases. In this paper, I will try to highlight only a few of Roberts-Austen's major contributions, and which show the various facets of these activities in his life.

Diffusion in Solid Metals and Modern Physical Metallurgy

Roberts-Austen was the first person to perform quantitative measurements on diffusivity in solid metals. In[14] he measured the concentration profile of Au in solid Pb (concentration vs. depth), used Fick's Laws and Stefan's (error function) solution to solve for diffusivity, using units of cm^2/day, and the values he published are within a few per cent of recently determined values.[1,2,6] He indicated that 'The diffusivity of Au in *solid* Pb is . . . slow when compared with diffusion in the *fluid* metal. In solid Pb . . . the diffusivity at 250°C . . . is 1/100th of the diffusivity at 500°C. . . . It is also clear that Au will diffuse into solid Pb at the very modest temperature of 100°C, which is 225°C below its melting point, a fact which must be considered to be remarkable, and one the existence of which has hitherto been unsuspected. The diffusivity is . . . only 1/100,000th of that which occurs in fluid Pb.' Roberts-Austen also compared the diffusion of gold in solid lead to that of gold in solid silver, and concluded that the rate of diffusion is a function of the melting point of the matrix metal.

Concerning this work, his colleague T. E. Thorpe writes: 'I remember . . . when he carried me off to see the first results of his inquiry into the diffusion of solid metals, and when be showed me the little beads of gold cupelled out of the several sections of the block of lead, which had been standing for days and weeks on a plate of the precious metal, all arranged at the proper intervals of the sections on a diagrammatic representation to actual scale of the leaden block.'[11]

In a follow-up paper[15] Roberts-Austen wrote that 'the amount of Au which would diffuse at the "ordinary temperature" in 1000 years is almost the same as that which would diffuse in molten Pb in a single day', and that he is 'trying to ascertain whether diffusion in the solid metal is, or is not, accelerated by the simultaneous passage of a strong electric current'. (See discussion in last paragraph of section: Roberts-Austen Ties Diffusion to Art Objects.)

He also performed the first quantitative work on the carburisation of iron[16] and indicated that 'If this curve (distance to which Fe has been penetrated by C, and the amount of C which has so penetrated) be compared with . . . the diffusion of Au or Pt in solid Pb, it will be found that the curve for the *carburisation* of iron resembles a true diffusion curve'. As a comment added to this same paper, J. E. Stead commented that other work he had performed together with Roberts-Austen on 'the MINUS cementation process' (=decarburisation) showed the reverse profile to that discussed by Roberts-Austen, except for a region towards the outside where carbon was absent.

Roberts-Austen tied his work on diffusion in solid metals together with his studies of the microstructural changes that he observed during heat treatment of steel. In[17] he wrote that 'there is abundant evidence of molecular unrest in solids . . . the molecular mean free path must be small . . ., but if it be granted that there is rapid molecular movements in solids, such movement may in time lead to disruption of masses . . .' With regard to the different microstructures observed in a 1.5 wt%C steel subjected to a variety of heat treatments and cooling conditions he indicated that 'Profound molecular changes take place . . . at each of these critical temperatures. . . . The point to insist upon is that all of them have been effected in the solid metal. . . . We are beginning to realise that the mechanical properties of a metal depend on the added elements, and, therefore, that "molecular unrest" and industrial applications are closely related.' This has to stand as the first real statement of what we today consider to be modern physical metallurgy, and the beginning of our understanding of diffusion controlled phase transformations.

The Colouring of Alloys and the Arts

In conjunction with his position at the Mint, he had always performed research with gold or silver-based alloys. When the cost of aluminium started to come down, it was natural for him to look at the effect of aluminium additions to gold. Roberts-Austen worked on melting and phase equilibria in the gold-aluminium system, and discovered the intermetallic compound $AuAl_2$.[18] This compound has a deep purple-black colour of great beauty, and Roberts-Austen was excited by its appearance. His colleague, T. E. Thorpe, records that, 'I shall never forget the manner in which he burst into my room, when at South Kensington, and showed me the first fragment of the beautiful rose-coloured alloy of gold and aluminium he had obtained. His delight was so real and unaffected – his joy almost infantile – as he turned and twisted the glittering fragment to the light to illustrate the depth and wonderful brilliancy of its purple. And, too, it was characteristic of him that, as I shared his admiration, he should, unasked, have seized a letter-weight and knocked off a portion of his prize and bade me take it.'[11]

In his textbook *Aluminium*,[19] J. W. Richards of Lehigh University, references (p. 501) the work of Roberts-Austen with Gold–Aluminium alloys. His book, first published in 1887, was the first, and for many decades, the only book on the modern industrial metallurgy of aluminium.

Concerning Roberts-Austen's interest in colouring and Japanese metalwork, Thorne writes: 'Much of his work was influenced by his strong artistic sense and by his passionate regard for beauty of form or colour. The secrets of oriental metallurgy had a singular fascination for him. He would literally gloat over some triumph of Japanese art, and the discovery of by what kind of "pickle", or by what kind of treatment, the lustre or colour or effect on a bronze had been obtained was a delight to him as intense as if he had lighted upon a new metal. The artistic side of his nature found frequent exercise in his work at the Mint, especially in medal-striking.' [11]

Japanese Traditional Metalwork

Roberts-Austen's fascination of the high level of workmanship and the use of colour in Japanese metalwork was a major factor in his influence on the arts in England. In order to set the stage so that we understand the nature of its appeal to him, it is appropriate here to review the state of awareness of the arts of Japan in the West during the last half of the nineteenth century.

W. C. Reed[20] describes the high state of metalworking technology in Japan in 1855: 'The art of manufacturing of steel is interesting, and I contend that they can manufacture better edge tools from their steel than any other nation. . . . I also visited their smith shops with interest, and became convinced that they were expert and skilful in metals. . . . The work turned out would be creditable to any of our blacksmiths or workmen in iron. . . . Their process of brazing copper was of much interest for it appeared to be done as easily as we could weld two pieces of iron together . . . succeeding in making six sheets of copper adhere together as if in one piece, astonishing me, and I do not think the thing can be accomplished by any of our workers in copper; hence in the nature and use of copper their knowledge exceeds ours.' As early as 1866, specific details of Japanese metalwork and its materials had been documented by Pumpelly.[21] He describes SHAKUDO as an alloy of copper and gold (ranging from 1 to 10%); and he states that SHAKUDO receives a beautiful bluish-black surface colour by boiling in a solution of copper sulphate, alum and verdigris. He tries to explain this colour 'by supposing that the superficial removal of the copper exposes a thin film of gold, and that the blue color produced is in some manner due to the action of light on this thin film of gold.' He notes that some Japanese objects made from copper have a bright red surface acquired by boiling in the same solution as used for SHAKUDO. He describes GIN-SHIBUICHI as an alloy of copper and silver (ranging from 30–50%); when subjected to the same patination process as SHAKUDO, this alloy develops a 'rich grey color much liked by the Japanese.' Most significantly, he indicates MOKUME (mixed metals having a woodgrain appearance; this technique is a major focus of this paper), as being made from 'several alloys and metals of different colors associated in such a manner as to produce an ornamental effect. Beautiful damask work is produced by soldering together, one over the other in alternate order, thirty or forty sheets of gold, SHAKDO, silver, rose copper, and GINSHIBUICHI, and then cutting deep into the thick plate thus formed with conical reamers, to produce concentric circles, and making troughs of triangular section to produce parallel, straight or contorted lines. The plate is then hammered out till the holes disappear, manufactured into the desired shape, scoured with ashes, polished, and boiled in the solution already mentioned. The boiling brings out the colors of the SHAKDO, GINSHIBUICHI, and rose copper.' (Spelling as in the original paper.)

Pumpelly[21] also gives the composition of Japanese brass alloys (SINCHU) and bronze alloys for use as bell metals (KARAKANE). Finally he lists the solder alloys used for joining silver, SHAKUDO, GINSHIBUICHI, and MOKUME, as all to be alloys of silver, copper and zinc. This description of Japanese alloys by Pumpelly predates the other references found in American publications by at least 8 to 12 years, and those in English publications by 16–20 years.

Edward Greey, a member of Commodore Perry's squadron during the opening of Japan and a collector of Japanese metalwork, later became an important New York dealer in Japanese goods.[22] New York City quickly became the centre of Japanese art goods retailing and marketing in the United States.[23] Francis Hall, helped found Walsh, Hall & Company, which became the leading American trading house in Japan. Hall returned to New York in 1866; the trading company he helped establish was sold to Mitsubishi in 1897. Hall reported[24] that prior to the Meiji restoration, the Japanese merchants were hesitant to sell objects of copper and brass, and that in late 1859 the Japanese government forbid further exportation of copper under the pretence that the Emperor needed it to rebuild his palace. As the demand for Japanese merchandise grew in the West, and as the economic pressure and desire for new technology grew in Japan, the new government that took power at the beginning of the Meiji era (1868–1912) sent its artisans overseas to study the market and learn modern techniques. Their products, often more Victorian than Japanese in style were then exported to international expositions, where they found immediate favour (in part because they looked so familiar) thereby earning foreign currency needed to finance the industrialisation of Meiji Japan.

The most significant booth exhibit of Japanese art in the 1860s, organised by the Japanese Government, was held in France in conjunction with the Paris Exposition in 1867. By this time, Japanese artistic influence was already strong both in England and in France. In the late 1860s and 1870s, Japanese stylistic influences on silver design in England grew and achieved peak popularity in the 1870s and 1880s. Elkington and Co., of Birmingham, and Hukin and Heath, of Sheffield, were two of the most significant companies who produced silver with Japanese-influenced designs. In France, similar metalwork was produced by the firms of Christofle and Barbedienne, and by individuals such as Alexis and Lucien Falize.[25]

In the United States, both Tiffany and Gorham also developed a strong interest in Japanese metalwork style, and it quickly developed as a major influence, especially under the guidance of Edward Moore at Tiffany. Tiffany's rapid growth, economically and technologically, was in large part due to their openness to new ideas and to the application of new technology and styles to develop new markets. In 1852 Tiffany became the first American firm to adopt the English sterling standard of 925 out of 1000 parts pure silver for their wares,[26,27] but they did not share the English attitude for basing value primarily on weight of the piece rather than on artistic aesthetic perception.

Christopher Dresser, a well-known author of three books on decorative design, was also opposed to the Victorian attitude and need for valuation of art by weight content of precious metals. He maintained that to use a solid metal such as silver was an unnecessary waste when the same effect could be obtained by using a technique such as electroplating.[28] In 1876, before going to Japan, he visited Philadelphia and attended the Centennial Exposition (at which Moore was also present) and he gave a series of three lectures at the Pennsylvania School of Industrial Art where he presented his theories concerning aesthetics and art valuation (which were certainly close to ideas held by Tiffany and Company). The objects that he brought back from Japan for Tiffany to sell in its shop when he returned late in 1877 included pieces of MOKUME and other metalwork which could have helped Tiffany perfect their work in Japanese style.[28] By the Paris Exhibition of 1878, Tiffany walked off with many major awards (Grand Prix for silver; one gold, one silver and four bronze medals for its co-workers; and Charles Tiffany became a Chevalier of the Legion of Honour). Tiffany's designs became the leading influence in metalwork worldwide.[26]

The relation between the Falize firm in Paris, and Tiffany is interesting in terms of how Japanese stylistic influences moved from Europe to America. A gold and cloisonné enamel locket designed by Alexis Falize in the Japanese taste and made between 1868 and 1870 is the earliest documented piece of that style known to have been in the United States shortly after it was made; the original presentation case that contains the locket indicates that it was originally sold by Tiffany & Co.[29]

The next great exposition after Paris (1867) and the first where the Japanese government organised a full formal display was in 1873 in Vienna. In that year, the first Japanese Manufacturing and Trading Company (KIRITSU KOSHO KAISHA)[30] was established in conjunction with Japan's participation in the International Exposition in Vienna. Its purpose was to promote the development of Japanese industry through the introduction of advanced science and technology, and to move Japanese production methods from emphasis on individual craftsmen to organised industries more able to deal efficiently with increased manufacturing needs; this policy was known as 'Increase Production and Promote Industry'. The New York Branch was opened in 1876 on the occasion of the International Centennial Exposition in Philadelphia. A major focus of this company was the production and export of traditional handicrafts such as fabrics, embroidery, ceramics, lacquerware, tortoiseshell work, leathergoods, folding fans, cloissonné, and metalcrafts, all under the general name of 'Art Industry'. Art industry flourished in Tokyo, Kyoto, Nagoya, Kanazawa and Arita, as well as other areas famous for traditional handicrafts. Vienna was important from an American perspective because it was the fair running when the United States was planning the exposition to commemorate its centennial in 1876 in Philadelphia. As mentioned above, Japan's Centennial Commission lobbied manufacturers in Japan to increase production of items suitable for sale and display in the United States. Within a year after agreeing to participate, Japan had opened an office in Philadelphia to coordinate operations in the United States. In addition to the 14,000 square feet of space in the main exhibition hall, Japan obtained an acre of land for the erection of Japanese houses to be used as a bazaar for the sale of Japanese goods, and another separate piece of land near the British buildings.[31] Japanese MOKUME was exhibited at the Philadelphia Centennial of 1876[32,33] and may have provided the impetus for experiments in the technique by Moore at Tiffany's. A newspaper report in 1878[32,34] describes the MOKUME process of Tiffany as producing 'the astonishing and most beautiful effects of veining'. Carpenter,[35] however, dates the experiments on mixed metals at Tiffany's to the 1880s. Interest in MOKUME as a matrix art form continued to grow. G. A. Audsley,[36] in *The Ornamental Arts of Japan*, describes 'The peculiar class of ornamental metalwork called by the Japanese MOKUBE . . . which produces a marbled effect'. (Spelling as in the original.)

The use of appliques and inlays of various metals and alloys by Tiffany in the Japanese style is first specified on drawings dated 1877.[32] These drawings include notations for yellow metal ('YM'), fine gold ('FG'), Japanese Gold ('JG') and red gold ('RG'), and the Japanese names SHAKUDO, and SHIBUICHI. Thus the firm was going well beyond the adoption of Japanese designs to the manufacture and use of Japanese alloys. Already in the mid-1870s published articles in trade journals contained formulas for the various Japanese alloys.[37,38] These coloured metals provided Tiffany a virtual palette of colours with which to represent tropical birds, ripen fruit and foliage, and the setting sun. An 1878 drawing shows a 'conglomerate' vase combining laminated metal inlays with applied and other decoration; the drawing includes a notation 'MM' indicating mixed metal, Tiffany's working term for MOKUME.[32] Edwin C. Taylor, writing in The National Repository in

1879 about Tiffany's Japanese silver, said that Tiffany had created a 'new school' of decorative art: 'The forms of some of the articles of daily use – the tea pots and the water jugs – follow nature very closely, and the gourd forms and modifications of vegetable life are used with admirable effect. These, too, are suggested by the familiar works of the Japanese, who show a rare sympathy with nature in their employment of decorative figures; but technically they have advanced far beyond their Japanese models, and they were regarded with astonishment by the clever Orientals at the Exposition, the Japanese commissioner purchasing characteristic specimens for his government'.

In the 1880s, Edward Moore experimented with mixed metals (not alloys) made from copper, silver and gold. The mixed metals were in the form of layers imitative of wood-grain texture (MOKUME, as above) which were fabricated by hammering the different metals together and heating them. Objects made from these mixed metals were exhibited at the 1889 Paris Exposition and were described by many writers of that time, including Samuel Bing.

There are two items in the files of Edward Moore that relate to the fabrication of MOKUME and to the colouring of metals. The first is a paper by W. Chandler Roberts-Austen[39] which contains figures and describes the principles of forming laminated metals and their colouring (Fig. 1); this pamphlet *postdates* the development of Tiffany's laminated metals and mentions the firm as a leader in Japanese-inspired metalwork.[32] There is no direct evidence that Moore had read Pumpelly's[21] paper, but the publication of similar descriptions in practical publications by 1875[37,38] could not have missed his attention and certainly suggests that a silversmith of Moore's stature would be aware of these alloys and their colouring. It is more likely that this paper by Roberts-Austen was in the Tiffany file because of its reference to Tiffany. The second file item is a note[35] (by Moore) explaining how the mixed metals were made: 'It consists of Fine Gold, Fine Silver, Pure Copper, SHAKUDO (Copper and Gold) SHIBUICHI (Copper and Silver). These five layers of metal are soldered together making a block about three-fourths of an inch thick. This is beaten and rolled out making it thinner. When it is reduced to about one-fourth of an inch in thickness, it is folded together, three parts making fifteen layers. This again is reduced and again folded double, making thirty layers. A sterling silver backing is now added and the process or variegation begins by cutting figures and spots through the outer metals and then by beating and forcing the under metals up through to the surface. This goes on until the desired fineness, effect and thinness is produced. It is now ready to be hammered into the shape designed, then smoothed and the colours of all the metals developed by chemicals producing the patina of each.'

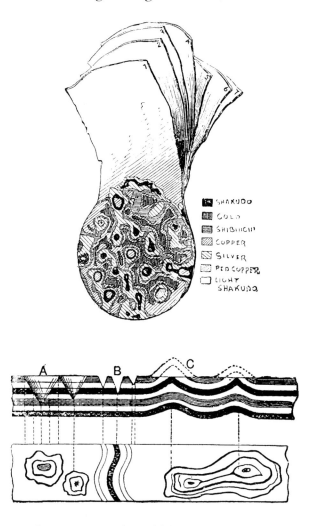

Fig. 1 Schematic showing materials and fabrication processing used in making MOKUME as described by Roberts-Austen.

The best known example of Moore's MOKUME work is a Tea Caddy[40] dated 1880. One Tiffany inlay piece (shown by Tiffany at the 1893 Columbian Exhibition) is particularly interesting. It is a silver bowl[41,42] inlaid with niello and mixed metal done in Southwest American Indian style; therefore, the bowl not only uses Japanese metalwork technique to reflect a different cultural style but also a different art medium, metal rather than ceramic. This is certainly reflective of the origin of Japanese mixed metal itself, believed to have developed from Chinese ceramics, and black and red lacquerware which certainly predate Japanese metal art forms.

In a recent study, Judy Rudoe[43] reports that an analysis performed by Susan LaNiece indicates the presence of platinum on black patinated spots present in an inlaid Japanese style Tiffany silver tray owned by the British Museum. If the platinum is in the alloy, it suggests that the blackened copper inlays are Tiffany's adoption of Japanese SHAKUDO alloys, with platinum replacing gold in the traditional alloy recipe. If the platinum is present in the patina, Tiffany may have used a platinum chloride solution to produce the black surface film. This method of achieving surface colour is described by Hiorns[44] in 1892, and also by Paul Wayland Bartlett[45] at about the same period of time. In any event, it appears that Moore at Tiffany was knowledgeable in the composition and traditional patination of Japanese alloys, as well as having experimented with alternative non-traditional methods for alloy preparation and or colouring.

The story of Japanese metal workers being brought over from Japan to work in Tiffany's Prince Street silver plant has been mentioned by a number of writers in recent years.[25,35,46] It was assumed that they were involved in the making of Tiffany's silver in the Japanese style. But the details of the story are elusive. The only contemporary accounts of such Japanese workmen are brief mentions in French publications of the 1880s and 1890s.

The first mention of Japanese workmen at Tiffany's was by Lucien Falize (writing under the pseudonym of M. Josse) in an article, 'L'Art Japonais', published in the *Revue des Arts Décoratifs in 1883*.[35,46,47] Falize stated that Tiffany & Co. had sent an associate to Japan who brought back a group of chasers (ciseleurs) to work in their silvermaking plant in New York where he said they were given relatively free rein except that they were restricted to conventional silver forms. G. de Leris,[48] writing in the *Revue des Arts Décoratifs* a few months later, also mentions, probably as a result of Falize's article, that Japanese workmen were brought over from Japan by Tiffany.

These French references have created a problem in that not a single reference to these Japanese workmen has been found in any contemporary American publication.[35] The volumes of Tiffany & Co.'s clipping files, which go back to the 1840s, appear to have no mention of the employment of Japanese workmen by Tiffany. None of the dozens of plant working drawings of the silver in the Japanese style have any Japanese names or Japanese notations of any kind on them, all pencil notes being in English. A promotional brochure issued in 1877 by Tiffany was rather emphatic about their silver being made by Americans. In describing the first floor of the store they noted: 'Passing still onward, the cases to left and right contain gold and silver plate, both in plain forms for ordinary table use, and also in every variety of costly pieces for decoration. Among these things it will be found that the spirit of Japanese Art has been largely utilised, while at the same time it has been so modified as to be brought into full harmony with the requirements of

American and European civilisation; and to afford, in the hands of American workmen, results of the most striking kind.'[35]

Samuel Bing in his book *Artistic America*,[49] first published in 1895, wrote: 'Moore had invited teams of Japanese craftsmen to America, under whose guidance tonalities of every kind were mixed with silver. The effect of these inlays, to which niello had already been added, was burnished still further by other colouring methods. Experiments with enamels followed, in which, by alternating muted, sombre tones with brilliant, translucent areas, a maximum degree of technical skill and ingenuity was attained. And finally, to be sure that every resource had been exhausted and to obtain the last work in splendour, precious stones were added.'

If Japanese workmen were employed in Tiffany's Prince Street works (and one assumes Falize and Bing had some basis for their statements) there is the question of when they worked in New York. It must have been after 1877 and possibly it was in the 1880s, since the objects employing techniques described by Bing were almost all made in the 1880s and 1890s.[35,46]

It therefore appears that no direct evidence of Japanese craftsmen working at Tiffany's has been found. The speculation that Japanese craftsmen might have come to the US could be based on the poor social situation of the craftsmen in Japan during this period. The Japanese civil war of 1871 had a tremendous influence on the metalworking craft. With the abolition of the feudal class, samurai were no longer permitted to use swords, thus putting a great strain on the metalcraft worker who had produced wares of outstanding artistic quality during the previous two centuries.

However, Japanese workmen may have been involved with the production of Tiffany and Co. wares even if they did not work in the New York shop.[25] An article published in 1878 indicated that objects produced in the US for sale at Tiffany were sent to Yokohama to be decorated 'with Japanese devices . . . so as to secure practical excellence and original and graphic Japanese Ornament.'[50]

As mentioned previously, in late 1877 Tiffany exhibited and sold a large collection of Japanese items that had been brought from Japan by Christopher Dresser, who had made a three-month trip to Japan earlier in 1877 in order to buy and make-up collections for both Tiffany in New York and Londros and Company of London.[46] Dresser may have directly or indirectly recruited workers for Tiffany from Japan,[46] but again there is no direct evidence that this is the case. Hawley[26] points out that Dresser also may have suggested to Tiffany that they could employ Japanese decorative techniques for metalwork which could not legally have been made and sold in England, as described below.

In addition to the work by Moore at Tiffany's, other American artists and firms such as Gorham[51–53] (mentioned previously); Whiting Manufacturing

Co.; Derby Silver Co.; James W. Tufts Co.; Pairpoint Manufacturing Co.; Meriden Britannia Co.; Reed and Barton; Simpson, Hall and Miller Co., were interested in and produced metal *objets d'art* using Japanese styles and techniques. The American sculptor Paul Wayland Bartlett also experimented with Japanese alloys and patinas; jewellery items produced by Bartlett were also exhibited at the Paris Exposition of 1889.[53]

To a significant degree, American success was related to production limitations which existed in England at that time. British laws regulating silvermaking prohibited the combination of silver and gold with base metals, such as copper.[25,26,54,55] In 1883 and 1884 English laws concerning the hallmarking of silver were modified; however there is evidence that a considerable quantity of silver items made by Tiffany and Co. in the Japanese style was sold to clients in England.

Concerning the lecture paper by Roberts-Austen found in the files of Edward Moore at Tiffany's, this paper was first published in 1886, and is in fact part of a series of papers that Roberts-Austen wrote on alloys and the colouring of metals. Roberts-Austen, in 1884, first reports[56] on the composition of Japanese alloys in conjunction with a reference to information from Charles Tookey of the Imperial Mint in Osaka. In his 1886 paper[39] he refers to analyses of SHAKUDO and SHIBUICHI by William Gowland (his former student at the Royal School of Mines) at the Imperial Mint in Osaka; this aspect is discussed in more detail later in this paper. In this same paper[39] he strongly praises the work of Tiffany and Co., hence the likelihood of it appearing in Tiffany's files. In a more extensive paper on colouring published by Roberts-Austen in 1890[60] he includes sketches of metalwork vases from 'the museums in Kensington' which show the use of mixed metal MOKUME (Fig. 2); in a still later paper, he includes a photographic figure of a vase showing the use of inlay of a variety of metals, among them SHAKUDO and SHIBUICHI. It was to my great and pleasurable surprise to find both of these same metalwork objects displayed for the first time in more than 100 years at a new exhibition at the Victoria and Albert Museum in London.[61] Toyojiro Hida[30] has attempted to locate items verified as being produced by Kiritsu Kosho Kaisha (the first Japanese Manufacturing and Trading Company), and he reports that 'I found only one bronze vase, on which flowers and birds are inlaid, in the collection of the Victoria and Albert Museum in London.' This vase, shown in figures 18 and 19 of his book, is the same as that shown on p. 202 of the Toshiba Gallery catalogue of the Victoria and Albert Museum, and similarly in the much earlier paper by Roberts-Austen.[60]

Because of his position as metallurgist at the Royal Mint in London, Roberts-Austen was well aware of developments in metal art technology in general, and of Japanese metalwork in particular. He bemoaned the fact[58,59] that 'so far as I know, no English artist has actually used this material (MOKUME) with the

Fig. 2 MOKUME vase described by Roberts-Austen.

exception of Mr Alfred Gilbert, ARA[62,63] (see below), who, I rejoice to say, has employed it in the now famous chain worn by the Mayor of Preston. But it is a beautiful material, that surely must be used by English art workmen if they hope to hold their own. The Americans have adopted it already and in the Paris exposition (of 1889) there was a very remarkable vase exhibited by Messrs. Tiffany, of New York. . . .'[60]

It is interesting that in 1888 Roberts-Austen suggested a solder for use with joining layered metals to make MOKUME-like materials; the content is given as 55.5 Ag–26 Zn–18.5 Cu by weight.[58] This is very close to the published composition of the ternary eutectic for this system,[64] and which melts at 665°C, about 100°C lower than the Ag–Cu binary eutectic at 773°C. Research into the melting behaviour and scientific understanding of binary and ternary eutectics were at their very infancy in Europe at this period of time (1880–1890) and this is a wonderful example of the rapid incorporation of these studies into industrial technology.

When Roberts-Austen was the metallurgist at the Royal Mint in London, his student, William Gowland, became the assayer at the Imperial Mint in Osaka. The two people maintained written dialogue both in official reports and in personal communication. Both Roberts-Austen and Gowland had a strong interest in Japanese art metalwork. Japanese style was being introduced in Europe just before Gowland departed for Japan. Upon his arrival in Japan in 1872, Gowland became actively involved in the study of the archaeology of Japan, and when he returned to England in 1888 he wrote numerous papers about traditional Japanese metalworking methods. Although the paper by Roberts-Austen written in 1884 mentions correspondence with William Tookey of the Osaka Mint,[56] the first mention in his work of Japanese metalwork in detail appears in 1886[39] and includes a reference to analysis of SHAKUDO and SHIBUICHI 'by Gowland.' Therefore these analyses must have been done by Gowland[60] while still in Japan and communicated to Roberts-Austen before returning to England. A later reference to Gowland by Roberts-Austen describes and shows sketches of Japanese metalwork provided to him by Gowland after returning to England. Although the personal correspondences between Roberts-Austen and Gowland (when Gowland was in Japan) have evidently been destroyed (see exception following), it may certainly be inferred that they had extensive dialogue about Japanese metalwork, both while Gowland was at the Osaka Mint, and after his return to England.

It does not appear that Roberts-Austen ever travelled to Japan, but by 1888 he did meet[58] with Japanese metalworkers in England, and by whom he was directly shown how to fabricate MOKUME. He indeed proceeded to make this material and provided it to others to make into repoussé metalwork.

In his textbook,[65] *An Introduction to the Study of Metallurgy*, Roberts-Austen juxtaposes figures showing flow lines in metals formed by a variety of mechanical deformation processes (as originally used and described by M. Tresca[66]) together with the figure describing the fabrication of mixed metal MOKUME that he had used in earlier publications. Comparison of the two figures in terms of deformation of flow lines in viscous flow is inevitable, and in the text Roberts-Austen combines the scientific principles demonstrated by these figures with many examples of real industrial technology to which they may be applied, i.e. the forging of iron rail, the coining of metals, and the interactions between a cutting tool and a work-piece during machining processes.

Interactions Between Roberts-Austen and Alfred Gilbert

Roberts-Austen's interest in colouring of metals ranged from the fundamental science, to the use of these colour effects in the medals whose manufacture he was associated with at the Royal Mint, to the colouring of bronzes for statuary, and to applications in the broad range of art metalwork. Roberts-Austen had a significant influence on the materials and techniques used by Alfred Gilbert, the leading English sculptor of his time.[62,63]

The first recorded use of aluminium for architectural application is the aluminium pyrimidal cap, cast by William Frishmuth at his foundry in Philadelphia, set atop the Washington Monument upon its completion in December 1884, and previously displayed in Tiffany's jewellery store in New York City in November of that year.[67] Frishmuth also had fabricated aluminium tubing for a surveyor's transit exhibited at the 1876 Centennial Exposition in Philadelphia. However, the statue of Eros, made by Sir Alfred Gilbert, commissioned in May 1886 (and completed in 1893), was the first and most significant aluminium public sculpture. Gilbert collaborated with William Chandler Roberts Austen on the casting of the various segments, and Gilbert kept close contact with the Society of Arts, which sponsored lectures on developments in aluminium and metallurgy. Penny notes that Gilbert attended lectures by William Anderson at the Society of Arts in March 1889 describing the Deville-Castner process developed in the previous year which made possible relatively cheap castings of the new metal.[62]

In reference 68 it is pointed out that Japanese metalwork such as the woodgrain texture MOKUME began to appear in Europe in the 1860s, and that JAPONISME was well entrenched in the statuette industry in Paris by 1880. Therefore Gilbert would have encountered these experiments during his Paris period, before being thoroughly schooled in Japanese bronze pickling techniques by Roberts-Austen as recorded by Dorment. Bluhm also points out[62, p.199] that Gilbert's statuette Charity (1899) reflects his experimentation with different alloys and patination methods to achieve colouristic effects, and that by 1888 he had met Roberts-Austen (the text describes SHAKUDO as a bronze with gold substituted for tin, and which could be patinated to a purplish colour by application of a caustic solution). Charity is a dark blackish-bronze except in the areas of flesh, where a pickling solution was applied.

Dorment gives a detailed description[62, p.83] of the Preston Mayoral Chain made by Gilbert (November 1888) and praised by Roberts-Austen, and indicates it to be of metal textured to imitate the graining effect found on Japanese bronzes. 'Particularly beautiful are the fish scales of the helmeted, double-tailed mermaid, and the imitation brick-work on the tiny castle and portcullis.', but[62, p.95] Dorment says Gilbert *did not* use Roberts-Austen's patination techniques for the Preston Mayoral Chain, as per Gilbert's own statements and in opposition to what Roberts-Austen reports in his papers (curious!). Dorment indicates[62, p.84] that 'Gilbert and Roberts-Austen certainly knew each other prior to December 1888', and that by the 1890s both worked side-by-side to obtain new and brilliant patinas in bronze. Dorment suggests[62, p.86] that before 1890 Gilbert had not used Roberts-Austen's methods, and[62, p.88] that the first use of these by Gilbert was the memorial panel for William Graham (1892). Dorment indicates that[62, p.95] at the request of Gilbert, Roberts-Austen read (1888) the paper 'On Certain Applications of Gold and Silver in Art Metal Work'[69] to the major meeting of the arts group in England. Finally, Dorment[60,62, p.86] quotes a letter of appreciation from Gilbert to Roberts-Austen which reads:

> My DEAR ROBERTS-AUSTEN – I can't say how truly sorry I am at being unable to support you at your lecture.
>
> * * *
>
> You know already how deeply interested I am in the subject of the 'Use of Alloys in Art Metalwork,' to understand that nothing short of necessity would prevent my being present on any occasion when there should be a chance of listening to you upon the subject. Your researches and discoveries connected with the use of precious alloys by the Japanese open up to me visions of an inexhaustible wealth of suggestions for the designer for metalwork in the future. I, myself, am longing most eagerly to employ some of the wonderful tints and actual information you have given me so generously.
> So much has been said of late years about texture and colour in works pertaining to the plastic art, but no suggestion, as far as I can see, for actually supplying these qualities for use in the arts in which they are most needed, viz., that of the goldsmith, have ever been made till now that you come forward with them. I jealously hope that it may fall to my lot to be the first to employ practically your discoveries, not for the sake of being first, but that by practical expression I may convey to you the best and most lasting form of gratitude for your efforts towards raising the beautiful art I love so much.
> Yours, very sincerely,
> ALFRED GILBERT.

The Avenue, 76, Fulham-road.
May 11, 1890

Roberts-Austen Ties Diffusion to Art Objects

Roberts-Austen's familiarity with Japanese methods of colouring of metals and of making mixed metal MOKUME also affected his scientific work. In his historic Bakerian Lecture[14] Roberts-Austen states that: 'It remains to be seen whether diffusion can be measured in solid lead at the ordinary temperature, and, with this object in view, cylinders have been prepared and set aside for future examination. In searching for evidence of diffusion in solid metals at the ordinary temperature, it will be well to examine certain alloys used in art metal-work by the Japanese, who often employ an alloy of copper containing a small proportion of gold (called SHAKUDO), which is soldered or welded in alternate layers with pure copper. The gold in the copper enables it to assume a beautiful purple patina when it is treated with suitable pickling solutions, which leave the pure copper of a red colour. In this way very singular banded effects are produced. Many of the specimens are centuries old, and I have attempted, by grinding away the existing patina and re-pickling the surface, to ascertain whether the widening of the coloured bands would show that, in the course of time, gold had diffused from the SHAKUDO layers and had passed into the copper. I believe that there is evidence that it does do so, but the enquiry is full of difficulty, and needs training in micrography, of which my friend, M. Osmond, is a master. We propose to study this part of the subject together, and I only allude to it here because, if diffusion occurs in copper, silver, and gold at the ordinary temperature, its results should be revealed in the products of this ancient oriental art.'

Despite extensive searching of the literature, and direct visit to the various institutions at which Roberts-Austen worked, no documentation of the completion of this experiment on diffusion in MOKUME has been found. Perhaps this can be explained by a later paper on diffusion by Roberts-Austen in 1900. In this later paper[15] Roberts-Austen writes that 'the amount of gold which would diffuse in solid lead at the ordinary temperature in 1000 years is almost the same as that which would diffuse in molten lead in a single day, provided no more gold is supplied in either case than can be held in solution. This will serve to show how important temperature is in relation to diffusion.' Furthermore, in his earlier paper,[14] Roberts-Austen also compared the diffusion of gold in lead, to that of gold in silver, and states that 'I am satisfied that the diffusivity of solid gold in solid silver at 800°C is the same order as that of gold in lead at 200°C. It would appear, therefore, that the melting points of the (matrix) metals have a dominating influence on the resistance offered to diffusion.' Thus I believe that Roberts-Austen came to understand the inability (at that time) to measure room temperature diffusion in MOKUME as he had originally proposed. However, in the spirit of the experiment proposed by Roberts-Austen, I would like to demonstrate the utility of measurement of interdiffusion in MOKUME materials, using modern analytical methods.[70]

Complex metal artifacts consisting of more than one material are often fabricated by either soldering or brazing, using a molten intermediary layer, or by solid state joining (diffusion bonding) without the use of a bonding aid. Traditional mixed-metal MOKUME is solid-state diffusion bonded and modern analysis (see below) shows that it most often does not use a solder intermediary material. In both types of fabrication, because the processes involved are controlled by the response of the particular materials to the temperature and time conditions during joining, it is often possible to determine major features of the fabrication history of such objects through examination with modern analytical tools.

Figure 3 shows the collar (FUCHI) that is found between the blade and the handle (TSUKA) of a Japanese sword. Although this piece of decorative metalwork is unsigned, it is typical of the styles fabricated during the early nineteenth century. This particular piece is crafted from thin alternating layers of copper and a copper–silver–gold alloy of the SHAKUDO type. Upon the formation of an oxide patina, the copper surface turns red, while the SHAKUDO turns a deep matte purple-black colour. The MOKUME pattern imitates a wood-grain or ink-floating-on-water texture, and in the specimen illustrated here it is used as the background to highlight a floating Paulonia plant fabricated from a solid piece of SHAKUDO and inlaid into the surface.

If a small area of the oxide patina is removed from the surface, or if a small metal sliver is extracted from an unobtrusive section, the sample may be examined using the focused electron beam of an electron microprobe analysis (EPMA) system. The electrons hit the specimen and produce X-rays which are characteristic of and can be used to identify the specific elements present in the solid material under the beam. The intensity of the X-ray signal is related to the concentration of the element detected and these intensities can be rapidly converted to wt% of that element, and a quantitative chemical analysis obtained. If the specimen is now moved at a fixed rate with respect to the focused electron beam by means of a motorised mechanical stage, the output intensities of the characteristic X-rays can then be used to produce a

Fig. 3 MOKUME FUCHI with Paulonia plant.

continuous record of composition vs. position as shown in Fig. 4a. For more precise work, the mechanical stage can be stopped in small increments and the output can be used to provide a profile obtained from discrete points along the specimen. In this manner, the SHAKUDO alloy was found[70] to contain 3.23% by weight (1.09 atom.%) of gold and 3.24% by weight (2.08 atom.%) of silver (the balance being copper), while the copper layers contained less than 0.1 atom.% silver as a residual element. It appears that these alternating layers are joined by solid state diffusion without the use of an intermediary joining material, as expected.

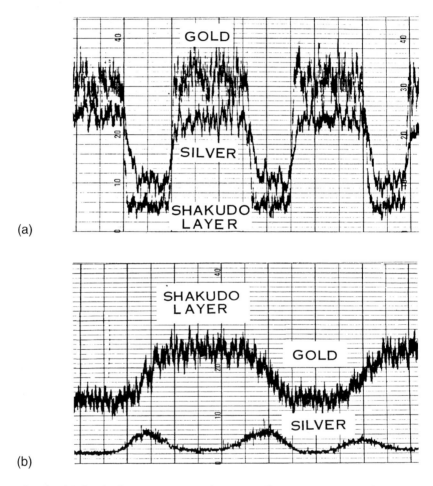

Fig. 4 (a) (top) Electron Microprobe trace of MOKUME piece shown in Fig. 3, showing characteristic gold and silver X-ray intensities as a function of position. (b) (bottom) Electron Microprobe trace of MOKUME dated late nineteenth century. Note peak in silver trace is at interface between copper and SHAKUDO. Specimen courtesy of C. S. Smith.

When two metals are joined together at an elevated temperature interdiffusion of atoms from both sides of the joint takes place such that the composition at any point depends upon the time and temperature, the geometric configuration and experimental constants for the two materials involved. The situation is more complicated if an intermediate compound forms between the joined metals, or if a ternary (three element) system rather than a binary system is considered. However, if the experimental diffusion constants have been predetermined, as they have been for the copper–gold–silver ternary system,[71] the composition profile for a given set of time-temperature conditions can be calculated and compared with the profile obtained by microprobe analysis. Moreover, for a situation where the two elements are dilute solutions in the third and where there is little effect of the presence of the third element upon diffusion of the other two (as is the case here), the two diffusing systems, i.e. copper–gold and copper–silver, can be considered independently. The experimentally determined profile is shown in Fig. 5 and is seen to match well with the computed profile at 800°C and 30 minutes for both the copper–gold and copper–silver systems. The calculated curves for 750 and 850°C are shown to be significantly above or below the experimental profiles, as would be the case with either shorter or longer times. The temperature (800°C) and time (30 minutes) appear to be consistent with good metallurgical practice for fabrication by solid-state diffusion.

The presence of a liquid intermediary layer could lower the temperature necessary for bonding. A composition profile for another MOKUME piece, dated late nineteenth century, is shown in Fig. 4b The presence of a thin silver-rich layer is apparent at the interface between each of the copper and SHAKUDO layers. This silver-rich layer could be produced by electroplating or by other coating methods and when heated would cause the local composition to reach that of the copper–silver eutectic, the lowest melting point composition in that binary system. A liquid phase would therefore appear at the joint at or below 770°C in the copper–silver–gold system causing rapid bonding; as the silver diffuses into the adjoining layers the local composition that corresponds to the eutectic composition melts, but further rapid diffusion causes the liquid layer to disappear after a short period of time. A sound joint could therefore be fabricated at lower temperatures and shorter times by using a 'disappearing liquid phase' rather than by relying on solid-state diffusion. This is common practice today.

With reference to Roberts-Austen's proposed experiments related to diffusion in MOKUME, the amount of interdiffusion caused by hundreds of years of contact at room temperature would be immeasurably small in comparison to the extent of interdiffusion measured at 800°C for 30 minutes. I believe that Roberts-Austen understood this as reflected by his comments,

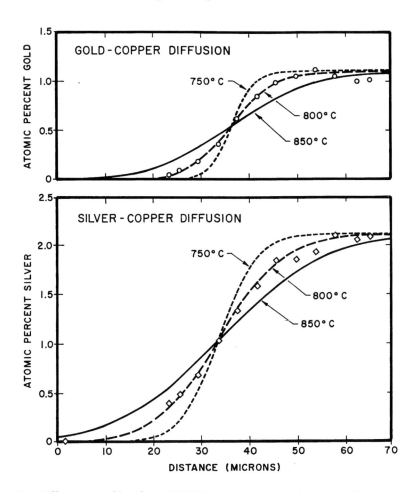

Fig. 5 Diffusion profiles for MOKUME piece shown in Fig. 3. Best fit curve (800°C) is shown superimposed on data; two other calculated profiles at different temperatures are shown for comparison. Time is fixed at 30 minutes.

and that he would be happy to see these present measurements performed in the context of a medium that he would enjoy.

Although Roberts-Austen performed his pioneering work on diffusion in solids at the very turn of the twentieth century, it was not until much later that his work was generally incorporated into the body of scientific literature. The view of solids in the early twentieth century questioned the permissability of a deviation of stoichiometry in alloy compounds and did not yet grasp the concepts of atomistic point defect formation. Thus a clear understanding of how solids could maintain continuously varying concentration gradients was not available. As pointed out by Mehl[72] it was not until 1915 that Von Ostrand

and Dewey[73] reconfirmed the results of Roberts-Austen. In 1932, Grube[74] demonstrated that the diffusivity in an alloy system varies with composition, and this then led Matano[75] to apply Boltzmann's solution to the case of varying diffusivity. It was not until the marker motion experiments of Kirkendall[76] that the point defect nature of diffusion was understood.

Roberts-Austen and the Royal School of Mines

As mentioned above, William Gowland and Roberts-Austen had many common interests focused on the mints, both in London and Osaka. The need for training of Japanese students in metallurgical skills was a critical need in the development of Japan during the Meiji Period. The Japanese government desired to send their best students abroad to study in countries with high quality institutions of learning, In the areas of mining and metallurgy, the focus schools became the German schools in Freiberg (with Ledebur at the Bergakademie) and Gottingen (with Tammann), and the Royal School of Mines in England.

Reeks[77] describes, in detail, the activities of Roberts-Austen at the School of Mines. She points out that Roberts-Austen succeeded John Percy, his professor, and that in turn he was succeeded by William Gowland, his student. This reference also indicates that the first student from Japan at the Scool of Mines was Nagayuki ASANO (1884–1889). In 1872 Roberts-Austen was instrumental in arranging for William Gowland to obtain the position of Assayer at the Imperial Mint in Osaka. Gowland and Roberts-Austen both encouraged students from Japan and other Asian countries to go to the Royal School of Mines.

In a letter[78] from Osaka, dated 1885, Gowland wrote:

> The Impl. Mint. Osaka. Japan
> 10th June 1885
>
> Dear Prof. Roberts,
>   Allow me to thank you very much for so kindly sending me your Cantor lectures. I have read them with great interest & they are now being eagerly devoured by my assistants & other mint officers. I am much pleased to hear your very favourable opinion of the Japanese students who have attended your lectures at the Royal School of Mines. I wish the government could be induced to send you some every session as there is the greatest need in Japan of men with a sound metallurgical training. . . .
>
>   Believe me to be
>   Yours very sincerely,
>   W. Gowland

It was obvious throughout his career that Roberts-Austen's professional life revolved around his attachment to the Royal School of Mines. Thorpe[11] writes that 'Roberts-Austen always cherished, as one of the most treasured memories of his life, the recollection of his early association with the Royal School of Mines. Although the Royal School of Mines is today incorporated with the Royal College of Science, a fusion of which Roberts-Austen entirely approved and which he loyally supported, his colleagues on the council of the school were more or less dimly conscious that deep down in his mind, "at the back of his head," as the saying goes, he was still apt to regard the school as a corporate entity with a separate existence, with all the power, privileges and prestige which it enjoyed as a separate entity in his old Jermyn Street days. There was probably no one position he coveted more than its chair of metallurgy, and no incident in his career which gave him a greater sense of pleasure and satisfaction than his appointment, in 1880, to that chair in succession to the late Dr. Percy. . . .

How loyal he was to the school, how affectionately he guarded its interests and how he studied to enhance its usefulness I, who was his colleague on the council of the Royal College of Science for upwards of nine years, desire now to bear testimony. It was the wish of his heart, had he been spared, that, after his retirement from the Mint, he might spend his remaining years, or so many of them as the regulations of the Department would have allowed him, to spend, in its service. It was possible that he cherished the hope that the erection of the new buildings on the other side of Exhibition Road might have afforded him the opportunity he had long desired, that of creating and equipping a metallurgical laboratory which should be worthy of this country and of an Empire whose sons are engaged in metallurgical work in almost every part of the globe. But if this was not to be, he has at least erected a monument to himself in the record of his past achievement; in the thoroughness and fullness of his teaching; in the scientific enthusiasm with which he sought to lay bare and illumine the problems of physical metallurgy. During the two-and-twenty years he held his chair, he trained a succession of men holding important positions at home and in many parts of the world, who are grateful to him for the stimulating influence of his teaching. . . .'

Summation

If there is such a persona as 'the metallurgical hero', Professor William Chandler Roberts-Austen is its epitome. He was a man of many facets and successfully linked together his interests in science with his aesthetic sensitivity in art. As just one example of the many areas of scientific research to which he made major contributions, is his work on the accurate quantitative measurement of solid state diffusion in metals, decades before anyone else. At a time when new art-styles were in vogue in Europe and America, and new metals such as aluminium were just coming into use, he played a significant role in fostering interaction between the sciences and the arts, and between the West and Asia. Roberts-Austen also had an abiding love for the Royal School of Mines, and he connected all aspects of his life to it as an institution.

References

1. M. Kiowa, 'Historical Development of Diffusion Studies', *Metals and Materials*, 1998, **4**(6), 1207–1212.
2. M. Kiowa, 'Diffusion in Materials-History and Recent Developments', *Trans. JIM*, 1998, **39**(12), 1169–1179.
3. F. X. Kayser and J. W. Patterson, 'Sir William Chandler Roberts-Austen: His Role in the Development of Binary Phase Diagrams and Modern Physical Metallurgy', *J. Phase Equilibria*, 1998, **19**(1), 11–18.
4. C. Tujin, 'On the History of Models for Solid-State Diffusion', in *Defect and Diffusion Forum*, 1997, **143–147**, 11–18.
5. L. W. Barr, 'The Origin of Quantitative Diffusion Measurements in Solids: A Centenary View', in *Defect and Diffusion Forum*, 1997, **143–147**, 3–10.
6. C. Tujin, 'On the History of Solid-State Diffusion', in *Defect and Diffusion Forum*, 1997, **141–142**, 1–48.
7. J. C. Chaston, 'Gold and the Beginnings of Physical Metallurgy – The Pioneer Work of Roberts-Austen', *Gold Bulletin*, 1977, **10**(1), 24–26.
8. S. W. Smith, 'Sir William Chandler Roberts-Austen', *Metallurgia*, 1943, **27**, 169–177.
9. S. W. Smith, *Roberts-Austen: A Record of His Work*, Charles Griffin, 1914.
10. T. E. Thorpe, Sir William Chandler Roberts-Austen (Obituary), *Nature*, 4 December 1902, **67**, 105–107.
11. T. E. Thorpe, 'Sir William Chandler Roberts-Austen' (Obituary), *Proceedings of the Royal Society*, 1905, **75**, 192–198.
12. H. Schliemann, *Ilios: The City and Country of the Trojans*, John Murray, 1880.
13. H. Schliemann, *Troja: Results of the Latest Researches and Discoveries on the Site of Homer's Troy*, John Murray, 1884.
14. Bakerian Lecture: 'On the Diffusion of Metals', in *Philosophical Transactions of the Royal Society of London*, Series A, 1896, **187**, 383–415.
15. 'On the Diffusion of Gold in Solid Lead at the Ordinary Temperature', in *Proceedings of the Royal Society of London*, 1900, **67**, 101–105.
16. 'On the Rate of Diffusion of Carbon in Iron', *Journal of the Iron and Steel Institute*, 1986, **XLIX**, 139–143, and Plate V.
17. 'On Molecular Unrest in Solids', *Proceedings of the Royal Philosophical Society of Glasgow*, 1900, **31**, 152–166.
18. 'On the Melting Points of the Gold-Aluminium Series of Alloys', *Proc. Royal Soc. (London)*, 1891–2, **50**, 367–368.

19. J. W. Richards, *Aluminium: Its History, Occurence, Metallurgy and Applications, Including its Alloys*, M. C. Baird, 1896, 501.
20. W. C. Reed, *The San Francisco Daily Herald*, December 2, 1855 (written April, 1855), referenced in Howard F. Van Zandt, *Pioneer American Merchants in Japan*, (Lotus Press Ltd, 1981, 130–133.
21. Raphael Pumpelly, Notes on Japanese Alloys, *American J. of Science*, 1866, **42**, 43–45.
22. William Hosley, *The Japan Idea: Art and Life in Victorian America*, Wadsworth Atheneum 1990, Footnote 92, p. 203. (See also *Japanese Bronzes in the Art Gallery of Edward Greey*, Privately printed, New York, 1888.)
23. *Ibid*, p. 43.
24. F. G. Notehelfer, *Japan Through American Eyes – The Journal of Francis Hall (1859–1866)*, Princeton University Press, 1992, 77, 85, and 91.
25. Gloria Ortiz de Cox, *Tiffany and Company Silver: The Japanese Influence 1869–1891*, Master's Thesis, Cooper-Hewitt Museum and the Parsons School of Design, May 1988, 4–5.
26. H. Hawley, Tiffany's Silver in the Japanese Taste, *Bulletin of the Cleveland Museum of Art*, October 1976, 236–245.
27. Daniel Cohen, *Charles Tiffany's 'Fancy Goods' Shop and How it Grew*, December 1987, Smithsonian, 53–60.
28. Isabelle Anscombe, Knowledge is Power, *The Connoisseur*, 1979, **201**(807), 54–59.
29. H. Hawley, A Falize Locket, *Bulletin of the Cleveland Museum of Art*, September 1979, 241–245.
30. Toyojiro Hida, *Kiritsu Kosho Kaisha: The First Japanese Manufacturing and Trading Co.*, Kyoto Shoin, 1987.
31. William Hosley, *The Japan Idea*, 1990, 32.
32. F. G. Safford and R. W. Caccavale, 'Japanese Silver by Tiffany and Company in the Metropolitan Museum of Art', *Antiques*, 1987, **132**(4), 808–819.
33. *New York Daily Tribune*, December 10, 1878.
34. *Montreal Herald*, November 6, 1878.
35. Charles H. Carpenter, Jr, *Tiffany Silver*, Dodd, Mead & Co., 1978, 180–211, and a note on Mixed Metals, 240.
36. G. A. Audsley, *The Ornamental Arts of Japan*, Charles Scribner, 1984.
37. *The Jewelers Circular and Horological Review*, December, 1874, **5**, 160; and January, 1875, 181–182.
38. *The Practical Magazine*, 1875, **5**, 101, 344; and 1876, **6**, 83.
39. W. C. Roberts-Austen, 'The Colours of Metals and Alloys', *Nature*, December 2 1886, **35**, 106–111.
40. Charles H. Carpenter, Jr and Janet Zapata, *The Silver of Tiffany & Co.: 1850–1987*, Exhibition Catalog, Museum of Fine Arts, Boston, 22, 1987.
41. *Ibid*, p. 40.
42. John Loring, *Tiffany's 150 Years*, 1987, Doubleday, 102.
43. Judy Rudoe, *Decorative Arts 1850–1950: A Catalogue of the British Museum Collection*, British Museum Press, 1991, p. 111–112.
44. Arthur H. Hiorns, *Metal Colouring and Bronzing*, Macmillan, London, 1892.

45. Carol P. Adil and H. A. DePhillips, Jr, *Paul Wayland Bartlett and The Art of Patination*, 1991.
46. Charles H. Carpenter, Jr, Tiffany Silver in the Japanese Style, *The Connoisseur*, January 1979, **200**(803), 42–47.
47. Lucien Falize, L'Art Japonais, *Revue des Arts Décoratifs*, **III**, 1883, 359 (writing under the pseudonym of M. Josse).
48. G. De Leris, *Revue des Arts Décoratifs*, 1883, **IV**, 119.
49. Samuel Bing, *Artistic America, Tiffany Glass and Art Nouveau*, 1970, MIT Press, 156.
50. *A Walk Through Tiffany's* National Repository, July 1878, 17.
51. Charles H. Carpenter, Jr, *Gorham Silver: 1831–1981*, 1982, Dodd, Mead.
52. Charles H. Carpenter, Jr, Gorham's Metalwares: The Japanese Influence, *Art & Antiques*, September/October 1982, 62–69.
53. William Hosley, *The Japan Idea*, 1990, 125–137.
54. Charles James Jackson, *English Goldsmiths and Their Marks*, 1921, London, 24.
55. Report from the Select Committee on Gold and Silver (Hall Marking) and Minutes of Evidence (1878), see paragraphs 379–380 in Minutes of Evidence. Also D. A. Hanks and J. Tother, 'Metalwork: and Eclectic Aesthetic' in *In Pursuit of Beauty*, Doreen Bolger Burke *et al.* eds. Metropolitan Museum of Art, New York, 1986..
56. W. C. Roberts-Austen, Alloys Used for Coinage–II, *J. Soc. Arts*, July 25 1884, **32**, 835–847.
57. W. C. Roberts-Austen, Alloys II, *J. Soc. Arts*, October 19 1888, **36**, 1125–1133.
58. W. C. Roberts-Austen, Alloys III, *J. Soc. Arts*, October 26 1888, **36**, 1137–1146.
59. W. C. Roberts-Austen, Alloys IV, *J. Soc. Arts*, November 10 1893, **41**, 1022–1030.
60. W. C. Roberts-Austen, The Use of alloys in Art Metalwork, *J. Soc. Arts*, June 13 1890, **38**, 690–701.
61. J. Earle, Japanese Art and Design (The Toshiba Gallery), *Victoria and Albert Museum*, 1986, 202, 203.
62. R. Dorment, *Alfred Gilbert*, Yale University Press, 1985, 111.
63. R. Dorment, *Alfred Gilbert-Sculptor and Goldsmith*, 1986, Royal Academy of Arts, London.
64. K. M. Weigert, Constitution and Properties of Ag–Cu–Zn Brazing Alloys, *Trans. AIME*, 1954, **200**, 233–237.
65. W. C. Roberts-Austen, *An Introduction to the Study of Metals*, 1904, 52–56.
66. M. Tresca, *Proc. Inst. Mech. Engineers*, 1867, 114.
67. G. J. Binczewski, The Point of a Monument: A History of the Aluminum Cap of the Washington Monument, *Journal of Materials (JOM)*, November 1995, **47**, 20–25.
68. A. Bluhm, *The Colour of Sculpture*, 1996, Van Gogh Museum, Amsterdam.
69. W. C. Roberts-Austen, On Certain Applications of Gold and Silver in Art Metal Work, *Transactions of the National Association for the Advancement of the Arts and its Application to Industry*, 1888, Liverpool Meeting, 119–121.
70. M. R. Notis, 'Study of Japanese MOKUME Techniques by Electron Microprobe Analysis', *MASCA Journal*, 1979, **1**(3), 67–69.
71. T. O. Ziebold and R. E. Ogilvie, Ternary Diffusion in Copper-Silver-Gold Alloys, *Trans. AIME*, 1967, **239**, 942–953.

72. Robert Franklin Mehl, 'A Brief History of the Science of Metals', *AIME*, 1948, 32.
73. C. E. Van Ostrand and F. P. Dewey, *US Geological Survey Professional Paper*, 1915, 95.
74. G. Grube, *Zeitschrift für Elektrochemie*, 1932, **38**, 797.
75. C. Matano, *Japanese Journal of Physics*, 1933, **8**, 109.
76. A. Smigelskas and E. Kirkendall, *Trans. AIME*, 1947, **171**, 130.
77. M. Reeks, *The Register of the Associates and Old Students of the Royal School of Mines and History of the Royal School of Mines*, 1920, 138–142.
78. Public Records Office, Kew, File: Japanese Mint, Gowland, 1885.

The Nuffield Research Group in Extraction Metallurgy (NRGEM): Founding and Evolution

C. B. ALCOCK
Emeritus Professor, Department of Metallurgy, University of Toronto, Canada

ABSTRACT
This account of the history of the origins of the Nuffield Research Group in Extraction Metallurgy, and its subsequent development into a world-class research activity in the Metallurgy Department of the RSM, covers the years 1950 to 1976. The group began life, under the direction of F. D. Richardson, as predominantly a laboratory researching high temperature thermodynamics, devoted to understanding metal-producing reactions. With the addition of the John Percy Group in 1965, it evolved to also consider the kinetics, as well as thermodynamics, of reactions. Parallel to this area, solid state research was inaugurated principally to supply information to the British nuclear power industry. These research studies provided a basis for the development of the Materials Chemistry contribution to the undergraduate curriculum of the Department. After Professor Richardson's retirement in 1976, the coherency and integrity of the group dissolved, and the research effort was carried on under a number of separate research supervisors. The work of the many research students in the group, and the contributions of a number of academic visitors, mainly from the Commonwealth, has had international impact and continues to influence the content of the Department of Materials undergraduate programme.

Introduction

Sir Charles Goodeve, at that time Director of The British Iron and Steel Research Association (BISRA), organised a Faraday Society Discussion Meeting in 1948 devoted to 'The Physical Chemistry of Process Metallurgy'.[1] The Head of the Chemistry Department at BISRA was Dr F. D. Richardson, a former post-graduate student of Sir Charles at University College London, and a member of Sir Charles' Weapons Development group of the Royal Navy during the Second World War. A paper presented at that conference, co-authored by Dr Ellingham of the Chemistry Department of the Royal College of Science and Professor Dannatt Head of the Department of Metallurgy of the Royal School of Mines, discussed the application of the

Ellingham Gibbs energy diagram to the selection of metal-producing processes. This was the first presentation of these concepts to an international audience.

In 1949 Dr Richardson (Fig. 1) won an award from the Nuffield Foundation, which appointed him as a Nuffield Fellow and enabled him to set up an academic research group at the Royal School of Mines. The Faraday Society meeting had shown the industrial value and strength of the research group in the thermodynamics of iron and steelmaking at MIT, headed by Professor John Chipman, of which there was no corresponding group in the UK. This US group had carried out a lot of fundamental work using large molten metal samples, of the order of a few kilograms, employing induction heating. Richardson's group had no funds for induction heating equipment at that time, and so a tradition was begun of using small samples of the order of a few grams for high temperature metal-making studies.

In describing the development of the group from 1950 to 1976, when Professor Richardson retired, I shall only have space to mention the new techniques and theories, which were developed by the group, and must apologise for any omissions of work, which came later. My aim here is to show when the group made the original contributions to its Science which established it as a major player in the field.

Fig. 1 Portrait of Professor F. D. Richardson (circa 1973).

(1950–1955)

When I, together with O. H. Gellner, arrived in 1950, there were five graduate students at work on PhD studies. Homemade equipment was the order of the day, winding furnace heating elements with asbestos insulation, and making glass gas-handling equipment was the job of everyone. Figure 2 shows the main laboratory of the Group during this early period.[2] Tutorials from the group leader on the construction of glassware using soda-glass, which was all that was available, showed Richardson's skill at experimental work. Mention should also be made of the help through the free supply of precious metals that was given by the Johnson-Matthey Company, not only then, but also over the years. By early 1953, the group consisted of Dr Richardson as Director, two members of academic staff (J. W. Tomlinson and myself) and eight research students.

These first studies included:

- The measurement of PbO activities in lead silicates using a gas transportation method involving PbO transfer through the gas phase to liquid lead samples, the oxygen content of which was determined by reduction in the solid state. A thin surface layer of the oxide, which formed on the

Fig. 2 General view of the main laboratory of the Nuffield Research Group in Extraction Metallurgy (circa 1953).

lead samples, was removed by surface reduction using a hydrogen plasma activated by a Tesla coil.
- The solubility of copper atoms in liquid silicates was determined in another study in which the corrosion patterns of the slag-gas and the slag-metal phase boundaries were clearly displayed. Subsequent studies of the solubilities of lead, silver and gold in the same slags showed that these increase as the metal vapour pressure increases at any given temperature, but decrease as the metal atoms radius increases at a given vapour pressure. Measurements of the solubility of lead in contact with lead alloys of a known activity coefficient showed a direct relationship between the solubility of lead in the slag and the lead activity in the alloy, and hence that the metal was dissolved in the atomic form.
- A classical counterpoint to the MIT studies was a measurement of the effect of chromium on the activity coefficient of carbon in solution in liquid iron. A graphite resistance furnace was designed and constructed in the laboratory, which for the first time, took the group's range of studies up to 1600°C.
- The first solid-state study carried out by a member of the group was of the oxidation of cobalt metal, which was a test of the oxidation theory of Carl Wagner. Using radioactive cobalt, the predominant diffusion of cobalt through the oxide and the predicted distribution of a thin radioactive cobalt layer on a sample of the metal through a subsequently formed CoO layer were confirmed.

These early days showed the significant changes, which the new research activity was instrumental in bringing about in the Royal School of Mines. One classic example of this was the tale of Bill Dennis and the Titfield Thunderbolt. Because it was traditional that the doors of the RSM were locked at 9 p.m., and no keys could be issued to research students, it was necessary for group members to find a less well-known path to leave their studies late at night, which was often necessary in long-duration equilibrium studies. One well-established route was through the basement of the then-existing Imperial Institute, which had a small gate leading out to Imperial Institute Road. Bill left his work at about 4 a.m. one morning and took this route out of the Department. On emerging from the Institute, he was staggered to find himself in a blaze of light, in which a group of men wearing Victorian clothes could be seen pulling a life-scale model of the Stephenson 'Puffing Billy' down the steps at the entrance to the Institute. To his credit, Bill appeared the next morning ready to regale us with his adventure which was taken as a sign of overwork and fatigue. The explanation for this experience was to be seen on the local cinema screens about a year later with the title 'The Titfield Thunderbolt', and was not a ghostly return to the early days of the Institute!

(1955–1960)

The next phase in the group's experimental development was aided by the availability of industrial equipment, such as furnaces and temperature controllers, together with homemade gas trains made in Pyrex glass. In 1955 further funds were obtained from the Nuffield Foundation and the Science Research Council, which enabled the group to acquire a high-frequency induction furnace and also to begin work on molten salt studies. The new furnace was not used for heating large metal samples, but to make small samples of master alloys, and to develop gas-liquid research using levitated metal drops.

In 1957, Richardson was appointed the Professor of Extraction Metallurgy, continuing as Director of the Group, and the revolution in this field at the RSM achieved considerably more momentum. The Nuffied Foundation continued to support a Nuffield Fellowship which was held by Dr J. W. Tomlinson. Some typical studies at this time were:

- The measurement of the sulphide and sulphate capacities of a range of molten silicate solutions, which followed Richardson's work at BISRA, and established a pattern of behaviour for many industrial slags. These capacities relate the uptake of sulphur by the slags from an environment with a defined sulphur and oxygen potential, and are thus of importance in metal-refining processes.
- The measurement of the solubility of calcium and magnesium in their molten halides was parallel to the work of Bredig in the Argonne National Laboratory on the solubilities of the alkali metals in their molten halides. This introduced the concept of electronic semiconduction in these ionic melts. From the extraction metallurgy point of view, it was important to establish the corresponding information for the Group IIA elements because of their important industrial role in the reduction of refractory metal halides. It was found that the solution of these metals into their chlorides led to the formation of complex Ca_2^{2+} and Mg_2^{2+} ions, which is quite different from the behaviour of the alkali metal solutions.
- Regular and quasichemical models of dilute ternary solutions were derived from experimental results and published data from the literature. These models were based on the assumption of composition-independent nearest-neighbour interactions only, leading to an equation for the concentrated alloys and the subsequent computer-aided calculations for ternary and more complex systems. Such calculations are now used in order to reduce the amount of experimental measurements required to establish the equilibrium behaviour of multicomponent industrial alloys.

- A number of studies in the solid state were carried out involving metal sulphide – oxide equilibria and sulphate decomposition, and sulphide ion diffusion in the cubic Co_9S_8, both of which required the development of new experimental methods.

(1960–1970)

This decade saw a considerable growth in the size of the group, and its scope of research. This was due in part to a building programme in the RSM, which provided new laboratory space, and enabled the introduction of an undergraduate practical laboratory based on the group's research advances, now called the Richardson Laboratory. Richardson also obtained further financial support from the Nuffield Foundation to form the John Percy group, which was named after the first Professor of Metallurgy at the RSM.

The John Percy Group Process Metallurgy was formed to extend studies to the kinetic aspects of metal-gas and metal-slag interactions, together with the science of the new pneumatic processes such as the Linz-Donawitz process which had revolutionised steel refining by the introduction of jets of oxygen to the technology, with the consequent formation of bubbles in the reacting phases. Much of this approach followed classical chemical engineering concepts in emphasising the importance of dimensionless numbers as a base for experimental design. Thus instead of working at 1600°C with liquid steel and silicate slags, an equally valid study of the process mechanism could be made at room temperature with liquid mercury and water-glycerine mixtures to represent the mechanical behaviour of these phases. These considerations of mass transfer in terms of diffusion- or interface-control augmented the analysis of metal-making processes when added to the already understood thermodynamic factors.

Another development in the Group's area of interest was influenced by the need for information and control techniques in the growing nuclear power industry, where there were a number of questions to be resolved concerning the future directions in the design of power reactors. This area led to an increase in the scope of the solid-state studies, which had previously been oriented towards solid-state metal-extraction processes. Representative of the studies which were made during this period are the following:

Extraction Metallurgy

The main thrust was in understanding the kinetics and mechanisms of metal deposition from molten salts. A distinction was drawn between the reversibility without activation overpotential for metal deposition on metal electrodes, as contrasted to gas ions, such as O^{2-}, which is accompanied by an activation overpotential. Chronopotentiometry was used to examine the extent of diffusion-control and the kinetics of transfer of metal ions from the electrolyte to the solid electrode surface as a function of morphology.

The liquid metal electrode behaves as a reversible, ideal electrode during deposition, and simple diffusion theory can be used to calculate the rate of deposition. Complications arise when a liquid metal is undergoing an exchange reaction with a molten salt, such as the sulphur-oxygen exchange, which occurs during the desulphurisation of metals by liquid slags. Here the surface tension of the liquid metal is significantly reduced by sorption of sulphur on the surface of the metal, and as the exchange occurs, the surface tension varies across the interface. This leads to interfacial turbulence, which was demonstrated in a model experiment in which manganese and iron were exchanged between a manganese-mercury liquid metal phase, and an aqueous solution of ferric nitrate. The interfacial turbulence was observed by Schlieren photography.

Experiments on gas-metal interaction kinetics provided information for the reaction of levitated metal drops with flowing gases in terms of a mass-transfer coefficient. These results could be directly applied to the analysis of pneumatic processes, which normally produce metal spray.

The bubbles that are formed in a metallic phase pass into a supernatant slag phase taking the form of spherical caps rather than the ideal spherical shape, due to the viscosity of the slag phase. A mass-transfer coefficient equation that was in the literature for mass transfer from a spherical cap-shaped bubble to a surrounding liquid was found to describe quite accurately the transfer of oxygen from pure oxygen bubbles to liquid silver. When a gas bubble emerges from a molten metal into a slag phase, the spherical cap bubble is initially covered with a thin layer of liquid metal. This film breaks as the bubble lifts into the slag phase, releasing small droplets of metal into the slag. The spherical-cap bubble carries a train of very much smaller spherical bubbles, which also take part in the transfer process. The John Percy Group obtained experimental data for this process.

Mass transfer coefficients for the transfer into a liquid metal surface across which a chemically active gaseous phase was directed were also measured as a function of the gas velocity, the chemical nature of the liquid metal, and the dimensions of the container, which determine the flow pattern in the liquid.

The structural nature of polymeric phosphate melts was obtained by dissolving the sodium salts in water and making a chromatographic analysis of the separated polymers. The calcium and zinc phosphates were analysed by adding a complexing agent to the water before dissolution. The size distribution was found to be smaller than that predicted by the Flory distribution theory. This is because the equilibrium constant for the concentrations of sequential polymers, e.g.

$$2P_n = P_{n-1} + P_{n+1}$$

is not unity as assumed in deriving the Flory equation, but has a finite value due to the heat of mixing of the polymers.

Materials Chemistry

Studies in this area were developed to some extent from the earlier work in the solid state, and led to the definition a new area of research, quite distinct from the mission of the extractive metallurgy group, in which ceramics now played a major role. A number of new experimental techniques were developed, which included the following:

- The parallel development of the gas transportation method at atmospheric pressure and the Knudsen effusion method in high vacuum for the measurement of vapour pressure made possible a comparison of the two methods in the same laboratory. It was found that the extrapolation of the gas transportation technique resulted to zero flow rate maximised the thermal diffusion effect which led to erroneous results for the vapour pressure. Most of the literature values for gas-solid equilibria were obtained from this extrapolation. After this was known, it was possible to get results for the vapour pressure of a substance by these two procedures, which were in agreement for pure metals such as copper and silver. Measurements were then made of the formation of volatile oxides by the platinum-group metals using the gas transportation method, and the stability of a number of uranium and thorium metallic compounds with elements in Groups III and IV of the Periodic Table using the Knudsen method.
- At MIT Kiukkola and Wagner (1957) used the results of much earlier work by Nernst to show that oxygen potentials could be measured by the use of zirconia-based solid electrolytes in an electrochemical cell. This work was followed in the RSM with a determination of the range of oxygen pressures over which these and the corresponding thoria-based solid solutions could be used as electrolytes. The zirconia electrolytes could be used successfully from atmospheric down to about 10^{-15} atmos. at 1000°C, and the thoria-based electrolytes were used to make a complete study of the Nb–O, Ta–O and the Mn–MnO systems at about 10^{-25} atmos. at the same temperature.
- An electrochemical method for the determination of the oxygen content of liquid metals using zirconia or thoria solid electrolytes to measure the

oxygen potential of the solution, and coulometric titration to vary precisely the oxygen content of a liquid metal sample was developed to measure the thermodynamics of oxygen dissolved in lead and tin. The results for liquid lead solutions confirmed the values originally obtained in the earlier vapour transport study. A zirconia electrolyte was used to measure the thermodynamics of oxygen dissolved in liquid copper, and the thoria electrolyte was used to measure and control the thermodynamics of oxygen solutions in liquid sodium, which was a necessary monitoring tool for the Dounreay fast nuclear reactor.

A number of subsequent studies were concerned with the application of the coulometric titration procedure to the determination of non-stoichiometry in solid oxides such as NbO_2, TiO_2 and MoO_2. This last oxide plays an important role in the microscopic determination of the degree of burn-up in nuclear fuels. Other applications included the control of the oxygen content of gases, as well as to some evaluation of other potential solid electrolytes in the perovskite and pyrochlore phases.

Data were needed for the evaporation rates of high-melting point ceramics which were candidate electrodes in gaseous MHD energy converters, and the construction of a tungsten-rod resistance furnace incorporating a torsion-balance apparatus which allowed measurements to be made up to 2300°C. A similar furnace was built into the beam source of a time-of-flight mass spectrometer, which was used in studies of alloy thermodynamics.

In order to understand the role of cation diffusion in nuclear fuel properties, measurements were made of the self-diffusion coefficient of uranium and thorium in their dioxides and solid solutions containing the natural distribution of isotopes. A solid state energy spectrometer was used to measure the energy spectrum of the alpha-ray emission of samples, after a diffusion anneal, which were originally coated with a very thin layer containing more radioactive uranium (U^{233}) or thorium (Th^{230}) isotopes before the diffusion anneal. Since the uranium oxide has a wide range of non-stoichiometry, the measurements were made over a range of compositions using an oxygen-potential controlled gas phase. It was possible from these measurements to separate the grain boundary from the volume diffusion coefficient. The inability to do this had vitiated the results of earlier measurements.

This technique was also used to measure the volume-controlled diffusion of thorium into nickel from the oxide, as a function of the oxygen potential of the gas phase. Subsequently an electron microscope was used to determine the Ostwald ripening of thoria dispersed in nickel, and the rate of this process was found to be slower than the diffusion rate of thorium or oxygen in the nickel matrix, indicating either interface control of transfer from the dispersed phase, or divacancy migration of thorium and oxygen atoms in the matrix.

Another technique was developed to measure oxygen transfer from a gaseous phase and the subsequent anionic diffusion of oxygen, into solid oxide electrolytes and potential electrode materials for solid oxide fuel cells. The exchange kinetics of oxygen-containing gases labelled with O^{18} can be separated into the interface exchange rate, and the subsequent volume diffusion within the solid oxide, showing the rate-determining step. This is important to know when assessing the potential performance of a fuel cell assembly. The technique was subsequently applied to SIMS analysis, which greatly enhanced the sensitivity of the method.

A mechanism for the results of measurement of the corrosion of nickel by SO_2/O_2 mixtures, in which sulphide particles were found at the metal/oxide interface, as opposed to the cases of copper, cobalt and iron, was elucidated involving sulphate formation and the grain-boundary diffusion of SO_2 through the oxide. As the gas approached the metal-oxide interface the decreasing oxygen potential led to an increased sulphur potential, and hence to sulphide formation.

Towards the end of this decade the original techniques, which the group had developed, were used in a wide range of studies covering many of the important high temperature chemical properties of materials, both thermodynamic and kinetic. There began to be a clear division of objectives between those involved in metal-extraction processes, the original mission, and the contributions to the rapidly developing field of Materials Science, in which the solid state studies principally were concerned. This new interest contributed to the introduction of an undergraduate Materials Science programme into the departmental curriculum to which members of the Nuffield group and the results of earlier research work supplied the materials chemistry content. The extractive metallurgy section of the group was instrumental in the introduction of an MSc course in the Department.

In 1973 an international conference was held in the Department to honour Professor Richardson, and the book published as *The Richardson Conference* recalls many aspects of the group's work, and the international friends, which the group had made over the years.[3] In 1974, Richardson's book *The Physical Chemistry of Melts in Metallurgy*, a two-volume work which contains references to much of the work of the NRGEM was published by Academic Press.[4] In 1976 Professor Richardson retired, ending a quarter century of outstanding contributions to the growth of the new RSM.

Conclusion Fifty Years On

This account of the nucleation and growth of a research group which placed British science in this field of High Temperature Chemistry in a leading position in the contemporary world, is dedicated to the memory of the group's leader and inspiration, Professor F. D. Richardson, and to the many scientists who based their future careers on what they learned, and what they contributed, in those days at the RSM. During this quarter century many students and already-established research workers and academics from all over the world, added to the scientific developments in the group.

It is also necessary to pay tribute to Government sources of funding during that time, but especially to the generous donations of the Nuffield Foundation, which demonstrated the importance of the application of private capital to someone of proven scientific ability with a new and untried concept, without the input of endless committees.

References

1. Report of Faraday Society Discussion Meeting on *The Physical Chemistry of Process Metallurgy*, 1948.
2. *Royal School of Mines Journal*, Plate VII, No. 2, 1953.
3. Jeffes and Tait (eds), *Proceedings of the Richardson Conference*, 1973.
4. F. D. Richardson, *The Physical Chemistry of Melts in Metallurgy*, Academic Press, 1974.

The Early Years of Transmission Electron Microscopy at RSM

PETER R. SWANN
Jumby Bay Island, Antigua

Introduction

Initially, the transmission electron microscope was not widely used as a research tool even though its resolution greatly exceeded that of the light microscope. This was because the ultra-thin specimens it required were so difficult to prepare that most researchers gave up trying to get results. This situation changed in the mid 1950s when the ultramicrotome and electro thinning techniques were developed to prepare biological and metal specimens respectively. Specimens of insulators and semiconductors could not be studied until about 1957 when the ion milling technique was invented.

One of the early researchers who worked on improving the electron microscope was Dennis Gabor. He emigrated to England from Hungary in 1933 and devised many new products for the electrical industry before using the electron microscope to invent holography in 1947. His technique has since become very important in other fields but has not attracted the attention it deserves in electron microscopy. Dennis Gabor joined the Imperial College of Science & Technology in London on 1 January 1949, first as a Reader in Electronics, and later as Professor of Applied Electron Physics, until his retirement in 1967. In spite of Gabor's presence at ICST only the Geology Department showed any interest in electron microscopy. This department had a Philips microscope that was used to search for microfossils in rocks.

The invention of the new specimen thinning techniques catalysed the production of commercial electron microscopes. Of these, the legendary Siemens Elmiskop 1 stood out above all others, because it excelled in the design of its electron gun and its double condenser lens which gave electron microscopists the intensity of illumination they required. It was this instrument that opened a window to physicists and metallurgists that can only be described as the 'golden era' of electron microscopy. This era lasted five years and started in 1956 when the Cambridge group, Hirsch, Horne and Whelan[1] and Bollman[2] saw features in aluminium and stainless steel specimens that they claimed were line defects in the crystal lattice known as dislocations (see Fig. 1).

54   Materials Science and Engineering: Its Nucleation and Growth

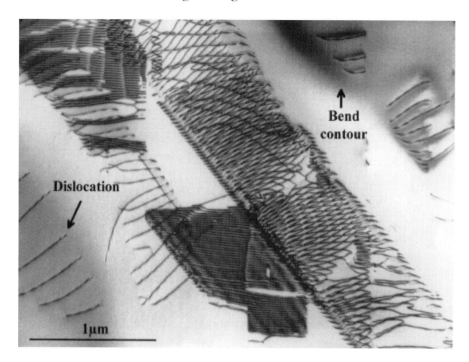

Fig. 1 Dislocations on slip planes in an Al 8%Cu alloy. Electron micrographs such as this convinced skeptics in the 1950s that dislocations do exist and that they are the fundamental unit of the plastic deformation process.

In 1956, the whole concept of single atom line defects as the fundamental vehicle for plastic deformation was not completely accepted. However, the actual observation of dislocations in the electron microscope left no doubt about their existence and this was a major breakthrough. It was not long before other types of crystal defects were observed such as stacking faults (see Fig. 2), atomic vacancy clusters and lattice strain fields around precipitates. Thus started a research revolution in which all kind of material properties were correlated with their electron microstructures. What surprised most researchers at the time was that the electron microscope could reveal atomic defects in metals when it did not have that level of resolution. However, the Cambridge group understood that the short deBroglie wavelength of high energy electrons makes their propagation through a crystal very sensitive to bending of the lattice planes and that it was the long range strain field around the dislocations that was being observed rather than the atomic disturbance at the dislocation core itself. It did not take long to produce a full understanding of most observed contrast effects in terms of the kinematical

Fig. 2 The planar crystal defects in this Fe 11%Al alloy were formed in the HVEM by the motion of partial dislocations. The fringe contrast associated with these defects was just one of the many contrast effects explained by the Cambridge group in 1956.

and dynamical theories of electron diffraction. The new ability to see and understand microstructure at near atomic dimensions in commercially important materials not only spawned a new equipment industry but altered the whole course of research in materials development.

The Arrival of the Million Volt Electron Microscope

In the 1950s, there was a tendency to look only at specimens that were easy to prepare. Every electron microscope specimen that was examined seemed to result in a paper submitted to the *Philosophical Magazine*. However, after five breathless years, the cream of the crop of 'easy' specimens had been skimmed and electron microscopists were looking for a new wave to ride upon. It came in 1961 when Gaston Dupouy announced the construction of the first million volt electron microscope at the University of Toulouse.

The Toulouse microscope, with its air-insulated, high voltage power supply was housed in a spectacular spherical building that looked more like an astronomical telescope than an electron microscope. The 1 MeV images obtained by Dupouy exhibited exceptional contrast and were clearly from much thicker specimens than those photographed in conventional 100 kV microscopes. Dupouy's results attracted worldwide attention and soon other laboratories started to build high voltage electron microscopes (HVEMs). In 1966 an experimental 750 kV microscope, designed by Smith, Considine and Cosslett, was completed at the Cavendish Laboratory. To protect operators from X-ray leakage this instrument was designed for remote operation.

At the time the Cambridge instrument was being built I was working at US Steel's Fundamental Research Laboratory in Pittsburgh, USA. Robert Fisher, the manager of the electron microscope facility there asked me to visit the Cambridge group to determine whether they would allow a copy of their instrument to be built for US Steel Corporation. Dr Smith and I had a meeting with the Managing Director of Cambridge Instrument Company but he was not as enthusiastic about the idea of building the HVEM as we expected. It was his view that the project would disrupt his company's core business and he suggested we looked elsewhere for a manufacturer.

US Steel eventually contracted RCA to build their high voltage microscope. It was a magnificent structure, housed in a special, four story building. It had an air-insulated accelerator surrounded by a stainless steel Faraday cage (see Fig. 3). Entering the accelerator room was like entering a cathedral and people would automatically lower their voices as a sign of reverence. The microscope soon became a public relations vehicle for US Steel and indeed, one of its first tasks was to reveal the structure of moon rock brought back to earth by *Apollo 11*. Robert Fisher, Scott Lally and Al Szirmae did a superb job on the electron microscopy of the moon rock and although this first project had nothing to do with steel it was great publicity for US Steel and no doubt this helped pay for the instrument. In the Autumn of 1966, before the RCA 1 MeV microscope was commissioned, I left US Steel to take up a Readership position at the Royal School of Mines. The plan was to bring RSM up to date in the field of transmission electron microscopy (TEM) and it was no secret that we would be applying to SRC for funds to build our own HVEM.

The Metallurgy Department at the Royal School of Mines was famous for its work in the field of Chemical Metallurgy under the leadership of Professors Richardson, Professor Alcock and their predecessors. However, with the appointment of Professor Ball and Professor Pratt the emphasis in the department shifted toward Physical Metallurgy and Materials Science. An AEI EM7 100 kV electron microscope had already been purchased in 1965

Fig. 3 The air-insulated, Cockroft–Walton generator and electron beam accelerator of the RCA million volt microscope at US Steel's Research Center. The actual microscope column is located on the floor below.

and was located in a small, stuffy room that opened onto a busy corridor on the sixth floor. When I took up my appointment, the microscope had not yet been used. Professor Ball, assigned to me a room next to the AEI EM7 so that I could set up a specimen preparation laboratory. The 'golden era' was over and my job was to apply TEM to a multitude of practical metallurgical and materials problems.

In those days, an electron microscope research project on a metallurgical problem would invariably start by searching for the correct electropolishing solution to suit the specimen being investigated. There were hundreds of recipes in the literature and almost all of them required us to mix a strong oxidising acid with an easily oxidisable organic liquid such as an alcohol. We did not know it at the time but rocket scientists were using the same protocol to make their rocket fuels. The main difference was that to get the best polishing results we also added a syrupy agent to increase the viscosity of our chosen mixtures. Stale Coca Cola turned out to have excellent qualities in this regard. My favourite starting recipe was a mixture of concentrated nitric acid, Coke and methanol – that is, until one day it exploded in front of me causing the loss of all my clothes and the laboratory windows. Miraculously, I survived but the experience made me search for better ways to prepare electron microscope specimens.

In September 1966, Professor Ball asked me to travel to Japan to evaluate the commercial HVEMs available there and obtain a quotation we could use for a Science Research application. I visited the JEOL and Hitachi facilities and examined specimens at both 650 kV and 1 MV. I was impressed by the performance of all the instruments but could not see any great benefit in raising the operating voltage above 650 kV. The Hitachi group was very co-operative and agreed to build a microscope for RSM to my specifications. In particular, the microscope would have a side entry specimen stage and an objective lens configuration that would accept an environmental cell and a range of specimen holders I had already designed. With this offer in hand and a very attractive price quotation I submitted a recommendation to Professor Ball that we buy the HU 650 kV instrument.

What I did not know, at the time, was that Professor Ball, was part of a nine member SRC panel studying the case for installing HVEMs at British Universities and that the Ministry of Technology, had already committed funding for two HVEMs based on the Cavendish design to be built by AEI. However, AEI needed at least one more order before it would commit its resources to building HVEMs. SRC was approached with a view to secure that order. Clearly, with this momentum for the AEI microscope my recommendation to buy an Hitachi instrument was not going to succeed. Eventually the SRC panel decided that not one but three HVEMs were needed and they should be placed at Imperial College (RSM), Birmingham and Oxford universities respectively. Later, SRC contributed a half share of a fourth microscope to be installed at British Steel Corporation in Rotherham. The universities of Leeds, Manchester and Sheffield had access to this microscope in exchange for SRC's half share in the purchase price.

Altogether, AEI received an initial order for 6 HVEMs. Of the nine members on the SRC panel, seven belonged to institutions that received a share of the 'HVEM pie'. The whole decision-making process was managed by Sir Peter Hirsch and is a brilliant example of how a carefully constructed and co-ordinated committee can extract large sums of money from the government. Even the main opponent of the HVEM investment, Professor Jack Nutting at Leeds University, was subdued when he was offered a half share in the microscope purchased by British Steel Corporation. The total investment (excluding the cost of buildings) was over £2 million, most of which probably came to the HVEM community at the expense of high energy physicis.

Each microscope was tested at AEI prior to delivery. The RSM microscope had its acceptance test in August 1971 (see Fig. 4) and was installed a month later. The investment in HVEMs provided a much needed boost to AEI which for years had been losing market share to Philips, JEOL and Hitachi.

What Did 1 Million Volt Electrons Have to Offer?

It is interesting to speculate why million volt electron microscopes excited so much attention. They certainly gave prestige to their owners and also a guaranty of research income for years to come. However, they did not offer any good prospect of a fundamental scientific breakthrough because they simply extended just one attribute of instruments already in existence i.e. a decrease in the deBroglie wavelength.

Fig. 4 The column of the AEI EM7 million volt microscope during its acceptance tests in 1971 immediately prior to its delivery to RSM.

In theory, a shorter wavelength should result in a better resolution, but in practice similar or better resolutions have always been obtained using microscopes operating at lower voltages. A shorter wavelength also reduces the probability of inelastic scattering and hence increases the electron transparency of specimens. However, there is no record of ia result from a 1 million volt microscope that could not also have been obtained at, say 300 kV.

A million is an attractive number in any field and could have a commercial advantage when selling an electron microscope. However, operating a microscope at 1 MV rather than say 300 kV introduces technical problems that are not justified by any corresponding scientific benefits. Not the least of these problems concerns X-ray shielding, building size (1 MV microscopes are too large to fit into conventional laboratory rooms) and difficulties in electromechanical stabilisation that prevent theoretical resolutions from being obtained.

A factor that may have swayed the decision of the SRC panel to opt for 1 MV (or even higher voltages) was the need to observe atomic displacement damage. For most researchers, such damage is an artefact to be avoided but for institutions like Harwell and CEGB there was the possibility that short exposures of specimens to a 1 MV electrons could simulate weeks of neutron exposure in a nuclear reactor. If this were true then a 1 MV electron microscope would pay for itself very quickly. Unfortunately, the types of damage caused by slow neutrons and fast electrons are not the same and with hindsight even this justification for 1 MV disappeared.

Some Highlights of the HVEM Work at RSM

In a meeting with Professor Ball in 1970 I emphasised the advantages of being able to construct special specimen holders to perform *in situ* experiments with the RSM high voltage microscope. He agreed and allocated one of the Department's young machinists, Mr Tony Lloyd, to work with me on the construction of these holders. To my surprise, we soon became the supplier of specimen holders to the whole HVEM community and the demand was so great that eventually, with the full support of RSM, we formed a small company, Gatan Ltd.

Reduction of Haematite

Our application to the Science Research Council emphasised RSM's strong background in Chemical Metallurgy and we proposed to use the HVEM to observe directly the solid state structural changes that occur during chemical reactions. Such changes are rarely studied by chemical metallurgists even though they play a significant role in the kinetics of gas-solid and gas-liquid reactions. To support our application we designed and built an environmental cell and a hot stage for use in the AEI microscope (see Fig. 5). The cell had four small apertures through which the electron beam had to pass. The space between the apertures was differentially pumped so that gas pressures up to one atmosphere could be maintained around a specimen with only a small deterioration in the microscope vacuum system. In one memorable afternoon, soon after the microscope was installed, we succeeded in firing up the world's smallest blast furnace and watching the reduction of a tiny crystal of haematite in 25 torr of carbon monoxide to form less than a microgram of the most expensive iron ever produced by man! However, for the first time, we were observing the world's most important chemical reaction at a resolution approaching the atomic scale and so we did not feel too guilty about the expense.

We soon discovered that at 300°C, which is about the temperature found near the top of a blast furnace, reduction starts by the formation of tunnels about 100 atoms in diameter in the haematite (see Fig. 6). We noted that the highly porous reaction product is similar to that formed when a base metal is leached from a noble metal alloy. In both cases, the highly curved reaction

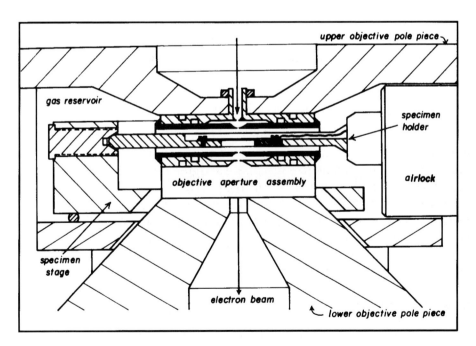

Fig. 5 A cross section through the environmental cell used to observe chemical reactions in the AEI EM7 at RSM.

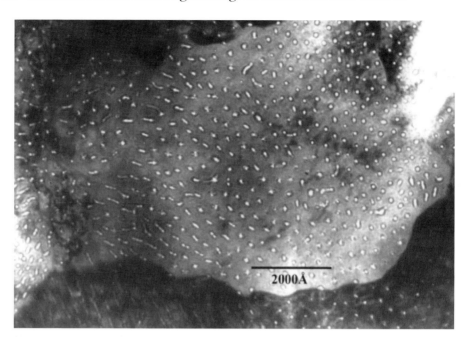

Fig. 6 Reduction tunnels formed in haematite when heated to 300°C in an atmosphere containing 40 torr of CO (courtesy of Nancy Tighe)

interface at the tips of the micro tunnels permits the two types of atom to undergo the required redistribution purely by surface diffusion. A planar reaction interface would soon become clogged by a layer of iron (or noble metal) thereby blocking access to the oxygen ions (or base metal) below the surface. The reaction rate would then be controlled by the rate of bulk diffusion which is much too slow for reduction to occur at 300°C. The creation of the porous structure involves significant swelling of the reaction product which is why the throat of a blast furnace is shaped like an inverted cone.

At higher temperatures, around 500°C, we saw that a new reduction morphology developed (see Fig. 7). Tunnels no longer formed but instead crystals of magnetite started to grow through the haematite. These crystals were partially coherent with the matrix and were surrounded by large strain fields. It was clear that at this higher temperature bulk diffusion of oxygen ions was starting to play a larger role in the reduction reaction and was forcing magnetite to nucleate in the bulk. This change in structure was accompanied by a decrease in reaction rate.

One disappointing feature of the environmental cell was that chemical reactions were found to be strongly influenced by the high energy electron beam. Usually, a chemical reaction would only nucleate in the area being

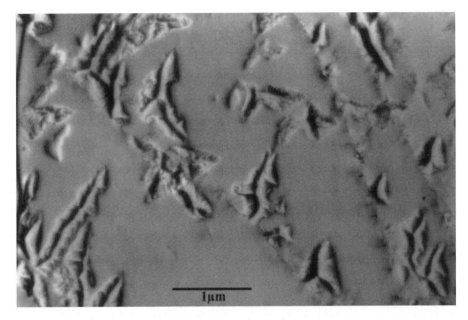

Fig. 7 Magnetite crystals nucleating at the surface of haematite at 500°C in an atmosphere containing 40 torr CO (courtesy of Nancy Tighe).

examined and so there was always the doubt that our observations were not typical of those that occurred in bulk material.

Corrosion of Silicon

In 1975 Professor Flower and I decided to use our environmental cell to study ice crystals.[3] We attempted to form the ice by passing helium gas, saturated with water vapour, over a single crystal of silicon mounted on a special cold stage in the environmental cell. However, instead of seeing ice crystals we saw bubbles of hydrated silicon oxide growing from pits in the silicon substrate (see Fig. 8). The bubbles would continue to grow as long as the electron beam was present but would stop if the beam or the gas supply was shut off. It was obvious that we were looking at an electron beam induced chemical reaction. We were able to correlate the size of the bubbles with the size of the pits and show that the driving force for bubble growth was the pressure of hydrogen gas within them. This was the first of many such beam induced artefacts we observed. Eventually, we discontinued using the environmental cell in the EM7 although we concluded that there could be a practical application in using an electron beam to define the location of a specific chemical reaction. Some years later I built an improved environmental

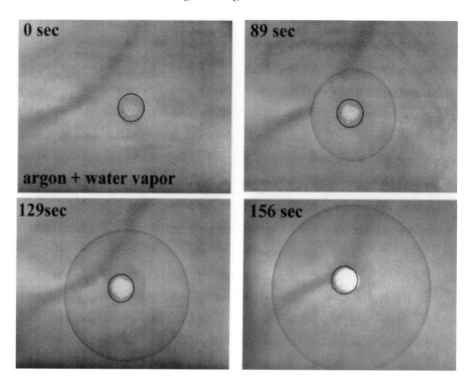

Fig. 8 A sequence of micrographs taken at 163 K showing a single, hydrogen-filled, silica bubble growing from a pit in a silicon substrate.

cell with a shorter electron beam path length to operate in a 300 kV microscope. This microscope had an excellent illumination system and to my surprise the specimen visibility was much better than with the EM7. Also, because of the lower operating voltage there were fewer radiation induced artefacts.

Stress Corrosion of Al–Zn–Mg Alloys

Al–Zn–Mg alloys have strength-to-weight ratios and weldabilty characteristics that make them commercially attractive. Unfortunately, they are susceptible to intergranular stress corrosion cracking in the presence of water vapour. To study this phenomenon we bent thin specimens of an Al 7% Zn 3% Mg alloy in air saturated with water vapour and then examined them in the microscope. We noticed that bubbles of hydrogen formed at certain grain boundaries (see Fig. 9). With prolonged exposure under stress the bubbles grew and eventually developed into well defined intergranular cracks (see Fig. 10). It is not

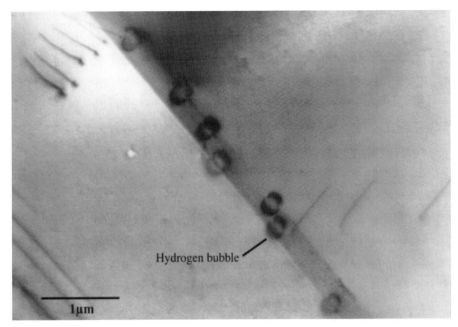

Fig. 9 Hydrogen bubbles formed at a grain boundary in a Al 7%Zn 3%Mg alloy exposed under stress to air saturated with water vapour (courtesy Reza Alani).

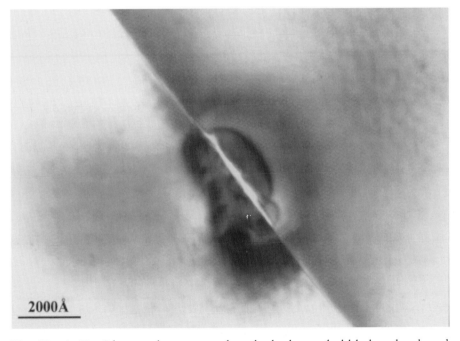

Fig. 10 As Fig. 9 but at a later stage when the hydrogen bubble has developed into a crack nucleus.

66   Materials Science and Engineering: Its Nucleation and Growth

clear from the observations what precise role is played by the hydrogen. It could be merely a source of localised stress or it could reduce the surface energy of the new fracture surfaces or it could do both.

Stress Corrosion of Stainless Steel

It is well know that austenitic stainless steels are susceptible to both transgranular and intergranular cracking in chloride containing environments. To study transgranular cracking we developed a miniature straining stage that could stress electron microscope specimens in the standard boiling magnesium chloride solution used for stress corrosion testing. When we examined the specimens in the HVEM it was immediately obvious that the initiation site for transgranular attack is the active slip step (see Fig. 11). When deformation occurs in electron transparent specimens each dislocation produces two slip steps – one on the top surface of the specimen and the other at the bottom surface. Although, the steps have the same height on each

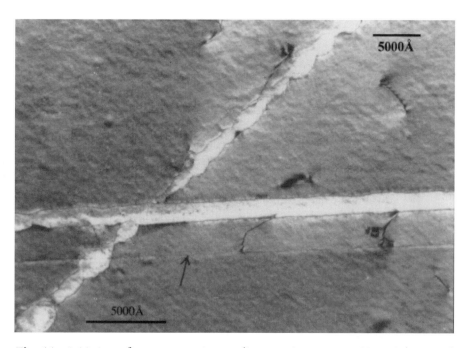

Fig. 11   Initiation of stress corrosion at slip steps in an austenitic stainless steel (micrograph courtesy of Geoffrey Scamans).

surface we noticed the amount of corrosion at each step is usually very different. This could be explained if the thickness of the passivating oxide layer that gives these steels their stainless quality is different on each side of the specimen. Corrosion only occurs if the oxide layer is broken and the underlying metal is exposed to the environment. Austenitic stainless steels have a low stacking fault energy which makes cross slip more difficult. As a result slip steps in these alloys are very large which makes penetration of the passivating layer more likely to occur. We considered this to be an important factor controlling the susceptibility of these steels to transgranular stress corrosion.

What Happened to the HVEMs?

Initially, the HVEMs were operated around the clock. There were great expectations that specimens would be much easier to examine because beam penetration would be greater. Actually, HVEMs were not as easy to use as conventional microscopes and local operators were needed to guide users especially those visiting from other departments. Furthermore, specimens were still hard to prepare because no attention had been given to the development of specimen thinning techniques since the late 1950s. It could take up to a week to prepare a specimen of a difficult, brittle material such as haematite and such specimens were so delicate that they would sometimes be destroyed during specimen loading.

Researchers who were not electron microscopists would get very discouraged when they discovered how difficult it is to prepare specimens. They did not know that electron microscopists generally pick out 'easy' specimen materials to work on and even then they publish less than 0.1% of all the micrographs they have obtained. The so-called 'typical' electron micrograph is usually anything but typical. This was the problem I faced when one of RSM's talented solid state chemists, Professor Brian Steele, came to me and described a potential HVEM project. I was sure he could eventually obtain very useful results if he put in the time required to learn how to use the instrument but I was equally sure he would soon give up because of specimen preparation difficulties.

In the early 1980s, new high resolution 200 kV, 300 kV and 400 kV microscopes appeared on the market and slowly the demand for HVEMs dropped off. Today HVEMs are only made at a huge cost to special order. By 1990 all but one of the AEI EM7s had been decommissioned. However, because of their size and the fact that they were housed in special buildings, they survived much longer than their useful lives.

	Commissioned	Decommissioned
NPL	1970	1979
AERE	1970	1986
Oxford	1971	1993
RSM	1971	1990
Birmingham	1972	1988
BSC	1972	1986

In 1969, even before the first AEI EM7 was delivered, RCA closed down its electron microscope business. AEI followed a few years later. This was a time of economic recession and big companies all over the world were tightening their belts by getting rid of unprofitable divisions.

What Went Wrong?

With hindsight it is clear that almost all the work that was done with million volt microscopes could have been done more easily and at lower cost with lower voltage instruments. The decision to build high voltage instruments in the 1970s created an elite and did great damage to the electron microscope industry by fragmenting the efforts of the microscope manufacturers. Only one manufacturer, Philips, resisted the temptation to enter the HVEM field. The order for six HVEMs had a political as well as a scientific dimension. It was widely believed that it would provide the vital infusion of cash that could save AEI's electron microscope division. In fact, it turned out to have the opposite effect. The basic fault in the SRC/MoT decision was that a committee took the unusual step of telling a manufacturer what it should build i.e. the tail wagged the dog. Generally, manufacturers make better decisions in this regard because they go out of business quickly if they make wrong decisions.

It is easy to see now that it would have been better for microscope manufacturers to have focussed on a basic electron microscope column geometry to suit a range of operating voltages up to say 300kV rather than to push the maximum operating voltage to such an extreme that interchangeability of column parts and attachments (such as specimen holders, cameras and analysers) would become impossible. Since the 1980s the column manufacturers have encouraged smaller specialist companies to develop attachments and government agencies generally supported the purchase of attachments if the scientific case warranted. In this way the column manufacturers spread their risk. A committee on the other hand is not concerned with spreading risk but rather with spreading responsibility. There is a lot to be said in favour of replacing committees with individuals so that responsibility is defined and reputation is risked.

What Happened to Gatan Ltd?

Gatan Ltd was forced to close soon after its formation when Mr Lloyd died suddenly of liver cancer. However, the company had a large backlog of orders and about a year later I decided to take those orders to Pittsburgh and begin a new career as an instrument designer and manufacturer. Wherever I went to visit electron microscope laboratories around the world I saw beautifully designed, expensive instruments sitting unused or badly used because researchers could not make good specimens. It therefore seemed worthwhile, that in my new career, I should make it my goal to design an instrument that could prepare electron microscope specimens easily and quickly. This goal was achieved in 1991[4] when we introduced a new type of precision ion milling machine with sufficient ion beam intensity to thin specimens at angles of incidence less than 5°. At these low angles ion beams polish rather than etch and ion implantation damage is greatly diminished. Figure 12 is an example of what the machine can do. The specimen was prepared by Dr Reza Alani and is a cross section through the surface of galvanised steel. Such a specimen, which is composed of materials having greatly differing ion

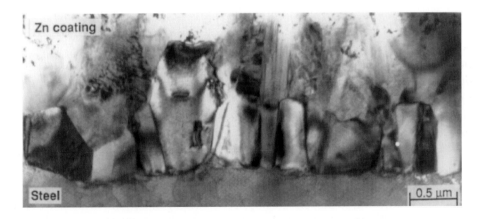

Fig. 12 A cross section through a zinc coating on the surface of steel. The large difference in thinning rates of zinc and iron makes the preparation of this specimen particularly challenging.

thinning rates, could not have been prepared in the 1970s. The specimen thickness problem that convinced us to invest in HVEMs was solved by learning how to make thinner specimens.

Acknowledgements

I have many fond memories of the HVEM research group and technical support staff at RSM and I thank them all for their support and enthusiasm. They will never let me forget my obsession with instrumentation, see Fig. 13. It was my particular pleasure to work with Harvey Flower, Nancy Tighe, Reza Alani and Paul Butler during my time at RSM. I also thank Ondrej Krivanek, Robert Fisher and Ronald Doole for providing historical information.

Fig. 13 The HVEM research group at RSM in 1976. From left to right: Ned Edwards, Harvey Flower, Sue Burrows (now Sue Ion), Graham Briers, Tony Lloyd, Howard Cheetham, Bill Bishop, Leo Christadoulou, John Woodhall and Paul Butler. The group is examining a double tilt heating holder which has been enlarged to about 100 times its natural size. Clamped in the jaws of the tilt mechanism at about 100th his natural size is the author who also designed the holder.

References

1. P. B. Hirsch, R. W. Horne and M. J. Whelan, *Phil Mag.*, 1956, **1**, 677.
2. W. Bollman, *Phys Rev.*, 1956, **103**, 1588.
3. H. M. Flower and P. R. Swann, *Corrosion Science*, 1977, **17**, 305.
4. R. Alani, R. J. Mitro and P. R. Swann, *MRS Symposium Proc.*, **480**, 263.

Transition from Metallurgy to Materials Science

ROBERT W. CAHN
Department of Metallurgy and Materials Science, Cambridge University, Pembroke Street, Cambridge CB2 3QZ, UK

ABSTRACT
The evolution of the discipline of materials science during the second half of the twentieth century is outlined. The concept emerged in the USA, almost simultaneously in an academic metallurgy department and in an avant-garde industrial research laboratory, and its development subsequently all over the world has been a joint enterprise involving universities, industrial laboratories and government establishments. The initial impetus came unambiguously from the well established discipline of physical metallurgy, but from the 1960s onwards, the input from solid-state physicists grew very rapidly, while materials chemistry is a later addition. Of all the many subdivisions of modern materials science, polymer science has been the slowest to fit under the umbrella of the broad discipline; its concepts are very different from those familiar to metallurgists. Two fields have contributed mightily to the creation of modern materials science: One is nuclear energy and, more specifically, the study of radiation damage, the other is the huge field of electronic and opto-electronic materials in which physics, chemistry and metallurgy are seamlessly combined.

The Beginnings

The English-language phrase *materials science and engineering* (MSE) emerged at an unknown time and place in the United States of America in the early 1950s. One suggestion is that it, and the concept it denotes, arose out of sustained discussions about research strategy between Herbert Hollomon, a notable research manager in the GE Research Laboratory in Schenectady, and some of his senior colleagues such as John Howe. Another is that the concept was first developed in academic circles in America, notably at the University of California Los Angeles and at Northwestern University. At all events, Daniel Rosenthal and Jack Frenkel at these two universities prepared pioneering lecture courses and Frankel published a book in 1957 entitled *Principles of the Properties of Materials*, the first book with a title focused on materials. (Since then, many dozens of texts have appeared under a wide variety of materials-centred titles.) Finally, at the end of 1958 the faculty of the Graduate Department of Metallurgy at Northwestern University at Evanston, near

Chicago, voted to change their department name to Graduate Department of Materials Science; in due course, the department embraced undergraduates as well, and the floodgates of change were opened. By 1985, there were three times as many MSE courses in American universities as residual metallurgy departments.

In this short overview, I do not propose to quote sources for my assertions. Extensive sources can be found listed in my recently published book, *The Coming of Materials Science* (Pergamon, Oxford, 2001), and I refer readers who wish for more background to that book.

Herbert Hollomon was a metallurgist, recruited by GE to form a new metallurgy research group. From an early stage, he incorporated ceramics into the same group, and as a consequence of that linkage, some extraordinary research was done on the nature of sintering of powder compacts. This in turn led to a major, wholly unforeseen, industrial development. Later, the ceramicists began to study electro-ceramics in depth. Many researchers of great ability joined this group – David Turnbull and John Cahn were examples – and spread the fame of GE around the world. At the same time, the Bell Laboratories also by degrees integrated the different constituent skills of materials science into mixed groups, and other laboratories followed suit later.

Apart from developments in university teaching departments and a few large industrial laboratories, the third change that firmly established MSE was the introduction of the interdisciplinary Materials Research Laboratories at a number of American universities. This arose from an anxiety felt by the ex-Hungarian mathematician, computer scientist and polymath, John von Neumann, when in 1954 he became a commissioner of the (American) Atomic Energy Commission: he was repeatedly told that what was holding up a number of developments in his area of responsibility was a shortage of properly trained materials scientists. So, in consultation with the great physicist Frederick Seitz, he conceived the idea of establishing buildings on selected university campuses with both offices and laboratories where metallurgists, physicists, chemists, chemical engineers, polymer scientists, electrical engineers, mineralogists and (indeed) mathematicians would work together and interact on a daily basis, and train new practitioners in the art of research. (The crucial idea was that the participating professors kept their departmental offices as well as their new offices in a MRL.) Soon after, von Neumann was taken ill and died. His imaginative project almost died with him, but it was revived by the psychological shock of the Russian sputnik satellite launched in 1957 and at last, in 1960, in the face of endless obstacles, the first three MRLs were established with support from the Atomic Energy Commission, the Air Force and other sources of finance. It was the creation of these MRLs which brought the solid-state physicists into materials science on a large scale. An increasing number of MRLs were established in the following years, and under new administrative and financial arrangements they are still going strong, a glory of American initiative and organisation. In my view, the MRLs were by far the most effective way of fostering the newly conceived discipline.

How the concepts of metallurgy fitted into this array of novel institutions and concepts will be discussed in the next Section.

The Role of Metallurgy in Launching Materials Science and Engineering

Many people are convinced that the true intellectual progenitor of MSE was the solid-state physicist. To counteract this impression, it must be remembered that until just before the Second World War, there scarcely was such a subject as solid-state physics. The heroes of pre-war physics, people such as Wolfgang Pauli, held the idea of applying physics to the solid state in near-contempt. In 1930, when Alan Wilson in Cambridge put forward the first formal quantum theory of semiconductor function, semiconductors were regarded as useless, 'dirty' materials, and Wilson was warned that in concerning himself with these deplorable substances, he risked professional suicide. Not everyone agreed, and the great books by Mott and Jones, and by Hume-Rothery, as well as the Bragg/Williams papers on the statistical mechanics of ordered alloys, in the 1930s helped to give birth to a proper solid-state physics, and in 1940, Seitz's great textbook of solid state physics marked a watershed. Soon after the War, Seitz brought out two classic papers on the interpretation of colour centres in ionic crystals (work which, in the fullness of time, led on to superionic conductors, modern storage batteries and fuel cells), and at about the same time, the first quantitative analyses of the dynamic properties of dislocations were published by physicists such as Nabarro and Mott.

All this prepared the ground for what I have called in my book the 'quantitative revolution' in metallurgy. Until about 1950, Hume-Rothery being the shining exception, metallurgists were apt to wave their hands and explain phenomena in qualitative terms, and even the inventor of physical metallurgy, Rosenhain, in the early years of last century indulged in the habit of proposing qualitative theories which could not be checked by experiment. A number of metallurgists, Cottrell at Birmingham University foremost among them, in 1949–1950 came up with detailed, strictly quantitative theories of dislocation behaviour in particular, as well as of alloy equilibria, which could be checked accurately by suitably designed experiments. This crucial development, which took place at a time when solid-state physics was still in its birth stage, forever changed physical metallurgy as a science. It was closely followed by the launching in 1953 of *Acta Metallurgica*, which focused researchers' attention on the new quantitative approach to metallurgy. Cyril Smith, in his preface to the first issue of the new journal, wrote: 'Now ... metallurgy is too broad to be encompassed by a single human mind: it is essential to enlist the interest of the "pure" scientists, under whatever name, and to increase the number of metallurgists whose connections with production and managerial problems are partially sacrificed in order that they may be more concerned with physics and physical chemistry as a framework for useful metallurgical advance. Even now much of the literature of a research metallurgist is to be found in the professional journals of the physicist and chemist and he can no longer depend on metallurgical journals alone.' He went on to argue that, in consequence, a new metallurgical journal leaning towards the pure physical sciences was urgently needed. The creation of *Acta Metallurgica* in 1953 was, in my view, the single key event in clearing the path to the emergence of materials science. I still recall the intellectual excitement which the first issues of that journal generated, and I became an enthusiastic contributor to it, and have remained so for almost 50 years.

At about the same time as the quantitative revolution and the launching of *Acta Metallurgica*, two other key developments were under way – civil nuclear power and the associated researches on radiation damage (which from the start was a joint venture of physicists, metallurgists and ceramicists), and the invention and rapid development of the transistor, with associated metallurgical developments such as single-crystal growth and zone-refining. These innovations fostered genuine mutual respect, almost for the first time, between physicists and metallurgists, and the latter increasingly involved themselves in team efforts to improve both semiconductor devices (including, eventually, integrated circuits) and nuclear fuels and related materials.

From the early 1960s, starting especially at MIT under the stimulus of people like David Kingery (as well as Coble at GE), a scientific approach to ceramics got under way; this was a big change, since until that time there was even less proper quantitative science in that field than there was in metallurgy. A little later, again, a proper science of fibre-reinforced composite materials began in a number of places, primarily under the impetus of metallurgists like Anthony Kelly. The big difficulty, for the new materials scientists, was to integrate the physics, chemistry and processing of polymers with these other topics. For a start, the idea (which for many years led to intensely hostile quarrels in the chemical community) that a substance can have a variable molecular weight was unacceptable to many, and such concepts as semicrystallinity, molecules extending beyond a single crystalline unit cell, rubberlike (entropic) elasticity, and crazing under mechanical action, were beyond the scope of most metallurgists. Today, after intense efforts, polymer physics and engineering are effectively taught in many materials science departments, but polymer (synthetic) chemistry has continued, and I believe will continue, on an entirely separate path. At any rate, polymers came to form part of materials science long after the new discipline began.

Solid-state chemistry and its recent derivative, materials chemistry, was essentially a postwar development that joined forces with materials science much more recently than did solid-state physics. A few remarkable scientists, Carl Wagner pre-eminent among them, made the link much earlier. A few early mineralogists and geochemists, Viktor Goldschmidt in Norway and Fritz Laves in Switzerland particularly, also contributed much to the concepts of materials science. In America particularly since the late 1980s, a rapidly escalating emphasis on improved processing of materials drew in a range of expert chemists (and chemical engineers also). Such concepts as self-assembly of nanosized particles, and the use of supercritical solvents in synthesis, were thereby injected into the body of materials science. So, just at a time when extractive metallurgy was progressively ejected into the world of the chemical engineers, chemists steadily improved the synthesis and treatment of advanced materials of many kinds.

If it be asked what concepts of physical metallurgy played a key role in the emergence of materials science, I would pick out the following as being the most important: the determination and exploitation of phase diagrams, the study of phase transformations, the management (as distinct from contempt for) impurities and trace elements, and above all, *the study, understanding, measurement and control of microstructure.* These were the concepts which underpin the claim of metallurgy to have acted as the father of modern materials science.

Is Materials Science a Proper Discipline?

This question is frequently raised, especially among protagonists of the most recent activities in the field, such as experts in biomimetics and in medical prostheses. The suggestion is sometimes made that materials science has become so immensely broad in its concerns that it has ceased to have any clear identity. In considering such criticisms, it is helpful to look at the history of other, older disciplines in their emergent stages. In my book, I devote an entire lengthy chapter to this analysis, with special attention to physical chemistry. It became clear to me that there are two main families of newer disciplines, those (like physical chemistry) that emerged by splitting from an older discipline (organic chemistry in that case), and those, like geology and materials science, that were formed gradually by the integration of distinct activities. (In the case of geology, topics such as stratigraphy, petrology, palaeontology, orography [including plate tectonics], mineralogy and geomorphology have all come together to generate what is sometimes simply called 'earth science'). 'Split disciplines' are open to the hostile criticism that they have destroyed a hardwon unity, 'integrated disciplines' to the objection that they are so broad that they consist of fuzzy generalities. Nobody says that about geology now, and in due course, nobody will say it about materials science.

This answer to the critics may seem rather feeble to some. Nevertheless, there is an identifiable science of materials, incorporating the key concepts listed at the end of the previous Section, and also the well-honed concept of relating all kinds of properties to composition (including the identity and location of trace elements), phase constitution and microstructure. That central concern, in turn, is linked to the extraordinary evolution of methods of characterising materials, including what I regard as three key inventions in the last century, transmission electron microscopy, electron microprobe analysis and differential scanning calorimetry. The evolution of materials science and that of characterisation have gone hand in hand. A few laboratories are large and varied enough to justify the establishment of a central group devoted to all methods of characterisation: one such, in particular, was established at GE, where the term 'characterisation' appears to have originated.

Should Metallurgy Still be Regarded as a Constituent of Materials Science and Engineering?

Numerous journals covering MSE are published on a regular basis – both those publishing original research and those focusing on reviews (those are especially important). They seem to me to fall into three categories: true broad-spectrum journals covering the whole of MSE; journals which began as metallurgical ones, sought to broaden their remit but have not yet conspicuously succeeded; and the many specialised journals which cover polymers, composites, semiconductors and, of course, metals. There is a tendency in the world of journals to subdivide the wide domain of materials into metals, and all other materials. This is not surprising: as a parent in his seventies, I am well aware that one's offspring, however loving they may be, emphasise the importance of pursuing their own, different path. But this split is not logically necessary, and some journals (such as *Journal of Materials Science* which I founded myself in 1966 – though I have long since laid down editorial control – and *Journal of Materials Research*, run by the Materials Research Society in America), have contrived to maintain real breadth in their coverage.

The Future of Research on Materials

This short paper is essentially about history, not about the future. Yet in conclusion a few sentences about the future seem necessary. Many in our profession are distressed by the difficulty of attracting enough young people to become materials scientists, and are dismayed by the slow but steady diminution of academic departments devoted to MSE, and of industrial research laboratories attached to many kinds of companies concerned with the manufacture or exploitation of materials. What John Ziman calls 'post-academic science' rules the roost, and long-range (fundamental, blue-skies, curiosity-driven) research has been almost wholly squeezed out of industrial laboratories. Companies expect government to fund such research, and governments appear to be increasingly resentful of this expectation. Some kind of organisational explosion is to be expected before too long. And yet, I often think that the gloom and doom are overdone. When something really novel and exciting appears on the scene, scores or even hundreds of small teams drop what they are doing and vigorously pursue the new glory – high-temperature superconductors, scanning tunnelling microscopy and the manipulation of single atoms, low-pressure diamond synthesis, carbon nitride superhard materials. There are still plenty of groups around – and they are not even exclusively in universities – who have the effective freedom and resources to pounce on something which catches their leaders' imaginations. So long as that remains possible, perhaps we need not worry too profoundly.

Structural Metals: Is there a Future?

JEFF W. EDINGTON
Consultant: Innovation with Technology

ABSTRACT
This paper outlines the business realities that all structural metals companies face in their long term, chronically oversupplied, slowly growing markets. Potential strategies that might be followed are outlined together with the implications for R&D, materials scientists and universities.

Metals manufacturing is a commodity business and should be managed as such. Developing the marketplace is different and the opportunities are now potentially more exciting than anything that has gone before. Innovation is key to being able to do business completely differently and so creating a new business model. This requires combining the revolution in telecoms and IT with all aspects of the company's knowledge base, as well as product development and systems engineering to provide new and valuable services for customers and consumers. A much greater, more sustainable growth in revenues and profits can be achieved in this way than has been achieved by most companies over the last two decades.

Done well, the impact can be the equivalent of the Toyota Manufacturing System that has driven Toyota's industry leadership position for the last 25 years.

Materials scientists have a key role to play here because they are taught to integrate, being at the junction of chemistry, physics, mathematics and engineering and are IT literate. Furthermore, to achieve success, knowledge of materials not just metals will be key. So will the ability to work with business developers, business information systems thinkers, designers, marketers and experts in other companies.

Universities have a key role to play in generating wealth by becoming a provider of new technologies and technology-based businesses. They need to make it as easy as it is in the USA. They also need to instil an open mind in their students and a recognition of a lifetime learning ethic.

Introduction

The science and engineering discovered and taught at The Royal School of Mines has been discussed elsewhere in this volume. They have been pivotal in the growth of the structural metals industries and have driven the huge rise in the standard of living over the last century in the UK, see Fig. 1.

Consequently my title may seem melodramatic because the worldwide structural metals industry provides the basic building blocks of civilisation. Without metals there is nothing – no houses, no transportation, no food industry, no computer industry, and the highly rated Microsoft would consist

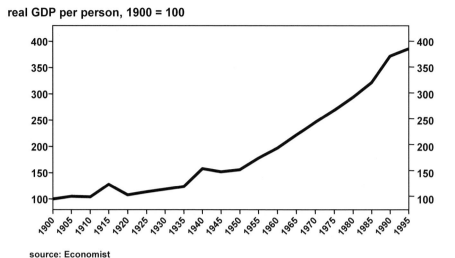

Fig. 1 Normalised growth in UK living standards.

of hundreds of software engineers wondering what to do while freezing to death in a field called Seattle.

However the title is not melodramatic. The metals industries are in for big change and for some companies it isn't going to be pleasant. However, for those companies that take a truly innovative approach and change, there is a good chance of success because a new range of tools is there, ready for exploitation.

Background

The metals industries are huge employers of people and have massive investment in plant. Some of their products demand a high price and millions of tonnes are used, see Table 1. Now if their products were in such demand you would expect that they would be highly profitable but they are not. Why? Because all metals are in chronic oversupply and have been for two decades.

Table 1 Metals Markets.

Metal	Global Demand ('000 tonnes)	Price ($ per tonne)	Market size ($ billion)
Steel	730,000	$400	$292.0
Aluminium	20,000	$2,500	$50.0
Stainless steel	13,000	$2,000	$26.0
Titanium	50	£35,000	$1.8

Consequently returns on operating assets are inadequate to create shareholder value. Indeed these industries have not even on average covered their cost of capital over the last decade, see Fig. 2. Worse, because of the high fixed costs, in some parts of the business cycle big losses are made. For example in the early 1990s Alcan had 13 consecutive losing quarters. Prior to that it only had three in the entire existence of the company (50+ years). British Steel lost £140 m in 1993, made £1 bn in 1996 and then lost almost £1 bn in 2000. These are typical examples that reflect the worldwide instability in these industries.

There are two messages here:

- First, the world needs healthy structural metals industries but it has not got them.
- Second, if we are to get them, the existing business model* has to be changed by management action.

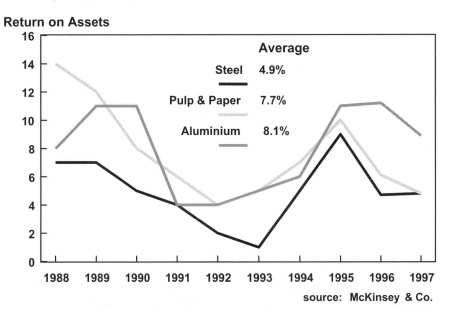

Fig. 2 Returns for basic materials industries.

*A business model is the holistic picture of what is delivered to customers and how it is done. For example, the mini-mill business model was to deliver basic standard steel products at the lowest possible price and in a completely new way by combining new, low capital cost technologies that were invented by suppliers. Very high levels of productivity were created by incentivising the people much more than ever before. Sites were built in the industrially underdeveloped Southern farming belt in the USA where wages were low to keep overall costs down. Management insisted on non-union operation to avoid the restrictive practices prevalent in the industry. Steel was delivered by barge on the extensive canal system, the cheapest possible way. The result was a low capital cost, incredibly productive, simple to run business that undercut its competitors, the incumbent integrated steel companies, by 40% for low end commodity steels. Over a decade, mini-mills took 70% of the market as upgrades in technology enabled it to make more and more sophisticated products.

The existing business model for most companies in the metals industries has been to be price competitive. Thus for decades companies have focused on cost cutting using easily measurable and predictable tools. These involve business downsizing, and mergers based on 'synergies' achieved by closures and workforce reduction. Unfortunately, in the words of Tom Peters 'You can't shrink your way to greatness' because the consequence of shrinkage is mass desertion.

First investors desert into more rewarding growth stocks so share prices are chronically depressed. Consequently equity cannot be used to buy a way out of trouble by acquisitions that could force change. Neither can sales of under-performing assets because they do not generate much cash. Mergers do not help either because they do not fundamentally change the business model. They usually just create a bigger, more unwieldy, less manageable organisation doing the same things less effectively.

Second, and more importantly, the best people desert to a more positive working environment so the human capital is depleted and the organisation becomes incapable of self-driven radical change.

The financial underperformance of most companies in the metals industry is a big issue: for the people employed there; for economies; for university teaching; and for Materials Scientists themselves. It will have to be dealt with in the working lifetime of many of the managers in the industries themselves and before most of the students graduating now are halfway through their careers.

The sky is not falling but action will have to be taken now because there is huge inertia in such capital-intensive industries and it takes time to execute the kind of change in business model that is necessary.

The Eight Change Drivers

Changing the business model requires companies to harness the major drivers of change in the world outside the business in a creative, innovative way. Many of these outside changes are revolutionary and have created a world of massive uncertainty where it is just as risky to behave normally as it is to act differently. Revolutionary change is not made up of subtle events and it is not tidy. The unexpected is the order of the day and detailed strategic plans have a 12-month lifetime or less. All this makes it very difficult for companies in the metals industries where decision-making is based on ideas of predictability implied by investment decisions that last for decades. Nevertheless these companies will have to find a way to deal with it.

Let us do a reality check. What is really going on that is shaping the future and is changing everyone's world?

First, and most importantly we live in a surplus society and it is everywhere. In Norway – population 4.5 million – you can choose from 200 different newspapers, 100 weekly magazines, and 29 TV channels. In Sweden – population 9 million – the number of beers to choose from has increased from around 50 to 350 in the last decade. Profusion abounds. Seiko sells more than 5000 separate watch models on the world market. In 1996 Sony launched 5000 new products worldwide, more than two per working hour. The ultimate is Disney, which developed, in 1998, a new product every five minutes.

Second, knowledge is exploding, particularly scientific and technical knowledge. Furthermore, 90% of the scientists who created it are alive today. So there is a huge opportunity to use this technical knowledge, which is a major driver of change and improvement, to create wealth. Unfortunately, in practice, it is often not done innovatively. As a result, the surplus society has a surplus of similar companies, employing similar people with similar educational backgrounds, working in similar jobs coming up with similar ideas, producing similar things at similar prices, warranties and qualities. This has arisen because of the focus in the 1980s and 1990s on quality, benchmarking, continuous improvement, sharing best practice, asking what the customer wants and other worthy activities. After two seconds thought it should be clear that this approach will obviously and inevitably produce similarity and convergent business models. Without innovation, everyone ends up competing on price. This is great for customers but not for employees' or executives' peace of mind and not for long-term profitability and viability.

Third, there is an integrated revolution in communications and information technology that is leading to almost cost-free communications, free access to knowledge and a completely different way of doing business for everyone. This is the equivalent of an intercontinental missile tipped with multiple nuclear warheads. It is unstoppable and it is coming faster than anyone imagines, and when it lands it is going to create hell. No one really understands this revolution because it is changing all the time. Even experts in the information technology and communications industry admit it. Witness the incredible prices paid for G3 telecoms licences in Europe that has left most telecoms companies with an unsustainable level of debt, massively reduced share prices and potentially unable to implement G3. However, understood or not it will have to be dealt with and the myriad of opportunities it generates because integrated telecoms technology and information technology are growing exponentially. Today there is more computing power in the average car than there was in the Apollo spacecraft that first took men to the moon. There is more computing power in a birthday card singing happy birthday than there was on the planet earth in 1950. The list of new information technology services beggars belief, from CDs, PCs, DVDs, totally computer simulated movies ('It's a Bugs Life', 'Ants') the internet, mobile telephones, palmtops, ubiquitous credit cards, worldwide ATM banking and on and on. All of these entered the market and took-off. Why? Because the more people who have them the more fun you can have. They spread as remorselessly as fertilised weeds. The old rules do not apply any more. Unlike Ferraris, country houses and Cartier jewellery where value arises from scarcity, for these products value arises from ubiquity. Mobile phones, access to e-mail, credit cards and belonging to the internet are only useful if everyone is involved. It is inevitable that such products and services will grow exponentially and will have a massive effect on every business.

Fourth, market mania rules. There are now markets for more things, covering a larger geographical area, than ever before. Deregulation and market liberalisation have unleashed market forces on virtually every human activity. There are markets in commodities and capital, body parts, children for adoption, every conceivable form of sex, any industrial component you can think of and any kind of service you can imagine.

Fifth, because consumers, you and I, drive the market, values of today are based on what we, as consumers really want to make our lives easier or more fun. Here is a list of products of the year in 1994 as identified by *Fortune* Magazine. Wonderbra, Mighty Morphin, Power Rangers, Oldsmobile Aurora, RCA DSS, Baby Think it Over, Snake Light, Mosaic, Svelt, Myst, the Lion King. Except for the Oldsmobile car, they are not metals intensive at all and it has probably got worse since.

Sixth is the youth effect[†] or what will tomorrow's people value? A study in the late 1990s concluded that children in Britain wanted, for Christmas; a bicycle, clothes, books, Nikes, computerised games, watch, Lego, computer, sports equipment, Nintendo. In Japan they wanted; electronic diary, mobile phone, word processor, CD/radio/cassette player, PC, Fax, CD/mini stereo. Only one of these is metal intensive. Of course you need metals to make them, but metals are not directly visible to the consumer in the product itself. They were for earlier generations – Bicycles, Dinky toys, Meccano, Hornby toy trains etc. The metals industries just don't have a high consumer profile any more and it matters, because consumers drive everything.

Seventh, the world is moving toward techno-economic parity. This means that there are very few commodities, technologies, products, services, insights,

---

[†] If you want to see the youth effect in action go to Offspring, an amazingly compelling Trainer store in Endell St, Covent Garden. Here, young people spend £120 on the season's cool Nikes. They end up with a product that is more coherently designed than any metals containing product, be it a car, house, office building or beverage can. They end up with a brand that scythes through all of the media fluff and hype and speaks to them directly. Their Nikes are all about status, about fitting in with the group. To this group, metals containing products are beyond tired. By the way, don't assume this will change when they 'grow up' and become parents. Their expectations have been formed.

knowledge areas or procedures in London, Paris, New York, Tokyo and Frankfurt that are not available to our friends in Bangalore, Shanghai, Prague, São Paulo, Bangkok and Singapore. This means that the basic prerequisites for doing business are becoming more available to everyone. The best business or person wins irrespective of where they come from. In the steel industry, for example, some of the most modern equipment is in the developing countries and in the software industry where comparatively little equipment is needed, some of the best people are too (India has 400,000 top software engineers).

Finally, eighth and most importantly, people are key. It is essential to make companies into places where the best people want to work. Management does not control people any more and the best people‡ know that and they know their value in the market place. The best people are key, because to take advantage of all of these trends innovation is essential and it's very difficult. Remember, the critical means of innovation is small, wet, grey and weighs about 1.3 kilograms. It is the human brain. The brain is controlled by its owner; not management, not shareholders, not investment funds, not any other body. Two legs can transport it out of one organisation into another and with it all its accumulated knowledge of the technology and the business. Most importantly those two legs take away all those great ideas on how things can be done differently to improve the business that were never explained to management because their owner was turned-off by poor leadership.

The Eight Change Drivers and the Metals Industries

What does all this mean for the metals industries? The surplus society combined with market mania and price and availability transparency due to the telecom and IT revolutions means that consumers are all powerful in a way that has not been possible before. Changing values of most people, combined with developing youth values, means consumers increasingly choose to spend on products and services that are not metals intensive. As a result, metals have become, at best, partial enablers to satisfy consumer needs and values. Metals are not the drivers or indeed critical factors in meeting consumers' aspirations that they were in the first 80 years of the twentieth century.

---

‡ Average people tend to produce similar results. What matters is the innovation that the people at the top end of the Maxwellian distribution curve can introduce. These people congregate. Without a critical mass they will go. Check if your prospective employer has them.

Although knowledge is exploding it is not often used innovatively to create new things that consumers value. In my view this is particularly the case for the metals industry whose last significant valuable innovation was the mini-mill over 20 years ago. Even that was a supply chain innovation not one visible to consumers. Since then the industries have been focused on benchmarking, introducing best practice thought up elsewhere, cost cutting, downsizing and developing similar products by listening to the same messages from their customers. It is obvious that this mechanical, clockwork universe approach to running a business will turn off the best people. It is also obvious that this would lead to largely similar business models and convergent strategies and a level of sameness that shows in the consistently low share price of most of the companies in the sector. Moreover, because of years of downsizing and cost cutting, most companies do not behave as if people are key so the best people have not been joining them for years and now many of their best people are leaving for a better working environment with better rewards. Most CEOs I know around the world recognise this and are extremely worried about it.

All of this, combined with worldwide techno-economic parity, means that companies must begin to compete on their own clearly differentiated business model. This will require a level of innovation and risk taking that has been absent for many years in most of the metals industry.

Successful management needs to focus on two things. First, business success is going to be driven primarily from the opportunities that the integrated revolution in communications and information technology makes possible. However, this is difficult and requires management commitment to change and enough excellent people properly rewarded. Second, without creating a culture that supports innovation and risk-taking, success cannot be achieved and key people will leave.

Changes Underway in R&D

Let us start by defining two types of R&D.

The first is incremental R&D, that is, invention in the field of existing products and processes. It involves the next version of an existing product or process. As such it is reasonably predictable and is amenable to the detailed budgetary processes and risk – reward analysis of large established companies. Its objective is to maintain competitive advantage and to protect or increase the company's market share. Its success rate is better than 50% on the most innovative projects and better than 90% on most of the evolutionary work.

The second is radical R&D, that is, invention and innovation associated with new products, new processes and new services that the company requires to seriously increase market share or move into new markets. This process is aimed at providing proprietary intellectual property rights (IPR) that forms the basis of growth that is difficult for competitors to copy. The fact is the success rate is extremely low and the activity is extremely risky.

The model for the second half of the twentieth century, when most of us did our R&D, was that large corporate labs did both of these. They did indeed invent much of the basis of the digital revolution the bio-technical revolution and the communications revolution. Think of Xerox PARC, Bell Labs. and GE Labs. However, the fact is that most of these discoveries were translated into the new industries that we have today outside these companies by small start-ups.

The model for today is different. Today, most large companies are very risk averse. To maintain and improve the company's position they know new IPR are needed. Efficient use of IPR is being looked upon as a balance sheet asset and is being treated as any other asset. IPR is being sold by major companies to those in other market sectors. Patent portfolios are being looked at very carefully and managed as never before. It is the IPR that is seen as the strategic asset not the R&D activity.

Smart companies are focusing internal R&D on incremental improvement and developing associated IPR. Here, their technologist's detailed knowledge of their own and their customers' manufacturing processes is an advantage. In contrast, they are effectively de-risking R&D that is aimed at radical improvement by buying in the necessary IPR. They are finding that this is the way to bypass those who are blinkered by the baggage of existing technology, products, processes and business models. Radical R&D is demonstrably done best by start-up companies and specialist technology companies rather than corporate R&D labs. mired in their cost centre culture. In these companies, the management and workforce are incentivised by stock options and the opportunity to float the company by developing the technology fast and demonstrating market acceptance. There is clear evidence for this in the US, where venture capital devoted to technology based start-ups now exceeds total company spending on R&D. Furthermore the money spent on technology transfer has increased by a factor of 10 in the last 15 years. This is real money being spent to create new wealth.

I believe that properly managed strategic activity to provide a mixture of internally generated and bought-in IPR can put many companies in a sustainable leadership position. It has worked spectacularly for Cisco Systems and most speciality chemicals companies as well as almost every pharmaceutical company.

Business Competition Today

Brainpower, innovation, breadth of business approach, market drive, openness to new ideas and the ability to implement fast all dominate successful, modern corporations. This integrated approach is their essence.

Manufacturing companies must compete on innovation and service as well as manufacturing. Manufacturing is not key to market success – it is only part of it. Car companies manufacture cars but their customer offering is about 30% service. Nokia manufactures mobile phones but its customer offering is more than 50% service and pure knowledge work. At Hewlett Packard and at IBM this figure is probably nearer 90%. Most successful manufacturing companies are being transformed from having limited service in their customer offering into service companies with a little manufacturing. In effect these companies are competing on business models.

Metals companies have to do the same. Most of them have a long way to go but there is no choice. Find a new business model or die.[§] Change or die. Innovate or die. Sweep the world for new ideas and approaches and implement them in your market, together with your own ideas, to create a new and different business model. Then keep moving it on or die. Nokia has Labs in Scandinavia, Japan, Hong Kong, Germany, Australia and the UK. It sends its people to Venice Beach in California and King's Road in London to pick up the latest fashion signals as the basis for innovation. The same is true, in their own context, for Hewlett Packard, IBM and all the automobile companies.

Smaller growth and technology intensive companies that I know personally, such as Ascot, AstroPower, Cobham, Scipher and Technology Partnerships all have these kinds of worldwide, ideas – sweeping activities through partnerships, joint ventures and marketing activities.

Business Competition in the Metals Industries

As pointed out earlier, most metals companies operate convergent business models based primarily on competition on price. This has largely been unsuccessful in attracting investors and good people.

However, some signals of what leads to success are there. Not all metals companies have failed to deliver shareholder value. In steel Fig. 3 shows that Nucor with a new mini-mill based business model consistently outperformed

---

[§] Death, being shunned by investors, happens. Between 1985 and 1990, 24 companies got pushed off the list of America's most valuable 100 companies, between 1990 and 1995 it was 26 and between 1995 and 2000 a further 41 bit the dust.

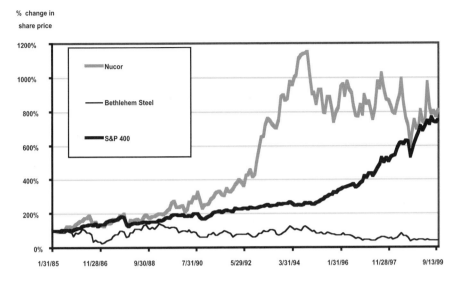

Fig. 3 Comparative share performance of Nucor steel.

the industry average and the S&P 400 average for ten years. It has not been the only successful steel company either. Figure 4 shows that several steel companies with new business models performed well over the period 1988 to 1997, i.e. reached the top right hand quadrant. Mini-mill leaders were Chaparral and Nucor. Speciality companies were Acerinox, Allegheny Teledyne, Carpenter Technology. Turnaround artist was Ispat International. Niche market segment focused were AK Steel, Rautaruukki and SSAB. Each had a unique, exploitable approach. In steel, size mattered. It made things worse. No large fully integrated company was anywhere near successful in this measurement. Similar success stories are there in the other metals industries but they are the exception not the norm.

I repeat, it is innovate or die. Innovate in process, in product in service and most importantly in business model. The ever-present question is 'how are we going to outsmart our competitors and continue to grow in revenues and profits'? It is also about perpetual innovation. Figure 3 showed that Nucor innovated once and has now stopped being rewarded for it as other companies copied their business model and the company did not innovate further. After you've done it once, you have to do it again to stay ahead.

By the way, the steel industry has already innovated massively and consecutively in technology four times (Bessemer, Open hearth, Basic Oxygen Furnace, Electric arc, steel making processes) and in business model twice (fully integrated steel mills and Mini-mills). There is no reason why it cannot be done again.

Options for the Metals Industry

In the previous section, in the discussion of Fig. 4, a few different business models that had been successful in the past were outlined. Clearly all companies in an industry cannot emulate these. The question is 'what other options might be available for business process innovation?'

1. By far the most powerful option available is to innovate with the integrated revolution in communications and information technology. The opportunity is to develop a new form of enterprise that is integrated both externally and internally in a completely new kind of way. Full integration enables management, customers and suppliers to challenge everything about the way business has been done before. Internally, a metals company will be integrated to greatly simplify and reduce the activities associated with order acceptance, manufacturing and delivery. Externally, its operations will be integrated with suppliers, customers and in some cases consumers so that there will be an efficient manufacturing chain with multiple owners but looking like a single enterprise. Externally, it will also be integrated with the design and marketing activities of its customers for truly innovative product and service

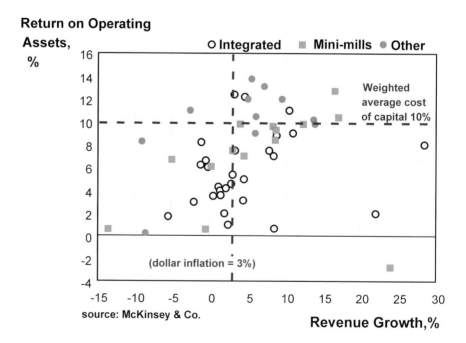

Fig. 4 Financial performance of steel companies 1988–1997.

development for consumers. Finally, it will be integrated with the knowledge base in Universities, consultants and other service providers. Most importantly it will share software and conduct service and product development activities in cyberspace, not just with its customers but also with other suppliers to produce better results quicker and more cheaply.

The key issue here is not to assume that creating the linkages is enough. Consistent, continued management action at a business model level is required. For example, it is essential to reduce the product complexity that has built up historically as a means of differentiating products just to give customers' purchasing departments a reason to place an order with a particular supplier. Indeed, the Scunthorpe works of Corus has 4000 products that it supplies in orders of 10 tons through a manufacturing system that produces steel in 200-ton batches and hundreds of combinations of section shapes and dimensions. This incurs gigantic costs in stocks, quality, order processing, logistics etc. In reality, many of these products are only variations on a theme that are produced to satisfy customers that they have a responsive supplier. They are not really different in performance terms. Consequently the first management action, before putting in expensive IT systems, must be to work with customers to simplify massively the product range and share the consequent savings.

I cannot emphasise enough how concerted action here can remove bureaucracy, speed up and simplify decision-making, and produce a level of service hitherto undreamed of. Overall, this is an opportunity to create an equivalent to the Toyota Manufacturing System that has, for more than two decades, clearly differentiated Toyota from its competitors in a way that is almost impossible for other car companies to copy. However, I want to emphasise that this is not a quick fix. It is a very serious commitment, like that to create the Toyota Manufacturing System. It took Toyota decades to develop it with massive intellectual effort. In today's world, I think it can be quicker but still will take a decade to execute although it will have many success milestones along the way. Even after a decade it will continue to be developed, as has the Toyota Manufacturing System, leading to a long-term defensible position. Such integration also provides the opportunity for some companies to create a brand position with consumers equivalent to 'Intel Inside' on this computer that I am using to write this paper.

Integration of the supply chain has been started in other industries. Boeing has integrated with suppliers and customers to design new commercial aircraft. Walmart and Benetton have integrated fully with their suppliers and Benetton has simplified its suppliers manufacturing processes. All three were dramatically successful. Why not a metals company and why not take it further? In this context it is worthwhile mentioning that large customers for metals, such as car companies, are already moving their supply chain to the internet. Their announced aim is to develop flexible manufacturing to enable them to build cars to order and massively reduce car stocks and distribution costs which is car companies largest remaining cost area. Suppliers who embrace this approach as part of their own new integrated enterprise business model will clearly win. Those who resist, because they see it only as a threat to squeeze their margins, will lose.

2. Another opportunity is to innovate in bundling the company's special knowledge with its products as a customer offering that is seen to be valuable. One knowledge area is the environment. Not only do metals companies understand the technology of how to reduce their own emissions, but they are also knowledgeable about the emissions associated with using their products, as well as the legal requirements in the different countries where they operate. This is valuable to some customers and can be bundled with the product as a service that can be enriched by joint ventures with environmental consultancy companies who can add their knowledge as required.

A second saleable knowledge area is logistics. Metals companies are used to managing a complex supply chain, both in manufacturing and distribution terms. This is transferable and potentially valuable to some customers, particularly when combined with the company integration system outlined above and enhanced by joint ventures with consultants with complementary expertise. One good example, in a related industry where this has been done successfully, is The British Oxygen Company. They built on their knowledge of logistics to create a subsidiary that supplied to the supermarket chain Tesco just in time delivery of all of their supplies to their shelves.

As a final point, financial knowledge can be bundled with the product. Advice can be given on areas such as debt management and raising capital. Furthermore, the financial strength of metals companies that comes from their cash generation can be used to underwrite product guarantees that might be required when a new product such as a steel-framed housing system is introduced to the market. Partnership with insurance companies may help here.

3. There is an opportunity to generate a new kind of materials company that provides a new integrated materials service. A company that had access to supplies of aluminium, stainless steel, plain carbon steel, magnesium, glass, plastic, titanium and concrete could sell an integration service to markets such as transport and construction. The service would provide design of components, systems and complete engineered

structures in any combination of materials. It would provide turnkey designs for low cost, lightweight structures designed for efficient assembly. This business could provide just in time delivery co-ordination from different suppliers and take responsibility for the quality and delivery of the final product to the customer. To do this it need not own any of the sources of materials. It simply needs supply contracts. This approach could provide another route to consumer recognised branding. 'Car body by Alcan' 'Building by Corus' could become the valuable and customer recognition equivalent of 'Intel Inside' mentioned earlier.

4. For some it may make sense to experiment with acquisitions as a route to radical transformation of the company like Preussag, Mannesmann and Nokia. This is not a route for all because the risk is high and some companies may not be structured for it in business and cultural terms. Furthermore the local attitude of stockmarket analysts may not be favourable. Clearly this is the most challenging route to the future but it is not impossible. Metals companies suffer from being caught in an interlocked combination of capital-intensive assets. In addition, they operate in markets with very large and powerful customers such as car companies and packaging companies. Directed acquisitions provide options that can be developed over the years into viable alternative directions for the company. Then the metals manufacturing business can be progressively off-loaded.

Preussag, Mannesmann and Nokia all started out as resource companies, the first two in steel and the latter in forest products. However, they either had, or acquired, other companies in different businesses that provided them with options and a mind set that change was possible. In all cases, the drive to move out of resource industries came because there were problems with them. The shift in strategic thinking took several decades to execute and mistakes were made but the companies regrouped and pressed on.

Mannesmann was already a conglomerate and well on the way to massive strategic change when an opportunity arose unexpectedly in the communications market. The German telecommunications industry was deregulated. Mannesmann bid for and won a license. This was the point of lift off when other assets were sold to fund growth. Eventually, the company was bought by Vodafone with considerable gains for the employees and shareholders. Nokia acquired interests in telecommunications and electronics and data systems in the 1970s. But the company was close to bankruptcy in the early 1990s. It chose a big bet route out of the problem and focused on telecommunications as a growth business it already understood because of its previous acquisitions in the communications area. The company sold almost everything else to fund growth. It worked and the company is now the premier mobile phone company in the world and had a market cap of $70 billion in 2000.

Preussag moved from being a commodities and raw materials company to become a dominant player in tourism, services and technology. The remaining metals interest is in trading rather than producing. The intermediate state was that of a conglomerate that provided the range of options available and the confidence that the company could manage the transition.

These three transformations were driven by a strategic intent and required the ability to ride several horses at once, at different paces and with different people. Outsiders played an important role in the process. Also the structure of the company allowed funding of the new model by progressive sale of businesses. The approach requires a management capable of emulating the ability of these companies to innovate continuously and to be opportunistic recognising an unexpected rapid growth opportunity and take it. The company must have a culture that enables it to remain determined to break free from the constraints of the metals industry norm through several management tenures. Finally, successfully finding rapid growth opportunities creates a forgiving climate.

5. Generation of a breakthrough technology for the marketplace and combining it with an innovative business model is another route that has been used. There are three examples from the aluminium industry that have had varying degrees of success, depending not so much on the technology but more on the business model. They involve different approaches to developing and benefiting from new technologies for making lightweight car bodies from aluminium. Three companies were involved, Alcoa, Alcan and Norsk-Hydro. All recognised the growth opportunity offered by car companies who were searching for way of reducing weight in cars but each had a different business strategy.

Alcoa developed a technology package based on extrusions and castings together with Audi as a partner. Initially, the partnership was so close that the two companies shared patents. Furthermore, Alcoa built a body assembly line for Audi to guarantee the quality of the car bodies for the first car, the Audi A8. It seemed that the Alcoa business model was to enter aluminium car body production by providing a complete car body production service to at least Audi. However, the joint investment went no further. Why is not clear. The aluminium body for the next car the A2 was built by Audi with different metal suppliers and so Alcoa's business model appeared to fail. The forthcoming A4 also follows this approach. However, Alcoa used the technology and business partnership activity in the initial stages to lever up its share price.

Alcan, which was strong in sheet manufacture developed a sheet based technology, initially by itself, then with Ford. Their business model was simply to increase their sales of sheet and this has been successful although the share price has not kept pace with that of Alcoa because it was not seen by investors to be a significant change in business model.

Norsk-Hydro who are strong in extrusions, developed an extrusion-based technology. It was different from that of Alcoa, was only sold to speciality car manufacturers such as Lotus, and the business model is currently not clear.

Other markets where breakthrough technologies are clearly needed include lightweight and high-speed rail rolling stock systems, fuel cell materials systems, battery materials systems, and pre-painted formable sheet for cars, white-goods and packaging. There are many others but all need to be associated with a partner and an original business model, and probably full integration between the companies to realise the full profits.

6. A difficult but potentially rewarding approach is to search growing markets for one where metals companies' in-house technical knowledge might be used. One such example that went unrecognised by the metals industry was the solar cell market for generating electricity from sunlight. All the price models showed a decade ago that growth of the market clearly required large price reductions of the order of a factor of 10. These could only be achieved by direct production of thin films rather than slicing up cast ingots. One company, AstroPower, developed a thin film direct casting technique based on the principles of the roll caster from the aluminium industry and direct strip casting being developed in the steel and aluminium industries. The solar cell industry is now on an exponential growth curve that one or two metals companies could have participated in, if they had recognised the opportunity. None did, because they were too inwardly focused.

7. The last approach I wanted to include is to restructure a selected market jointly with other players. The most obvious is construction, the single biggest market for metals. It is also the most fragmented, low technology, low quality and low profitability market with the highest level of customer dissatisfaction. For a large building, which is the best starting point for change, the conventional approach is sequential. It starts with the client, for example a hotel chain or British Airports Authority, who raises the finance and carries the risk. Then an architect is engaged to design the building then a consulting engineer is hired to write the tender document. After a competitive bidding process, a main contractor is

selected who hires contractors, subcontractors and suppliers, also via competitive bids, and the building is constructed. Because the low bid always wins, margins are wafer thin and performance is poor because everyone does everything on the cheap. Consequently the profits are made on the variations from the original specifications and the relationships are confrontational and often end up in court. The first of two alternative models is for the client to manage the whole project creating a team of the same actors as before but acting in unison not confrontation. The second is for the main contractor to manage the team but with the client retaining design control. Suppliers such as metals companies have the opportunity to help or even drive this change to provide a quality product, a more efficient design and construction process and a more profitable activity for all.

This is clearly an incomplete list. Creative and innovative people will develop more opportunities over the years if the leadership of companies allows it.

Implications for the Future of R&D

Competition, from now on, will be between business models that generate growth in revenues and profits. In metals companies that do decide to follow this route, management of IPR for business success will be one key technical and management activity. The IPR will come from an increasingly broad range of sources, internal R&D will just be one source, in some cases a minor one.

Of course not all companies will take this route but those that do not will not have much R&D activity. They will not be worth working for in any case because they will become steadily more emaciated as cost reduction is their only driver because they compete only on price. Assuming companies decide to change and recognise the importance of IPR, let us follow the sequence of the previous section.

1. If companies are to develop a fully integrated business model based on the IT and communication revolution, the process will be business led. However, the role of R&D will be key. It will have to organise and staff-up to become a key node in the seamless link with universities and consultants in the process of generating the best incremental R&D and associated IPR. It must be capable of integrating this with IT and telecoms and delivering it to customers. I want to emphasise that it is an extremely challenging activity to innovate in industries that are in many ways on the flat part of the learning curve. Done well, it will be

extremely rewarding for all concerned. Metals companies will add strong systems management capability to ensure the correct development of IT to support the fully integrated business model. Metals companies will also add systems engineering capability to enable them to fully integrate with customers since design and engineering of systems is the common language with customers. It will be extremely stimulating for R&D to develop a seamless integration with these activities. The combination will determine the success or failure of the enterprise
2. The knowledge in R&D is key in developing bundled knowledge with products as a valuable service. Again an integrated delivery network is key because it is an integrated package that needs to be delivered.
3. The knowledge in R&D is also a key component of a materials service company. Such things as materials compatibility, corrosion protection, paint compatibility, joining, formability and finishing are key parts of the knowledge base that design engineers need, to provide the most efficient in service and the most efficiently manufactured multi-material engineered systems. R&D will have to contain a much broader range of materials experts than it does in a conventional one metal company and the interaction between these experts will lead to new technologies for getting the best out of materials combinations and design.
4. If the company attempts radical transformation like Preussag, Mannesmann or Nokia, R&D will become integrated with the new activities and radically changed by the process. Some of it may be sold-off, with key businesses, to fund the transformation.
5. If generating breakthrough technologies for the existing business is the chosen route, there is a strong case for outsourcing this to an organisation unencumbered by the company history and mindset. In-house R&D would only be peripherally involved. Otherwise it would try to prove that it would have been better to do it in R&D all along.
6. The search for growing markets, where in-house technology might be used, is extremely difficult and highly people dependent. In my experience, R&D is not very good at it because of the lack of business skills. In fact, very few people are capable of doing it. If companies have any they will probably be found in unexpected parts of the organisation. This is because they are self selected out of most companies who reject apparent misfits. Outside help is key here since we are really talking about radical thinking that is best imported, as discussed earlier. R&D is probably not key in finding the opportunity but can help in mating in-house technology with the opportunity.
7. The R&D role in restructuring the market is minimal.

In summary, in a well led metals company that is changing for the better as outlined above, good management of IPR for business success will be key. The IPR will come from an increasingly broad range of sources. Internal R&D will be a key node in the network and will be a knowledge source for incremental technology development. External sources, business start-ups and technology commercialisation businesses will also be key nodes in the network. This IPR activity will be either corporate or business driven and led depending on the scale of the strategic impact. R&D and the internal and external materials knowledge base will become a major, shared information base in a network of relationships designed to change the company. Because of its position in the network R&Ds relationship with IT, communications activities and product systems engineering activities in the company will be critical. So will a seamless porous interface with business development activities. More conventional engineers will be involved in R&D and the activity will be led by a technically qualified businessman not an R&D specialist.

Implications for the Future of Materials Scientists

Materials scientists are valuable people in this new world. Material scientists already have a broad quantitative education encompassing physics, chemistry, mathematics and engineering. Key additional skills and experience that will need to be added during continuing professional development will be in IT and communications. This can either be toward becoming a technical expert or towards making effective use of IT and communications services. For others, a business qualification may be added for those who want to enter the new business start-up world or the technology development company.

Each person must take a broad view of the value of their knowledge base and intellectual property in the environment that they find themselves and exploit it fully. This means that one route to a good career for the best materials scientists and engineers will be to become specialist in branches of the discipline that are key to the future of the company. They need to be prepared to change that specialisation when the company's needs change, or move on when they choose. Another route may be to become hybrid materials generalists/engineers/business developers as they move through the company or move on outside. Most importantly, materials scientists must, like everyone else, manage their own careers and be ready to add to their skills or move on when the time is right.

Selection of companies to work in is critical. Prospective companies must show signs of changing as outlined in the section on options for the metal industry or they come off the list. Success in business enterprises requires the organisation to break the people free to develop and implement change increasingly quickly. I am firmly convinced that there are more good ideas and implementation plans in the heads of many people that never see the light of day in organisations today because the management has not created the right atmosphere for people to feel confident about voicing them. If you're not in such an environment, move on. Success is not forever. Many companies have a finite life, particularly those at the top. Very few stay there forever particularly those based on rapidly developing technologies. When you see the writing on the wall move on to one that is like that.

Implications for Universities

I believe that a major objective of universities is to create wealth. In this context university research will become a key provider of new options for successful metals companies. The best faculty will become networked into the IPR management activity I have described and have the opportunity to contribute greatly. Some faculty may choose to become part of the environment of technology-based new business start-ups. Others may lead technology commercialisation businesses that start from a university base. The universities need to make this process easy, as it is in the USA.

From an education point of view, the only thing I would emphasise would be to try and instil an open mind with regard to career in students and recognition of the lifetime learning ethic. The basic education programme in most universities does not need radical overhaul. Perhaps, just provide a few more optional courses for students to taste but not in business subjects. This comes later with an MBA when people are more experienced.

Conclusion

Yes, some structural metals companies have a bright future, but only if they are truly innovative in defining it.

Further Reading

D. Coupland, *Microserfs*, 1995, Pub. Flamingo ISBN 0-00-225311-9.
P. F. Drucker, *Innovation and Entrepreneurship*, 1994, Pub. Butterworth–Heinemann Ltd. ISBN 0-7506-1908-2.
R. N. Foster, *Innovation: The Attackers Advantage*, 1986, Pub. Macmillan London Ltd. ISBN 0-333-43511-7.
T. L. Friedman, *The Lexus and the Olive Tree*, 1999, Pub. Farrar, Straus and Giroux NY. ISBN 0-374-19203-0.

G. Hamel, *Leading the Revolution*, 2000, Pub. Harvard Business School Press, ISBN 1-57851-189-5.

G. Hamel and C. K. Prahaled, *Competing for the Future*, 1994, Pub. Harvard Business School Press, ISBN 0-87584-416-2.

A. Hartman, J. Sifonis with J. Kador, *Net Ready*, 2000, Pub. McGraw Hill, ISBN 0-07-135242-2.

S. Johnson, *Who Moved My Cheese*, 1998, Pub. Putnam NY, ISBN 0-399-14446-3.

R. Jonash and T. Somerlatte, *The Innovation Premium*, 2000, Pub. Random House Business Books, ISBN 0-7126-8428-X.

J. Kaplan, *Start-up*, 1995, Pub. Houghton Mifflin Company NY ISBN 0-395-71133-9.

R. Leifer, C. M. McDermott, G. C. O'Connor, L. Peters, M. P. Rice and R. W. Veryzer, *Radical Innovation*, 2000, Pub. Harvard Business School Press, ISBN 1-57851-903-2.

R. N. Nayak and J. M. Ketteringham, Breakthroughs, 1986, Pub. Rawson Associates NY, ISBN 0-89256-294-3.

T. Peters, *The Circle of Innovation*, 1997, Pub. Alfred A. Knopf, ISBN 0-375-40157-1.

J. B. Quinn, J. J. Baruch and K. A. Zein, *Innovation Explosion*, 1997, Pub. The Free Press NY, ISBN 0-684-83394-8.

T. R. Reid, *The Chip*, 1984, Pub. Simon and Schuster, ISBN 0-671-45393-9.

J. Ridderstrale and K. Nordstrom, *Funky Business*, 2000, Pub. Book House Publishing, ISBN 0-273-64591-9.

J. M. Utterback, *Mastering the Dynamics of Innovation*, 1994, Pub. Harvard Business School Press, ISBN 0-87584-342-5.

F. Wiersema, *Customer Intimacy*, 1997, Pub. Harper Collins, ISBN 0-00-255821-1.

Role of Advanced Materials in Modern Aero Engines

PETER PRICE
Engineering-Defence (Europe) Rolls-Royce PLC, UK

ABSTRACT
Advanced materials have been a key technology throughout the history of the aero engine, continually contributing to its increasing effectiveness. The constant drive for increased performance necessitates materials that demonstrate not just ever-increasing high-temperature strength but also decreasing density for lower weight and higher thrust-to-weight ratios. As current materials reach the limits of their potential, new materials are emerging that offer significant improvements in capability, or, to a military operator, the potential edge in combat situations.

The composition of the aero engine has changed radically in the last 40 years. Until around 1960, steels accounted for about 60% by weight of the engine. They now represent little more than 10%. The presence of aluminium alloys has also vastly diminished. With the latest generation of high bypass turbofan aero engines, titanium alloys and nickel superalloys account for up to 75% of the engine weight. Titanium based alloys offer mechanical strength and light weight, and are used in fan and compressor blade and disc manufacture. Nickel superalloys have superb high-temperature capability, and are heavily used in the turbine 'hot-end' section of the engine.

Composite materials have also steadily increased in use and, while not yet fulfilling eaarly predictions, they have overtaken aluminium in overall usage and make up a large proportion of the engine's static structures. New materials, such as titanium and nickel aluminides and advanced composites, can offer a step change in capability over today's alloys. However, the aero engine market is cost driven, so development of existing alloys remains vital, as the challenge of demonstrating performance improvements and simultaneous cost reduction via new advanced materials is undertaken.

The Past, Present and Future of Magnetic Data Storage

WILLIAM O'KANE
Seagate Technology (Ireland) Ltd, 1 Disc Drive, Derry, N Ireland BT48 0BF
e-mail: William_J_O'Kane@Seagate.com

ABSTRACT
The rigid disc drive industry is intensely competitive, with manufacturers competing for a limited number of major customers. The principal competitive factors in the rigid disc drive market include product quality and reliability, form factor, storage capacity, price per unit, price per megabyte, product performance, production volume capability, and responsiveness to customers. It is generally recognised that the most important factor that gives the industry the ability to leverage these factors is increasing the areal density and access time of the disc drive.

In the early 1990s areal density grew at a relatively modest 30–40% per annum. However, with the advent of anisotropic magneto-resistive (AMR) and more recently giant magneto-resistive (GMR) head technology, areal density has since grown at an amazing 100–120% per annum (Fig. 1).

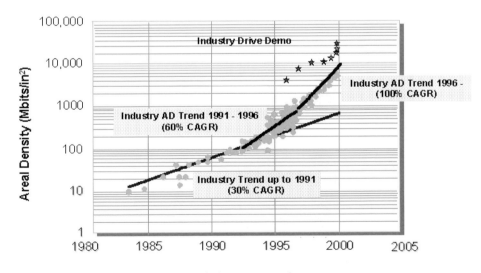

Fig. 1 Areal density growth vs. time.

It is unrealistic to assume that this rate of increase will occur indefinitely. However, due to current consumer appetite for increasing storage capacity, coupled with the competitiveness of the disc drive industry, it is clear that these areal density advances will have to be maintained for the next few years. The challenges required in order to sustain these advances are tremendous, and it is the aim of this paper to outline some of the key challenges facing the industry. An overview of how the industry has evolved to this point will also be provided.

Introduction

A schematic illustration of a modern disc drive is shown in Fig. 2 and can be seen to consist of three main components, these being:

1. The media which stores the desired information,
2. the recording head which writes the information to the disc, and subsequently reads it back, and
3. the electronics sub-system which processes the data.

The challenges required in order to sustain current areal density advances that will be discussed in this paper are divided into two main categories; namely those related to performance requirements of (1) the recording head, and (2) the recording media. The effect of the drive electronics in determining disc drive performance is beyond the scope of this paper.

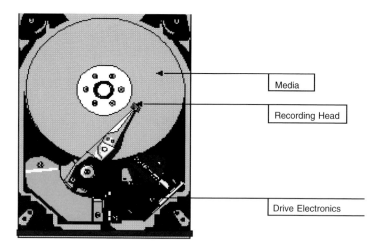

Fig. 2 Schematic illustration of disc drive.

Recording Head Overview

In a hard disc drive, information is stored in a series of concentric tracks with each track holding many millions of magnetic transitions. The areal density of a disc drive is defined as the product of the number of tracks of information per inch (tpi) and the number of transitions per inch along the track (bpi), expressed in bits per square inch (Fig. 3).

From the above illustration it is clear that areal density can be increased by either increasing the number of tracks per inch of disc or by increasing the number of bits stored within a given track. Each of these results in the bits that store the information becoming smaller. Development of recording media that will store these bits reliably constitutes one of the main areas of media development that will be discussed in the next section.

Smaller recorded bits are more difficult to form and subsequently read. More sensitive heads are required in order to maintain acceptable drive performance. In addition, ever increasing data transfer rates, with associated bandwidth and noise implications, make the need for amplitude even more acute. The need for improved write and read performance will be discussed in following sections.

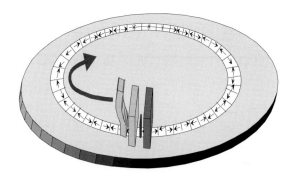

Fig. 3 Schematic illustration of magnetic recording system.

Media Challenges

As mentioned above, increases in areal density results in magnetic bits of information becoming smaller. As bits become smaller, their thermal stability decreases. The thermal stability can be improved by increasing the crystalline anisotropy of the media.

Media noise is proportional to N, the number of grains in a bit. Thus as areal density increases and the bit size falls correspondingly, the grain size needs to decrease. The following is a list of desirable media attributes:

108   The Past, Present and Future of Magnetic Data Storage

- High coercivity media, H_c
- High remanence, M_r
- Thin squareness, S
- Thin media layer, $M_r t$

Increases in media coercivity and remanance are obtained by transitioning media from binary and ternary alloys, to quaternary and quinternary alloys, as well as potential shifts from the CoPtX based systems used in present generation disc drives.

The main challenge for media to support ever increasing areal densities is related to the super-paramagnetic limit, which results in spontaneous demagnetisation of the media. The exact areal density where conventional longitudinal recording can no longer be employed is still the subject of debate; however, there is general consensus that it is in the 50–100 Gb/in^2 regime. In order to suppress demagnetising effects of the media, the aspect ratio of recorded bits is changing from a 20:1 aspect ratio today, to a projected aspect ratio of 4–6:1 at areal densities around 100 Gb/in^2 as shown in Fig. 4 below.

As well as advances in materials, alternative-recording mechanisms will most likely have to be employed. Perpendicular recording offers a means of extending areal density recording beyond the 40–60 Gb/in^2 regime due to the increased thermal stability of the recorded transitions. However there are many unresolved issues associated with perpendicular recording media, including the ability of the media to support high data rates, the effective coercivity of the media at typical recording frequencies, and incorporation of the media into a working product.

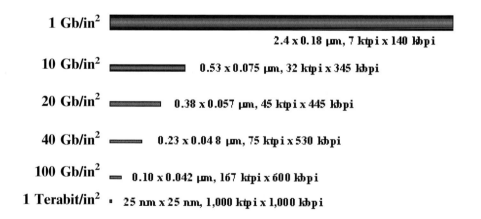

Fig. 4   Relationship between areal density and magnetic bit size.

Beyond perpendicular recording, patterned media may have to be employed. This should offer extremely high density recording, however, research into this is still in early stages, and volume production issues have yet to be addressed.

Recording Head Challenges

Inductive Write Head

There are two main challenges to overcome in the design and manufacture of future generation writer structures, these being (i) the ability to write media layers with ever increasing coercivity, and (ii) writing data to the disc at ever higher rates.

The need to write data to higher coercivity media results in the need for pole materials with increasing saturation moment, $4\pi M_s$. As media coercivity increases from a current value of 3000–4000 Oe to values in excess of 6000 Oe, the saturation moment of the poles will have to increase correspondingly from a current value of 14–16 kG to values approaching 24–28 kG for areal densities in the range 80–100 Gb/in^2.

Developing pole materials with a higher saturation moment is critical to the success of future generation recording heads. Currently most commercial disc drives employ electroplated Ni_xFe_{1-x} alloys where $x = 45$–81. However, the highest saturation moment that can be attained from such an alloy is in the region of 16 kG. Recently there has been considerable focus on the development of electroplated ternary CoFeX alloys where $X = Ni$, Cu etc., and moments as high as 21 kG have been reported.

Sputtering deposition is also being investigated as a means of developing high saturation moment alloys. The most common material under investigation is based on the FeXN family of alloys (where $X = Al$, Ta, Ti, W etc.) and can be deposited with a saturation moment approaching that of pure iron. $Co_{40}Fe_{60}$ has the highest known moment of any alloy. In bulk form, this alloy has a saturation moment of approximately 24 kG. However, due to the absence of Ni, corrosion issues may preclude the use of such a material. Finally there have been some reports indicating that $Fe_{16}N_2$ has a giant moment, with reports indicating various values of saturation induction in the range 28–29 kG. However it is not yet clear what the exact moment of this material is, and if production of such a material is scalable to a manufacturing environment.

As well as being capable of writing transitions to higher coercivity media, the materials must also be capable of performing at higher frequencies. Most leading edge drives currently operate at frequencies in the 200–250 MHz range, with this value likely to increase to 500 MHz within the next 2–3 years. As a result, the pole materials must retain their properties at high frequencies. Loss in performance at frequency is usually due to internal eddy current losses within the pole materials at high frequencies. This can be overcome by laminating materials, thus reducing the conducting thickness of each individual lamination period and extending the frequency accordingly. Alternatively the use of high resistance alloys (such as the amorphous Co family) may circumvent the need for lamination.

Read Sensor Challenges

As areal density increases, the magnetic bit size becomes much smaller. In addition the recording media itself is expected to become thinner. Both of these result in the sensor seeing a lower signal from the media. Thus to ensure that the information can be read back from the media repeatedly, the sensitivity of the reader element must increase. Recently the industry has transitioned from inductive playback heads (with a sensitivity of 100–150 $\mu V/\mu m$) to AMR sensors that display a MR ratio of 2–2.5% and have sensitivities in the region of 300–500 $\mu V/\mu m$. These in turn are currently being superseded by spin valve sensors with MR ratios of 6–10% and a sensitivity approaching 2 $\mu V/\mu m$. In order to sustain the required signal-to-noise ratio as areal density increases, the sensitivity of the read element must be further increased. Increasing the sensitivity of devices is a key challenge to the industry, and further work must be undertaken in one or all of the following areas in order to realise these advances:

- Improved understanding/control of deposition processes.
- Alternative deposition techniques.
- Improved understanding/control of interfaces.
- Materials with improved thermal stability.
- Control/manipulation of grain size.

It is not clear how much amplitude can be obtained from spin valve/GMR sensors, and research is currently underway to evaluate alternative sensor structures. The most promising of these is based on spin tunnelling and MR ratios as high as 25% have been reported. However, there are considerable challenges associated with reliable fabrication of the tunnel junctions. Additionally, because of the inherent insulating nature of the junction, these devices typically have very high resistances with the result that the drive electronics will have to adapt accordingly.

Summary of Challenges

In summary, the magnetic recording industry is evolving at a tremendous pace that is resulting in areal density capability doubling every 12–15 months. However, to sustain areal density advances there are several challenges that must be overcome. For future generation recording media the challenges (and thus areas for development) may be summarised as follows:

- High coercivity, small grain media.
- Media with improved grain size distribution.
- Means of overcoming issues associated with super-paramagnetic limit.
- Development of low noise perpendicular recording media.

With respect to recording heads, the main challenges can be summarised as:

- Sensor materials to support areal densities in excess of 200 Gb/in^2.
- Ultra high moment materials for high density recording.
- Reader structures for extremely high density recording.
- High frequency materials for both writer structures and read sensors.
- Novel process technologies to support fabrication of sub micron devices.

On the Horizon – Some Material Advances in the Nuclear Industry

SUE ION
Director of Technology and Operations, British Nuclear Fuels plc

ABSTRACT
This paper considers some of the challenges facing the nuclear industry today and how they may be overcome in future decades. Examples will be given of where materials engineering is already playing a central role in these advancing technologies and some broad-ranging conclusions are drawn.

The Early Days

Few industries raise as much passionate debate as the nuclear industry. Still fewer play such a large role in society as the nuclear industry.

Born out of military necessity in the early 1940s, the UK's fledgling civil nuclear power programme was quickly established. Research and administrative centres were created, notably at Harwell near Oxford, and nuclear reactors together with their fuel cycle support facilities were designed and constructed in short timescales that seem almost unbelievable by today's standards (Appendix 1). Materials science, of course, was intimately associated with these early developments – what is the best cladding? What is the best form of the fuel itself? What to use as a moderator? Or as a coolant? Harwell, particularly, became a national and indeed an international centre of excellence in all matters to do with the nuclear fuel cycle, but its contribution to materials science is probably one of its most famous achievements.

Of course, famous achievements are made possible by famous people. To attempt a list of materials technologists who have made significant contributions through their roots in the nuclear industry would guarantee to offend the multitude of their colleagues who shared those pioneering times. However, few could argue with a short list of eminent professionals whose names have stood the test of time:

Professor Sir Alan Cottrell
Professor Brian Eyre
Professor Sir Peter Hirsch
Professor Marshall Stoneham
Professor Derek Hull

Professor Geoff Ball
Sir Monty Finniston
Professor Robert Cahn
Professor Ray Smallman
Professor Geoffrey Greenwood

But innovation was not limited to materials science. Indeed, as Lord Penney remarked, Harwell was '... bursting with ideas about possible reactor systems'. Following the first experimental reactors GLEEP and BEP0, prototype fast reactors ZEPHYR and ZEUS were constructed, as well as a toroidal fusion reactor ZETA. Fields of research covered:

- Magnox reactor systems, such as Calder Hall.
- Liquid sodium-cooled Fast Reactors.
- Fusion.
- Ceramic fuels, ultimately used for Advanced Gas Reactors.
- Pressurised Water Reactors, from 1954, in which zirconium was identified as the best cladding material, but was expensive.
- Organic liquid moderated reactors.
- Heavy water reactors – based on the Chalk River design.
- Uranyl sulphate dissolved in Heavy Water reactors.

However, despite an intensive recruitment campaign in the early 1950s, the intake of new staff did no more than keep pace with wastage. Reactor Schools were established at Harwell (1954) and Calder Hall (1957) and concerted efforts were made to introduce nuclear physics and engineering into Technical College courses – all in an effort to boost skilled manpower. So serious was the shortage of manpower, coupled with the need to retrain contractors in new standards of welding and higher safety standards, that the UK's nuclear programme was being delayed. This led to the recommendations of the Strath Report (1956) which proposed that the Advanced Gas Reactor (and Steam Generating Heavy Water Reactor as a lower priority) should be adopted as the reactor systems of the future – the Pressurised Water Reactor was to be abandoned because the UK simply did not have the people to lead its development (Fig. 1).

And Today

Water-cooled and moderated reactors have become dominant. The Pressurised Water Reactor (PWR), born out of submarine propulsion systems, was pioneered by Westinghouse in the US and now forms the basic design for

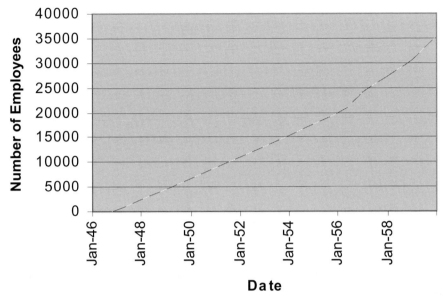

Fig. 1 UKAEA employees.

nearly 50% of the world's reactors. The Boiling Water Reactor (BWR) accounts for 22%. Gas-cooled graphite-moderated reactors were deployed in the UK (the Magnox reactors) and in France, with the second-generation Advanced Gas Reactors being developed in the UK. The CANDU reactor utilises Heavy Water (D_2O) as both coolant and moderator and has found favour in Canada, Korea, India and Pakistan. In the Russian Federation, the light-water cooled, graphite moderated RBMK reactor has been constructed and the BN350 and BN600 sodium-cooled fast reactors are in commercial operation.

In some countries, notably those with poor indigenous energy sources, the proportion of electricity generated by nuclear means is very high (Table 1). Within the OECD countries, the proportion of nuclear-generated electricity is around 24% (36% in Europe), whilst worldwide the proportion is 17%, equivalent to 350 GWe, produced by over 430 reactors. Within the UK, some 27% of electricity is produced by reactors owned and operated by British Energy and BNFL.

Into the Future

All credible predictions suggest that World energy demand will grow over the next fifty years. Such demand needs to be balanced against environmental constraints and the availability of raw materials; meeting the World's energy needs in this sustainable manner is a huge challenge to mankind.

Table 1 Proportion of electricity generated from nuclear energy.[3]

Country	% Nuclear
Lithuania	77
France	76
Belgium	55
Sweden	45
Ukraine	45
Japan	36
UK	27

Many reviews of world energy demand have been published.[4] Electricity generation makes a significant impact,[5] not just because power generation has a great influence on the environment, but also because increasing industrialisation and quality of life in the developing nations leads to much increased electricity demand. It is projected that maintaining nuclear generation at 15% of the World electricity generation will require some 1000 GWe of capacity by 2050 – roughly tripling today's nuclear capability.

Japan is continuing to plan for new nuclear build – Tomari-3 and Shimane-3 were added to the Basic Power Development Plan in 2000 and should be operational by 2008. In the UK, the Trade and Industry Select Committee looking at future energy policy recommended in 1999 that '. . . a formal presumption be made for the purpose of long-term planning that new nuclear plant may be required in the course of the next two decades'. Similarly, the Royal Society concluded that decisions on future energy policy should be taken early enough to enable nuclear to play its full, long-term role in national energy policy.

However, considerations of energy generation and of environmental consequences must go hand-in-hand. Most leading climate scientists now agree that human pollution – mainly from fossil fuels – has added substantially to global warming in the past 50 years. Furthermore, the Earth is likely to become far hotter than previously predicted, with immense consequences for people and wildlife. Average temperatures today are 5°C higher than they were at the end of the last Ice Age; a worst case prediction suggests that average temperatures will rise by 6°C in the next 100 years (the rise estimated in 1995 was 3°C).[6] Certainly, renewable energy sources have their place – wind power, wave power, hydro-electricity and solar cells. However, renewables cannot be the complete answer. For example, the British Isles are the windiest area of Europe so wind farms offer a good opportunity. However, if the UK is to meet its Kyoto target of 10% fewer emissions by 2010 through windpower alone, then it is said that three offshore windfarms would be needed to be built *every day for the next 10 years*. Even then, windpower is unreliable and the voltage and frequencies fluctuate to the extent that it is doubtful if the grid could sustain the high quality electricity supply that modern society (and their computers!) has come to expect.

Nuclear generation goes a long way towards providing the solution. The European Commission recently concluded that '... nuclear may be the only non-fossil fuel-burning source of electricity that would be technically capable of filling a substantial part of the gap in electricity supply that would be created if fossil fuel electricity generation were to be drastically reduced as a response to Kyoto'. Lifetime extensions of existing reactors is seen as a first step, but beyond 2010, it is essential to maintain long-term research if the nuclear option is to be kept open for future generations.[7] Following such a responsible path suggests[5] the need for about 2500 GWe of nuclear generation by 2050 i.e. about 15 times today's capacity.

So what will the nuclear industry look like in the future?

Reactors

In descending order of priority, the criteria for reactor systems in the future are likely to be:[8]

- Economics – with a particular drive to reduce construction times and hence financing costs.
- Safety – a drive for *passive* safety.
- Management of fuel cycle waste.
- Proliferation resistance.

These criteria suggest an evolutionary approach to reactor designs:

Now Existing Light Water Reactors (LWR)
$\downarrow$
Advanced LWRs e.g. ABWR, AP600 for which detailed designs have already been completed.
$\downarrow$
Around 2020 Novel designs such as ABB PIUS system, the Eskom Pebble Bed Modular Reactor or the General Atomics Advanced Thermal Reactor.
$\downarrow$
2040–2050 Liquid-sodium-cooled Fast Breeder Reactors.

Enrichment[9]

Two-thirds of today's enrichment needs are met by obsolete, energy-intensive diffusion plants. By 2015–2020, these will require replacement. Further evolution of the gas centrifuge is likely, though there may be competition from one or more of the laser-based isotope separation technologies such as AVLIS.

Fuel[9]

The economics of the thermal reactor fuel cycle depend on both the 'front end' (pre-irradiation) and the 'back-end' (post-irradiation). In the front end, the economics improve with burn-up to about 50–60 GWd/t. Beyond this level, the separative work function rises steeply and the corresponding irradiation times imply a greater discounting penalty, both of which restrict the likely economic burn-up. Back-end costs will increase with burn-up. Hence the economic optimum burn-up for thermal reactor fuels are not much different from today's technological capabilities up to about 60 GWd/t.

If the fuel is to be reprocessed, 60 GWd/t is likely to be about the limit for a PUREX-type flowsheet. However, for a once-through cycle, 80–100 GWd/t may be attractive. However, radical fuel designs and cladding will be necessary to achieve such levels of fuel burn-up. The Pebble Bed Reactor, using fuel particles coated in silica or silicon carbide/nitride should present few barriers to achieving high burn-up.

Reprocessing[9]

Developments of the current PUREX-type reprocessing technology are being directed towards reducing waste volumes and activities as a means of reducing costs. After 2015, it is possible that non-aqueous fuel reprocessing technologies, such as those based on molten salts, may offer competition. The benefits of molten salt reprocessing are said to include:

- Performed in one unit, leading to lower costs.
- Minimised waste volumes through concentration of fission products from the molten salts.
- Final products ready for fuel manufacture.
- Suitable for a wide range of fuel types.
- Improved proliferation resistance.

Disposal[9]

Disposal of wastes from historic and from current operations is likely to involve surface storage in a treated (e.g. grouted or vitrified) form followed by sub-surface disposal in geological formations. New technical solutions are unlikely to be necessary, though it will certainly be a challenge to treat large volumes of historic waste whilst dealing with current arisings at a time when discharge authorisations are tightening. Public acceptance will be important, particularly if the concept of an international repository is to make progress.

Disposal of spent fuel, as an alternative to reprocessing and recycling, will also be necessary to deal with the 100,000 tonnes of used fuel currently stored at nuclear power plants and interim facilities around the world. Disposal in Sweden is expected to commence in 2008–2012 and use a 400–500 m deep tunnel system. The spent fuel will be isolated from the biosphere by a multi-barrier approach:

- The fuel, almost impossible to dissolve in water.
- Copper canister, to encapsulate the fuel in a very corrosion resistant material.
- Emplacement in watertight swelling bentonite clay.
- The bedrock offering 'lasting' conditions and acting as an efficient filter.

Again, winning public acceptance will be crucial.

The Materials Challenge

We will now turn to some specific examples of the 'materials challenge' faced when turning today's vision into tomorrow's practise.

Fabrication of Advanced Nuclear Fuels

Any commercial-scale reactors ordered in the near future are likely to be based on today's proven designs. Advanced light water reactors, incorporating passively safe designs such as the AP600 which has already been licensed, will be a natural evolution of reliable reactors. The fuel for these designs of reactors will be similar to that in use today, consisting of pressed and sintered ceramic oxide pellets containing fissionable materials (usually enriched uranium) contained in zirconium alloy sheaths.

Beyond the advanced LWRs, a number of novel designs are under development. Amongst these is the direct cycle Pebble Bed Modular Reactor designed to operate at high temperature (around 900°C gas exit temperature) using helium coolant. The successful operation of the reactor will rely heavily on the fuel attaining quality standards achieved in past High Temperature Reactor programmes and which have well-understood quality performance characteristics. Despite this proven design and manufacturing basis, there are still many aspects of the performance and physical properties of the coated fuel particle that present significant materials challenges. Many of these challenges relate to understanding how the changes in the manufacturing process affect the performance of the fuel in reactor.

The fuel elements, or fuel spheres, for the Pebble Bed Modular Reactor (PBMR) can be described as 60 mm diameter spheres of graphite-rich matrix in which are embedded approximately 15,000 coated particles containing a total of 9 g of the fissile uranium inventory. Within the 110 MWe concept PBMR, there will be approximately 375,000 such fuel spheres which are transported through the core in a slow, but regular, removal, inspection and replacement operating cycle. The graphite-rich matrix provides the necessary moderator material content of the reactor core but is also a key structural support material for the coated particles during their transport through the reactor core. Accompanying the fuel spheres in the current design of the PBMR are fuel-free pebbles of identical appearance whose only functions are to provide additional moderation and ensure a specific configuration of fuel spheres is achieved.

Key to the PBMR design concept is the coated fuel particle consisting of multiple layers of pyrocarbon coatings and a single layer of ceramic silicon carbide deposited using a chemical vapour deposition technique (CVD). It is the integrity of the coated particle that determines the fission gas retention characteristics of the overall fuel. Key design objectives for the fuel will have to be demonstrated by modern production quality assurance practices and by the irradiation testing of the initial product (Figs 2 and 3).

Fig. 2 Sintered UO_2 kernel particles.

Fig. 3 Coated particle.

The silicon carbide (SiC) coating layer is both the main load-bearing layer and fission product diffusion barrier in a TRISO particle but the other pyrocarbon (PyC) layers, the buffer layer, the inner-PyC layer and the outer-PyC layer have important roles. For example, the buffer layer, which immediately surrounds the UO_2 kernel, is of low density, has a thickness of 90–95μm and has sufficient voidage to accommodate fission product gas build-up during the life of the fuel and also serves to protect the inner Py–C layer from recoiling fission product fragments. The inner Py-C layer has the function of protecting the UO_2 kernel from the corrosive gases produced during CVD-deposition of the SiC layer and, along with the outer Py–C layer, provides additional structural support to the silicon carbide layer. This structural support comes from compressive stresses exerted on the SiC layer during radiation-induced shrinkage and which counteract the hoop stresses generated within the particle as a result of fission gas accumulation (Fig. 4).

Obviously, such key performance features rely on the particles being manufactured in a uniform, reproducible manner and with the correct thickness, density and microstructure characteristics. Bearing in mind the very large number of individual particles in the fuel inventory of a PBMR, then the roles of quality assurance and advanced materials characterisation systems are evident. These roles will become even more important when consideration is given to further improving the generating capability of the reactor system – by raising operating temperature limits or by increasing burn-up levels. Understanding how all the various layers interact and how effective each layer is in the duty it has to carry out will be key to improving the economy of the reactor and improving the utilisation of the uranium resource.

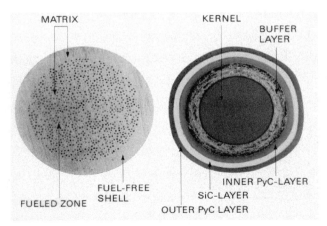

Fig. 4 Sections through a fuel sphere (60 mm diameter) and an individual coated particle.

Pushing the Limits with Advanced Cladding

'Cladding' is the term given to the protective envelope that covers the heat-generating fuel in the nuclear reactor and prevents interaction between the fuel (usually uranium dioxide ceramic pellets) and the coolant (often water or gas).

The desirable properties of cladding include:

- Low cost material.
- Ease and safety of processing.
- Low neutron absorption characteristics.
- Suitability for both pre-reactor, in-reactor and post-reactor handling and treatment (including reprocessing or long-term storage).
- Optimum mechanical and chemical performance in the reactor.

The more significant mechanical and chemical properties are:

- Strength.
- Creep rate.
- Fuel side corrosion.
- Coolant-side corrosion.
- Growth e.g. radiation-induced and hydride-induced.

Zirconium was recognised as a candidate for water-cooled reactors like the Pressured Water Reactor. Indeed Zircaloy-2 and Zircaloy-4 were developed in the early 1950s[10] where the combined addition of tin, iron and chromium resulted in a marked improvement in corrosion behaviour relative to sponge-zirconium. Since that time, many other zirconium-based alloys have been evaluated, but only a few compositions (such as the Zr-1% Nb, Zr-2.5% Nb) had achieved commercial exploitation. Quenching and tempering the Zr-2.5% Nb alloy to completely precipitate out the Nb in the form of very fine uniformly distributed β-Nb particles led to further improvements in water-side corrosion rate, particularly when tested in-reactor in oxygenated boiling water coolant.[11] Similar effects, benefiting from both the β-Nb and from $Zr_{0.5}Nb_{0.3}Fe_{0.2}$ precipitates were expected in the Zr-1% Nb alloy containing iron and tin additions.[12]

From the early 1970s, the drive towards improved efficiencies of PWR systems led to higher fuel burn-ups, extended cycles with higher lithium levels in the coolant, higher power peaking factors and high coolant temperatures. Such changes led to more severe fuel duties and, when coupled with in-reactor enhanced corrosion of Zircaloy-4 cladding at higher burn-ups, resulted in waterside corrosion being one of the limiting factors that restricted the deployment of high-burnup fuel designs. The search for cladding materials exhibiting greater waterside corrosion resistance was led by Westinghouse and focused on the Zr–Nb and particularly the Zr–Nb, Sn Fe alloys, which form the basis for the cladding ZIRLO.[13] Typical compositions and properties are given in Table 2.

Autoclave corrosion tests were carried out in pure water, steam and lithiated water. The low Nb (Zr-0.5% Nb) alloys showed accelerated corrosion over the temperature range 589 K to 727 K and all the binary alloys showed a high sensitivity to corrosion. ZIRLO was superior to Zircaloy-4 in such environment (Fig. 5).[14]

In reactor trials, the binary alloys showed low uniform corrosion rates, but localised modular-like corrosion above an apparent burn-up threshold of about 55 GWd/t. This tendency to form locally thick corrosion oxide increased with increasing power and burn-up. In contrast, neither Zircaloy-4 nor ZIRLO showed any localised corrosion, with the oxide being thinner on ZIRLO samples than on Zircaloy-4 samples. The enhanced corrosion resistance of ZIRLO was also demonstrated in commercial reactors (Fig. 6).[14]

Dimensional stability is also important. Excessive or unexpected dimensional changes of a fuel assembly can result in operational issues such as incomplete control rod insertion or potential fuel assembly interactions and handling concerns due to increases in the fuel assembly envelope.

Table 2 Typical characteristics of Zircaloy 4 and ZIRLO.

	Zircaloy 4	ZIRLO
Nb, wt%	<50 ppm	1.02–1.04
Sn, wt%	1.48–1.52	0.96–0.98
Fe, wt%	0.22–0.23	0.094–0.105
Cr, wt%	0.11–0.12	79–83 ppm
Zr	Balance	Balance
Precipitate size, nm	160	79
Room temperature tensile		
UTS, MPa	84	81
Elongation, %	15	16.5
658 K tensile		
UTS, MPa	47	49
Elongation, %	18.5	15
Post transition corrosion rates, mg/dm^2/day		
633 K	0.6	0.4
633 K in 70 ppm Li	2.9	0.5
633 K in 210 ppm Li	30	13
727 K	50	5
Hydrogen pickup, μg/dm^2/day		
633 K	11	3

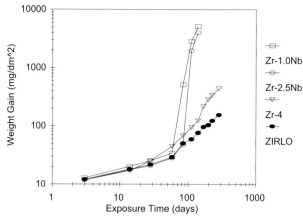

Fig. 5 Corrosion weight gain vs. time for zirconium alloys in 633 K water containing 70 ppm lithium.

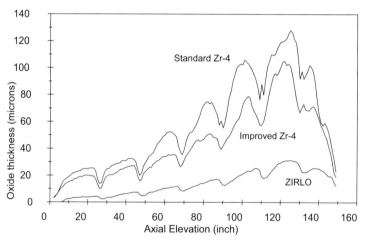

Fig. 6 Oxide thicknesses as a function of position from bottom of fuel assembly, for both Zircaloy and ZIRLO cladding (52.5 GWd/t).

Zirconium alloys, in common with many other materials, expand with increasing neutron flux as the neutrons damage the crystal structure. However, for zirconium alloys, hydrogen pickup during the waterside corrosion process is believed to lead to the precipitation of zirconium hydrides which cause a volumetric expansion equating to approximately 0.3% length increase per 1000 ppm hydrogen content. This effect is largely independent of texture or prior history and has been supported by laboratory and commercial reactor data (Figs 7 and 8).[15]

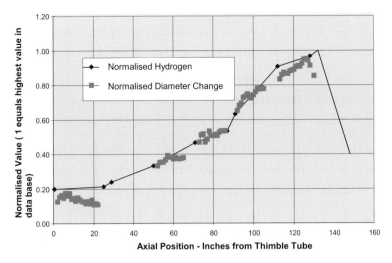

Fig. 7 Axial profiles of diameter change and hydrogen levels (relative values).

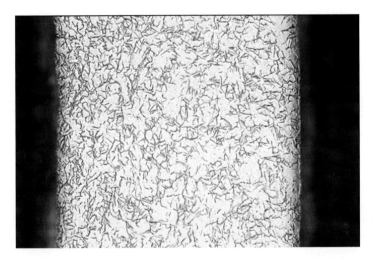

Fig. 8 Transverse view of zirconium hydrides, 760 ppm hydrogen (50 GWd/t).

The choice and development of cladding alloys has been a triumph for materials technology. But there is more to do; engineers are working on further proprietary developments in an effort to further improve performance and reduce costs. What's the limit? Are ceramic-clad fuels viable? Or have we pushed conventional designs to the limit and now need to turn to more radical fuel and reactor designs, such as the Pebble Bed Modular Reactor?

Structural Integrity and Surveillance

Most commercial reactor designs operate under pressure. It is therefore vitally important that the reactors are correctly designed and flawlessly constructed if leakage of radioactive material is to be avoided. Guaranteeing the integrity of the reactor and its associated coolant circuits over the lifetime of the installation is a challenge, but one which will enable the reactor to continue operating, with a happy regulator and a tolerant public. Similarly, the chemical plants that support the nuclear reactor itself often employ high temperatures and/or corrosive chemicals; they too must be safe *and proven to be safe*. Structural integrity surveillance and assessment, often in extremely hazardous environments, has therefore become an integral component of plant life management.[16]

An example of the development work taking place in the Structural Surveillance field is remote ultrasonic inspection. In this technique, laser energy is transmitted into a hazardous environment and delivered to the target directly or by means of a fibre optic cable. Ultrasonic waves can be created within the target, particularly if the incident power is sufficient to locally ablate the target surface. Remote *detection* of the induced ultrasound and its reflections is more difficult. Trials have shown that induced ultrasound can be detected by confocal Fabry-Perot interferometers and by photo-emf detector systems. The latter, which can match and often out-perform the interferometer also have the benefits that they are compact, robust detection systems suitable for industrial application.

With direct application of laser light, Rayleigh, Lamb and compression waves can be detected in plate specimens. Thickness measurements in 35 mm aluminium or 30 mm steel plate are possible to ±0.1 mm and planar defects >2 mm are detectable. Lamb and Rayleigh waves can also be detected in 3 mm plate samples with stand-off distances of over 2 m (Fig. 9).

Beam delivery via fibre-optic cable has many advantages. However, use of such cables restricts the beam power that can be applied in the creation of the ultrasound. So far Rayleigh and Lamb waves have been detected, and thickness mapping and defect detection has been proven in the laboratory.

Further advances in signal processing and improved coupling of laser delivery and detection systems to optical fibres are expected to open up opportunities for:

- Defect detection.
  - Corrosion, weld defects, surface cracking.
  - Suitable for curved and inaccessible surfaces.
- Thickness gauging.
- Determination of elastic constants.

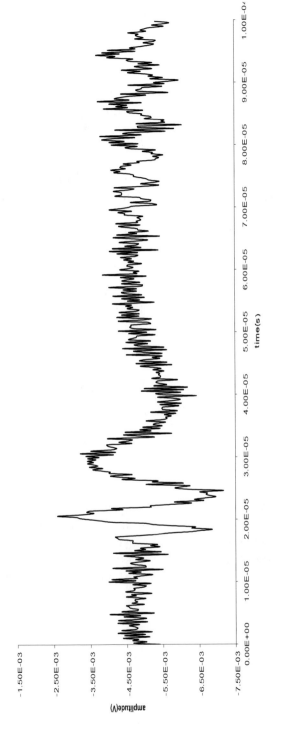

Fig. 9 Laser generated Lamb waves in 1.56 mm Al plate measured using an optical fibre coupled photo-emf system.

Such a system would be particularly suited to high temperature or radioactive environments or confined systems or with rotating objects. Development work continues.

Next Generation Reprocessing Options

Many utilities choose to have their used fuel reprocessed. In today's reprocessing plants, the fuel assemblies are mechanically guillotined into short (typically 25 mm long) sections before being dissolved in hot nitric acid. The liquor is chemically separated into uranium, plutonium – both of which can be reused as fuel – and wastes. Notwithstanding the development of radically different technologies such as molten salt reprocessing, there are considerable advantages to be gained by integrating process steps and reducing mechanical handling. One such development is the solution contact electrochemical dissolution of nuclear fuel.

Solution contact electrochemical dissolution of fuel avoids direct electrical contact with the fuel assembly. The assembly is lowered into a stepwise pyramid dissolver and dissolution occurs in an impressed anodic region of the fuel assembly that lies close to the cell cathode (Fig. 10). As dissolution of the outer fuel pins proceeds, the assembly descends into the dissolver where the now exposed inner fuel pins are dissolved via a lower electrode pair.

Experiments have been conducted in a 3 m high pilot rig using half-wave rectified AC power supplies and 8 M nitric acid at 85–90°C (Fig. 11).[17] Dissolution rates increased with voltage in the range 12–20 V rms, and feed rates of 40 cm/hr could be achieved. This dissolution rate would be further increased by adopting a full-wave rectified power supply.

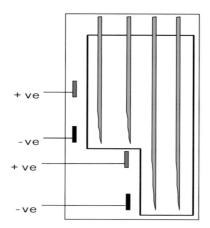

Fig. 10 Schematic of pyramid dissolver.

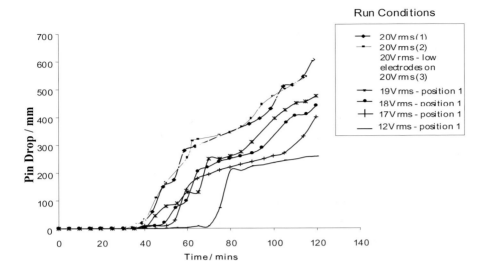

Fig. 11 Effect of potential on pin dissolution rate.

Interaction between the electrode pairs has no significant effect on the dissolution rate. However, fuel pins towards the centre of an assembly are shielded from induced field and dissolve more slowly. This is convenient as it prevents premature severance of the inner fuel pins.

Because the zircaloy components are passivated by an oxide layer during their service in the reactor, it is necessary to reverse the polarity of the electrodes for about 10 minutes prior to commencing dissolution in order to activate the surface. For continuous dissolution, this depassivation process needs to be repeated on a regular basis if undissolved shards of fuel are to be avoided.

If this compact, low-maintenance electrochemical dissolution technique is deployed on full size irradiated nuclear fuel, the solution can be fed directly to the chemical separation stage. Zirconium bearing liquors, together with ZrO_2 sludges from the process have the potential to be converted into stable solid wasteforms for encapsulating the high level radioactive wastes.

What to Do with the Waste?

Like all other industrial processes, the generation of electricity by nuclear means produces waste. Wastes may be in the form of treated gaseous or liquid effluents which are discharged to the environment in accordance with strict Discharge Authorisations issued and monitored by Government Departments. Solid wastes are traditionally classed as Low Level (with radioactivity less than a specified limit), High Level (self-heating wastes of high activity) and Intermediate Level.

In the UK, Low Level solid wastes such as towels, clothing and mildly contaminated plant are drummed, compressed where possible and grouted into containers before being consigned for shallow land burial at Drigg in Cumbria. Intermediate Level wastes include various sludges from the reprocessing operation, ion-exchange resins and activated fuel cladding that has been stripped off the irradiated fuel. These items are size-reduced where possible, then remotely loaded into containers and immobilised in grout. The containers are stored in passively safe buildings pending the availability of a deep geological repository. High Level wastes are the used fuel itself and the liquids containing the fission product wastes separated from the re-usable components in the reprocessing operation. This waste amounts to about 3% of the fuel and is currently evaporated, mixed with a glass frit and melted into stainless steel containers which will be surface-stored for many years in accordance with Government policy. Long term disposal is likely to be in a deep repository.

The required characteristics for a High Level wasteform include:

- Chemical stability, in container and ultimately in surrounding rock formations.
- High temperature stability.
- Radiolytically stable.
- Low leach rates.
- Incompressible solid.

Borosilicate glasses are likely to remain the materials of choice for immobilising historic wastes and wastes arising from today's reprocessing operations. However, developments of the process to deal with certain high sulphur and high zirconium-content wastes, such as the 55 million gallons stored at Hanford in the US have been carried out. Trials using a Joule Ceramic Melter have shown that sulphurous and refractory wastes may be processed at twice the normal throughput by increasing the operating temperature by 200°C, tailoring the waste glass composition and closely controlling the melting process. The challenge is one of minimising the wear and tear on the melter materials and avoiding the volatilisation of active species.

Wastes generated by potential advanced reprocessing technologies, such as electrochemical dissolution (above), will present new challenges and new opportunities. Such wastes are rich in zirconium, iron and chromium as well as the fission products and non-recycled actinides. Titanate ceramics derived from Synroc are being developed for such wastes. These wasteforms can use the fuel assembly components as functional constituents of the phase assemblage enabling waste loadings of up to 50% and leach test performances exceeding those currently achieved for vitrified wasteforms. Furthermore, the final waste volume is reduced by a factor of four (to around 0.1 m^3/t of fuel) in comparison with a vitreous wasteform and thus makes the ceramic waste much more economically attractive.[18]

The materials challenges to develop, understand, optimise and deploy waste management technologies are considerable.

Conclusions

From the very earliest days, the UK's nuclear power programmes have benefited from the understanding and advances in materials technology. Likewise, as the Harwell experience amply demonstrates, materials technology has benefited from the nucleation and growth of the nuclear power industry. The science and the practical application have gone hand-in-hand.

Even today, in a nuclear power industry that many regard as 'mature', there are significant challenges for the materials technologist. Improving processes to reduce costs or optimise products often requires the introduction of new materials either by evolution or by a step-change. Adopting commonly-used materials from outside the industry can also demand careful choice, design and implementation if they are to perform in the unique environments found in the nuclear industry. Structural surveillance and integrity assessments are also a vital component of our licence to operate – expected by the public, the politicians, the regulators and ourselves in the industry.

In the future, the challenges will be just as great. Guaranteeing integrity and justifying life extension; new reactor designs as the world increasingly recognises the environmental benefits of nuclear power; the fuel, fuel treatment and wasteforms which support such advanced reactors – all these will rely on the materials technologists of this generation and the next.

Some would say that the nuclear industry is currently sleeping. But when the environmental wake-up call is sounded, the industry must be in a position to respond. Never again must the UK be in the same position as in the early 1950s where lack of skilled resources inhibited the growth of nuclear power. We need a healthy supply of good graduates and technicians, both now and into the future. Furthermore, we must retain the laboratories, test reactors and other facilities that support the industry.

Academia has a role to play here. We are members of an exciting profession, working in an exciting environment. We must help the school pupils of today to understand engineering, particularly materials engineering; help them to recognise the benefits to UK wealth creation; encourage them to take up the materials challenge and offer exciting and enticing courses that meet industry's needs. This requires an adequate number of high quality academic institutions working in close collaboration with the 'end-users'.

With its foundations 150 years ago as the Royal School of Mines, Imperial College is well placed to lead in this particular 'materials challenge'.

References

1. R. F. Pocock, Nuclear Power: Its Development in the UK, *I. Nuc. E.*, 1977.
2. *The Development of Atomic Energy: 1939–1984*, UKAEA.
3. *2000 World Nuclear Industry Handbook*, Nuclear Engineering International, 2000.
4. J. Houghton, *Energy for Tomorrow's World – The Realities, the Real Options and the Agenda for Achievement*, World Energy Council, 1993, Kogan Page.
5. B. L. Eyre, Prospects and Challenges for Nuclear Power in the 21st Century, *Proc. KAIF/ANS Conf.*, 1997, 45–59.
6. Draft report of new research by Intergovernmental Panel on Climate Change, October 2000 (final report expected May 2001).
7. Technical Background Document to Commission Green Paper on 'Security of Energy Supply', October 2000.
8. S. E. Ion, Criteria for the Reactors of 2020, *Proc. ICONE* **8**, 2000.
9. L. N. Chamberlain, S. E. Ion and J. Patterson, Fuel Cycle Technologies – the next 50 years, *Proc. IAEA Symp. on 'Nuclear Fuel Cycle and Reactor Strategies: adjusting to new realities'*, June 1997.
10. S. Kass, The Development of the Zircaloys, *Proc. Corrosion in Zirconium Alloys*, ASTM, 1964.
11. J. E. Lesurf, The Corrosion Behaviour of 2.5% Nb Alloy, *Proc. Applications-Related Phenomena in Zirconium and Its Alloys*, ASTM, 1969.
12. Amaev, *et al.*, Corrosion Behaviour of Zirconium Alloys in Boiling Water Under Irradiation, *Proc. 4th UN Int. Conf. of PUAE*, 1971, **10**, 537.
13. Sabol, *et al.*, Development of a Cladding Alloy for High Burn-up, *Proc. Zirconium in the Nuclear Industry: 8th Int. Symp.*, ASTM, 1989.
14. Sabol *et al.*, *In Reactor Fuel Cladding Corrosion Performance at High Burn-ups and Higher Coolant Temperatures.*
15. Kesterton *et al.*, *Impact of Hydrogen on Dimensional Stability of Fuel Assemblies.*
16. P. E. J. Flewitt, Structural Integrity Assessment of High Integrity Structures and Components: User Experience, in *Mechanical Behaviour of Materials*, A. Bakker (ed.), Delft, 1995, 143.
17. J. D. Wright, P. W. J. Rance, S. Banner and T. Tinsley, Solution Contact Electrochemical Dissolution of Nuclear Fuel – Pilot Plant Studies, *Proc. Atalante 2000 Conf.*, Avignon, Oct. 2000.
18. Titanate Ceramics for the Immobilisation of High Level Wastes arising from Advanced Purex Reprocessing Technology, in *Environmental Issues and Waste Management Technologies in the Ceramic and Nuclear Industries*, **VI**, Am. .Ceram. Soc., 2000 (in press).

Appendix

Civil Nuclear Power in the UK – Chronology of the Early Years[1,2]

1938	December	Hahn and Strassman in Berlin discovered nuclear fission by bombarding uranium with neutrons.
1940	February	Peierls/Frisch memorandum 'On the Properties of a Radioactive Super bomb'.
1940	April	Prompted by the Peierls/Frisch memorandum, the War Cabinet set up a Sub-Committee on the Atomic Bomb under Prof. Sir George Thomson. Later renamed the Maude Committee, Maude being a code word.
1941	July	Two Maude Committee reports submitted to the then sponsoring Ministry for Aircraft Production, 'Use of Uranium for a Bomb' and 'Use of Uranium as a Source of Power'.
1941	October	Directorate of Tube Alloys (code name for atomic energy R&D) set up within Department of Science and Industrial Research.
1944	September	Technical Sub-Committee of Tube Alloys Consultative Committee urges the need for a UK experimental establishment.
1945	April	UK Government decision to undertake a broad programme of R&D in atomic energy, including building a research establishment. Harwell site announced in October.
1946	January	Attlee appoints Lord Portal as Controller of Atomic Energy. Reporting to him were Prof. John Cockroft as Director of Research and Christopher Hinton (formerly of ICI) as deputy controller and responsible for fissile materials. Cockroft's group established themselves at Harwell and commenced theoretical studies for the production of plutonium, which had been shown to be more effective than uranium in explosive devices. Hinton '. . . with no staff, no office accommodation and practically no information' set up headquarters at Risley on 4 February 1946.
1946	March	Construction of fuel plant at Springfields commenced.
1946	June	Construction of British Experimental Pile 0 (BEP0) commenced at Harwell.

1947	July	Establishment of Windscale announced; site clearance in September.
1947	August	GLEEP, Britain's first successful research reactor went critical, followed by BEP0 a year later. BEP0 used to observe the behaviour of uranium, graphite and other materials in radioactive fields.
1948	January	First uranium cast at Springfields – 22 months after construction started.
1950	September	Windscale piles, being enlarged BEP0 piles, went critical and the first UK bomb was detonated on 3 September 1952.
1953	January	Further supplies of plutonium were needed. Prompted by the very harsh winter of 1947/8 and worries that the National Coal Board were failing to meet their output targets, the Calder Hall reactors were approved. Calder Hall was designed as combined heat and plutonium-producing reactors based on Harwell designs. The contract was let on 22 July 1953 and Reactor 1 went critical at 7.15 pm on 22 May 1956. The contract value was £$2\frac{1}{2}$m.
1956	October	'Future generations will judge us, above all else, by the way in which we use these limitless opportunities which Provenance has given us and to which we have unlocked the door. They offer us a vital and timely addition to the industrial resources of our nation and to our material welfare . . .'. The Queen opens Calder Hall.
1956	November	Contracts for Berkeley and Bradwell Power Stations placed. The civil nuclear industry in the UK had arrived!

Materials in Medicine

R. C. BIELBY, L. D. K. BUTTERY and J. M. POLAK
Tissue Engineering Centre, Imperial College School of Medicine,
Chelsea and Westminster Hospital, 369 Fulham Road, London, SW10 9NH

ABSTRACT
The human body has a limited capacity for regeneration and repair. For hundreds of years one of the principal goals of medicine has been to find artificial ways of overcoming the debilitating and disabling effects of tissue and organ damage. The very earliest attempts may date back to prehistory: remains at Palaeolithic sites and cave paintings depicting amputees suggest that early humans did attempt to tackle the problems that came about through disease or trauma.[1] Pliny and other authors describe the design and use of prostheses.[2] With the arrival of the Renaissance and the re-birth of science, medicine began to make important advances in understanding the functioning of organs and how, in certain cases, artificial substitutes might be used to treat patients. Although these early attempts might appear rather crude, many prosthetics from mediaeval times onwards show high degrees of skill in both their design, use of materials and manufacture.[2] From these first attempts arose a set of principles which still apply today,[3] although advances in science and engineering allow them to be applied in a greatly more sophisticated way.

Introduction

The earliest use of synthetic organ replacements was in prosthetics. Both functional and cosmetic prostheses have been widely used under many different circumstances utilising the whole breadth of extant materials. Before surgery had the benefits of antiseptics and antibiotics, amputation was a commonly practised form of life saving surgery – although mortality rates ran as high as 70%.[4] The aftermath of the American Civil War saw the first widespread use of modern design and materials in prosthetics,[5] although the claims made for many artificial limbs were exaggerated. Earlier designs were updated and incorporated the very latest materials. Prosthetics have always sought to mimic as closely as possible the material strength and form of the tissue being replaced, assuming that millions of years of evolution have already achieved the most efficient design. Compared to those of 140 years ago, modern prosthetics are much lighter, stronger, resilient, more easily adapted to by the patient and cosmetically superior.[6] Increasing sophistication in electronics, micro-motors, interface design, computer software and processing power has allowed ever more superior control by the user and even the capacity for sensory feedback. However, they all have one thing in common – they are synthetic, non-living replacements for biological tissue. Function can be restored, but not the capacity for growth or adaptation to physical requirements – though they can be designed and altered to meet different needs. This is a minor criticism however when one considers the profound impact of prosthetics on the lives of those who have become disabled.

Medical Implants

In more recent times, a wide number of applications have opened up for a different use of materials – medical implants. Successful interventions have been developed where synthetic materials are used to restore some degree of normal functioning to a diseased or damaged tissue. There is a steadily rising requirement for treatments for patients suffering from traumatic or degenerative conditions as the proportion of elderly individuals in the population increases. Improvements in surgical techniques and post-operative care has meant that operations that were highly unlikely to succeed 50 years ago are now commonplace. As the potential to intervene has increased, so has the need for implants.

Principles of Implant Design

A number of general criteria apply in the design and testing of medical implants for all conditions. In order to develop a successful treatment, knowledge of normal tissue physiology, anatomy, biochemistry and biomechanics are necessary, as well as changes that occur in disease. In order for a synthetic implant to successfully substitute, its chemical and mechanical properties must closely match those of the tissue it replaces. Various materials (both organic and inorganic) have been used since the earliest recorded civilisations to replace tissues lost to disease, injury and (more recently) old age. Even the most rudimentary use of materials such as the suture has been dramatically changed by the discovery and use of new materials types and the potential for tailoring materials properties to suit different applications. This has led to the two principal groups of implants – those composed of polymers for soft tissue replacement and those utilising metals and ceramics for hard tissue replacement. Due to the complex structure of biological tissues, often no single material can satisfy all the requirements and composites involving combinations of materials are becoming increasingly common.

Synthetic polymers come in two varieties – condensation polymers (e.g. polyester, polyurethane) and addition polymers (polyethylene, polyvinylchloride). They are in use in most types of implant where flexibility is required. Metals and alloys are used where mechanical strength is the principal requirement. Some, such as titanium, steel and aluminium are used as the base metal and are often found in implants alloyed with other elements. Common alloying elements include nickel, vanadium and chromium. Depending upon the proportion of constituent metals within the alloy, the material properties can be manipulated to achieve desirable characteristics such as low corrosion.

Ceramics are finding increasing use in medical implants because of their high degree of resistance to chemical and mechanical degradation, even if they may not be quite as physically strong as metals.

Many of the materials used in implants have not been designed and developed specifically for medical use, but have been adapted from other fields where high performance materials are required. Unfortunately, the requirements within a biological environment can be very different to those in an aeroplane wing or a combustion engine even if superficially they appear similar. This has led to a trial-and-error development process not always based on first scientific principles. The basic scientific reasons behind the success (and failure) of synthetic implants is still relatively poorly understood from both a materials and biological viewpoint. It is hoped that by addressing these basic questions, the next generation of implants can be greatly improved. Biocompatibility is one area where basic research has a significant contribution to make to the improvement of implants. A wide range of testing can now be conducted *in vitro* using cell cultures to assess the suitability of materials for implantation into the body. The proliferation of cells, their production of extracellular matrix and the cytotoxicity of breakdown products released by implant materials can all be examined in tissue culture before moving on to animal models.[7] Although some testing on animal models may always be necessary, earlier stages of testing conducted in tissue culture will hopefully lead to a reduction in the scope of these procedures.

Reaction of the body's immune system to the implant as foreign is a frequent cause of implant failure. Events at the implant-tissue interface and wound healing responses post-implantation, which are mediated by migratory inflammatory cells, determine whether or not implant integrity (both structural and functional) can be maintained.[8] As well as direct effects on the implant material itself through active radicals (e.g. superoxide anions), the local release of cytokines and growth factors can cause chronic inflammation and fibrous capsule formation. Techniques for improving the biocompatibility of implant materials are being developed. Coatings can be applied to implants which reduce their inflammatory effects. A recent innovation, using compounds originally developed for systemic use, is to modify the inflammatory response by chemically incorporating synzymes (or enzyme mimics) onto the surface of the implant. These act on the inflammatory cells or their secreted products to reduce the amount of damage done to the implant or to dampen down the local inflammatory effects. This concept has already been demonstrated using superoxide dismutase mimics in a subcutaneous implant model.[9] By catalysing the breakdown of superoxide radicals, the enzyme mimics were able to reduce local inflammation and fibrous capsule formation.

Medical Uses of Implants

The success of implants can be seen by their use to treat a wide range of conditions, where they have made substantial contributions to patient health and survival rates.

Skin Grafts

Skin is the largest organ in the body, and also one of the most important as it serves as the interface between the body and the environment. Unlike many organs, the skin has a capacity for self-repair which produces mechanically and functionally acceptable, if not optimal, tissue. Light wounds will close by themselves and binding or sewing wound edges closed helps to facilitate natural healing. More serious damage caused by burns or trauma that remove large areas of skin or cause damage to a significant depth are harder to address. Without a protective layer of skin, the underlying tissue cannot properly regulate its temperature, control water loss or prevent the entry of micro-organisms. A wound dressing is required to replace the functions of the skin either temporarily while it heals or permanently where damage is so great as to prevent natural healing. A number of synthetic barrier dressings have been developed using polymers.[10] Polyurethane films with adhesive backings have proved successful as wound dressings. Varying the porosity of the film has an effect on the progression of healing and various types are produced tailored to different applications. Apart from treatments involving dressings based on synthetic components alone, new strategies have been developed using a combination of synthetic and biological materials. Dermal substitutes utilise xenografts (i.e. tissue from a different species) of cell free skin.[11] Chemical treatment of the graft is sometimes carried out in order to mask antigenic proteins which would trigger a rejection response from the recipient.[12] In ideal circumstances, a biodegradable dermal substitute is used, and once the graft has bonded to the patient's tissues, it is slowly replaced from beneath, acting as a barrier in the meantime. Xenografts may also be paired with a synthetic polymer layer to improve their characteristics.[13] 'Artificial skin' is the first clinical success of tissue engineering – the process of creating functionally and architecturally natural biological tissues in the laboratory. It has been possible to grow sufficient epidermal cells for grafting in culture for some time,[14,15] but the approach adopted by tissue engineering – combining cell biology with materials science – led to the development of growing cells on biodegradable substrates.[16] This created a wound dressing containing living cells with all the characteristics of normal epidermis that bonds to the tissue and is gradually remodelled by the body's own healing mechanisms.

Skeletal Tissues

The hard tissues of the body which comprise the skeletal tissues bear the weight of the body and absorb the stresses placed upon it. They are therefore placed under high mechanical demands and are at risk of failure. Attempts to replace hard tissues go back to the beginnings of the twentieth century with the first joint replacement procedures, but have only really begun to find widespread application in the last three decades.

The three most common orthopaedic procedures in which synthetic implants are used are cruciate ligament repair, total hip and total knee replacement surgery. All three are now considered routine, with a good probability of successfully rehabilitating the patient to normal levels of mobility. In the case of anterior cruciate ligament (ACL) repair, synthetic replacements have proved less effective than hoped. Carbon fibre, braided polyethylene and polytetrafluoroethylene have all been used in attempts to re-stabilise the knee joint following ACL injury.[17] While the short-term improvements are good, long-term they are not as effective as grafts of ligaments either from other sites in the patient or from donated sources.[18] Much more success has been achieved using biodegradable polymeric materials which provide a mechanical support while the damaged ligament repairs itself. Polyglycolic acid and trimethylene carbonate have both been used in this role.[19,20]

Total Joint Replacement

Hip replacement is one of the most commonly thought of surgical procedures when discussing implants. The increasing incidence of osteoporosis and arthritic diseases has led to ever greater demand for joint replacement in the elderly. Although joint replacement was attempted early in the twentieth century, failure was frequent due to poor material qualities, inadequate design and implant loosening. The breakthrough came in the 1960s with the use of poly(methyl methacrylate) – PMMA – as a cement for fixation of the implant.[21] Most modern hip prostheses are composed of metal (cobalt-chromium or titanium-aluminium-vanadium alloys) or plastic (high molecular weight polyethylene) plus PMMA cement. However, implant fixation is still the problem that causes most hip replacements to require revision. Femoral component loosening is often seen at 7–8 years and acetabular cup loosening at 10–15 years.[22] A number of possible causes of this problem have been investigated. Mechanical signals are important for maintenance of bone matrix and mineralisation and the presence of a metal stem may lead to localised unloading of bone, causing resorption at the tissue-implant interface and associated loosening of the implant.[23] There is also a large body of work on the effects of cement wear debris.[24,25] Small particles of cement are able to infiltrate the tissue-implant interface where they cause a local inflammatory response leading to bone resorption and loosening. Various modifications to the way the cement is prepared and administered have brought some improvements in fixation by strengthening the implant-cement-bone interface.[26,27] Surface roughness of the implant may be a mechanism for increasing the effectiveness of cement fixation and pre-coating of the implant surface with PMMA increases the strength of the implant-cement interface.[28]

Although the conditions requiring replacement of the knee joint are often similar to those of hip replacement (i.e. damage or wear of the articular surface), a different strategy is employed for this procedure.[29] Rather than replacing all the components of the joint, the tibial plateau is resurfaced using plates manufactured of various materials. Metallic plates on bone have proved to be stable for long periods. In cases where resurfacing is not a suitable or adequate procedure, it has taken longer to find successful strategies. Metallic hinges with stems which inserted into the bone had lots of early problems with limited motion, pain and wear and loosening problems similar to those seen with hip replacements, as did early types of cemented metal-plastic replacements. Improved fixation has been achieved by combining a tibial coating element synthesised from plastic which was then enclosed in a metal tray, providing a combination of good surface characteristics with improved fixation. The most commonly used formulation is cobalt-base alloy combined with ultra high weight poly(ethylene) (UHMWPE) with fixation via PMMA cement.

The lifetime of a knee replacement is principally determined by loosening of the tibial surface element. Wear debris from PMMA and UHMWPE is often implicated in a phagocytic response which is associated with loosening. Other problems often found in total knee replacement are wear of the tibial surface by deposition of particulates (bone or PMMA wear debris) between the gliding surfaces and fatigue of UHMWPE due to repeated loading.[30]

Cardiovascular Implants

Heart disease is a major killer in the Western world. Diseases of the cardiovascular system contribute to approximately 20% of fatalities in people between the ages of 36 and 74. The most common conditions are atherosclerosis (which causes narrowing of the artery lumen) and leakage around the valves of the heart which results in poor cardiac output. As is the case with many types of synthetic replacement, materials are used due to a shortage of suitable biological sources for donor tissues.

Vascular graft or bypass surgery has found widespread application for the treatment of coronary heart disease. Obstruction of the arteries which supply the cardiac muscle by atherosclerotic plaques can lead to a lack of oxygen reaching the heart. Coronary surgeons can either try to unblock the artery to restore the blood supply or graft on an extra piece of vessel to bypass the blockage. The first type of procedure usually involves angioplasty. A tiny inflatable balloon is passed though an endoscope to the site of the narrowing before it is inflated, which causes localised distension of the artery wall. This widened vessel lumen is then held open by the placing of a stent, a small tubular lattice usually made from PTFE or Dacron, via an endoscope to the angioplasty site where it is manoeuvred into place and expanded. It presses itself against the artery wall and forces it outward, so maintaining the diameter of the lumen and allowing more blood to flow through. Materials science has played a crucial role in the development of this procedure by providing materials out of which such revolutionary surgical tools can be manufactured.

Angioplasty is a relatively fast and minimally invasive procedure when compared to traditional heart surgery and it can make significant improvements in the patient's condition. In some cases however, often where multiple coronary arteries are blocked, angioplasty cannot provide sufficient improvement and a bypass must be considered if the patient is to stand a good chance of recovery. Wherever possible, autografts, i.e. a vessel from elsewhere in the patient's own body (usually from the saphenous vein in the leg), are used for bypass operations. Although there are well known problems associated with using veins to perform arterial bypass, living tissue is usually at least as good as, if not better than, synthetic alternatives. Synthetic vascular grafts are usually classified according to their size large (12–40 mm in diameter), medium (5–10 mm in diameter) and small (<4 mm in diameter). Different materials and different fabrication techniques are used in each case. Poly(ethylene terephthalate) is a standard material in clinical practice for large vessel replacement. For medium size vessels, poly(tetrafluoroethylene) is the predominant product in current use.

These synthetic polymers can be woven, knitted, crimped and even branched for treating aneurysms which arise at bifurcations in the circulation.

Many grafts are supplied in porous and non-porous forms for different applications. These range from coronary bypass and bypasses in other parts of the body (e.g. in the femoral arteries) to replacing the aorta and other major blood vessels which may rupture after aneurysm formation. Good success rates have been achieved using these artificial vessels.[31] The problems associated with them are not as significant as with other prostheses. Early difficulties with weaknesses at the suture line between patient tissue and the graft were overcome. Blood compatibility remains an important consideration, and the likelihood that plaques can form or attach on the walls of the artificial vessel, thereby narrowing it. Infection rates are low, but can occur a long time after the surgery has been performed, and graft infection is associated with a high risk of graft failure.[32]

Replacing small diameter vessels has proved to be a much greater challenge. There are problems with achieving sufficient flow rates and preventing narrowing of the lumen by clot formation. Expanded, porous poly(tetrafluoroethylene) is widely used with varying pore sizes: 20–30 microns has been reported to be optimal.[33] One method for trying to reduce these problems has been seeding artificial grafts with autologous endothelial cells.[34] By augmenting the physical properties of the graft with the biological properties of an endothelial layer, it may be possible to reduce some of the problems found in small vessel replacement, though there is still much scope for improvement

Cardiac valves control the flow of blood between the chambers of the heart and between the heart and the two major arteries of the body – the aorta and the pulmonary artery. In order for the circulation to perform efficiently it is important that these valves do not leak, otherwise the unidirectional flow of blood into, through and out of the heart becomes disrupted. During an average lifetime, these valves will open and close several billion times. Any synthetic replacement must therefore have a high capacity to withstand fatigue as well as being biocompatible and able to cope with the high pressures and flow rates found within the heart. Two strategies have been adopted to deal with valve failures. One uses totally synthetic, mechanical devices of the caged ball or disc design. The other type of valve might be termed a 'bioprosthetic' where actual valves from a donor are implanted, but must be attached to a synthetic ring in order to stabilise their insertion into the tissue.

The first prototype mechanical valve was described in the 1960s[35] and was composed of a silicon rubber ball enclosed within a cage constructed from cobalt-chromium-molybdenum alloy. Various modifications have been made to this early design to lessen clotting, the chief problem facing mechanical valves of this type. Patients receiving mechanical valves have to be maintained on anti-coagulant drugs to prevent dangerous blood clots forming which might cause fatal embolism or stroke. Despite these problems, mechanical valves have reasonably good survival rates, though the anti-coagulant therapy obviously brings other associated risks.

Bioprosthetic valves are preferred over mechanical ones because of the lower risks of clot formation (which is approximately the same for someone receiving a mechanical valve and anti-coagulant treatment). Donor sources are usually porcine or bovine valves or pericardium.[36] Chemical treatment with glutaraldehyde is used to cross link the collagen fibres in the tissue to make it more resistant to enzyme degradation and help mask antigenic sites (i.e. reduce the risk of rejection). The synthetic part of this valve prosthesis is the sewing ring, onto which the valve is attached. The ring is usually constructed from polymers such as PTFE or poly(propylene). The performance characteristics of bioprosthetic valves varies widely between designs and from those of natural valves. Therefore their performance can be variable, but a compromise is made because of their better biocompatibility. Bioprosthetic valves usually fail due to mechanical fatigue, leading to degeneration of the organic component and calcification. Their survival rate is better than mechanical valves for the first ten years[37,38] but thereafter they are as likely to fail as a completely synthetic implant.

Ophthalmological Implants

The eye is one of the most remarkable organs in the body and provides one of the primary human senses. Like other parts of the body though, there is a clinical need to repair damage and correct defects. Huge advances have been made in the treatment of vision defects, many of which have benefited greatly from new types of materials – the contact lens being a striking example. The earliest contact lenses were hard, impermeable lenses manufactured from PMMA and other materials.[39] These afforded good optical qualities, but were relatively impermeable to oxygen and wearers experienced problems such as corneal abrasion when using them for extended periods of time, which made them rather impractical. The next generation of 'soft' lenses were more permeable, giving improved oxygen delivery to the cornea, but these were not problem free either. Soft lenses are vulnerable to accumulating deposits from tear fluid and growth of bacteria.[40] The most modern lenses are made from rigid, gas permeable materials such as silicones and fluorocarbons. Silicone was thought to be a suitable material with high oxygen permeability, but it also had problems due to reactions with components of tear fluid and lens solutions. Modification of silicones with methylmethacrylate can alter the surface chemistry towards a more favourable hydrophilic state without compromising gas permeability, and similar achievements were made with fluorinated silicone acrylates, but both lacked sufficient oxygen delivery to allow extended wear. Extended wear, rigid gas-permeable lenses were finally produced by using fluorinated methacrylate or pyrrolidone,[41] although another valid approach under examination to increase the permeability of hard acrylic lenses by using lasers to create microscopic channels through the lens material. Problems associated with extended use have not yet been completely solved and research continues to highlight that the problems associated with increased frequency of use of contact lenses.[42] Risks of developing such problems as corneal abrasion, infection and hypersensitivity are higher for soft lenses than hard lenses.[43]

Corneal transplants have restored sight to many patients – approximately 2500 such operations are performed each year in the UK. As with all transplant procedures, there is a shortage of donors even though the UK has a well-established network of Eye Banks. The research effort to produce an artificial cornea for transplantation has shown some promise using poly(2-hydroxy-ethyl-methacrylate) (PHEMA), though its optical and mechanical properties require further study and improvement. Work with PHEMA and other candidate materials such as PTFE have highlighted the importance of porosity for successful integration of corneal implants.[44] Materials which are hydrophobic and opaque at implantation can, over time, become wettable and translucent, if they are porous. Cells can invade the implant, resulting in deposition of extracellular matrix components such as collagen and hyaluronan and, in the case of PTFE, epithelial differentiation can occur over the surface of the implant.

A less obvious, though very important, use of materials has been in the use of viscoelastic solutions in ophthalmic surgery. The vitreous body of the eye has specific mechanical properties that are important to the functioning of the eye.[30] Development of artificial vitreous substitutes has expanded the range and scope of surgical procedures that can be performed whilst maintaining the shape of the tissue and protecting the delicate structures which are essential for vision. The chief components of the natural vitreous body are type II collagen and hyaluronan, so solutions of hyaluronic acid are a logical choice for an artificial vitreous substitute. The chief obstacle to this concept has been the need to purify suitable material that was non-pyrogenic and non-inflammatory for use in human patients. Since this was achieved, HA has been the material of choice for viscoelastic substitutes. A number of commercial forms exist, varying in the percentage composition and molecular weight of the hyaluronan molecules. The viscoelastic properties of HA solutions are influenced by the capacity of HA molecules to form networks and high molecular weight HA chains perform best in this respect. Besides HA, a variety of other large molecules have been used as viscoelastics including chondroitin sulphate, hydroxypropylmethyl cellulose, collagen and polyacrylamide.[45]

Novel Materials

One of the major factors in the rapid growth of research on advanced medical implants has been the variety of new materials becoming available. Advanced synthesis techniques allow materials to be produced in which parameters such as surface area, porosity, surface charge, surface chemistry and dissolution kinetics are carefully controlled. Therefore, particularly desirable properties (in addition to those which are already well known as prerequisites for successful implants) can be incorporated to provide materials which are non-toxic, resorbed by the body and bio-active. Although this technology is at an early stage and clinical studies will be needed to clear the way for their therapeutic use, bio-active materials present an important opportunity. The use of dynamic materials which assist or accelerate the healing process can radically alter the outlook for patients suffering from degenerative diseases.

Tissue Engineering

The availability of advanced materials which support cell attachment, cell growth and formation of living tissues has further broadened the options for treatment in the future. The combination of bio-active materials with cells allows living implants to be made which can go on to perform an active biological function as opposed to an inert structural function. The development of scaffolds for tissue engineering has progressed rapidly in the fields of soft and hard tissue medicine. Cells from articular cartilage have been grown on a variety of materials[46–48] and can synthesise an extracellular matrix of similar composition to natural cartilage. The use of three-dimensional scaffolds has been particularly important in the development of cartilage repair systems, as chondrocytes lose their specialised characteristics when grown for long periods on a flat surface, leading to poor quality tissue formation.

As already discussed, repair of hard tissues has utilised ceramics for some time, both alone and in composites with metals. Newer types of ceramics such as synthetic hydroxyapatite and bioactive glasses have already found applications as bone graft substitutes for oral and crano-facial surgery,[49,50] but more ambitious applications are now under development. Collaborative work utilising expertise in both materials and cell biology has begun to establish the mechanisms by which these ceramic materials interact with osteoblasts to stimulate cell growth and tissue formation. Human osteoblasts grown on Bioglass 45S5™ exhibit increased cellular proliferation and expression of markers of active bone formation.[51] Three-dimensional tissue structures are formed with cells embedded in organic matrix which has strong bonds to the underlying ceramic material (Figs 1 and 2).

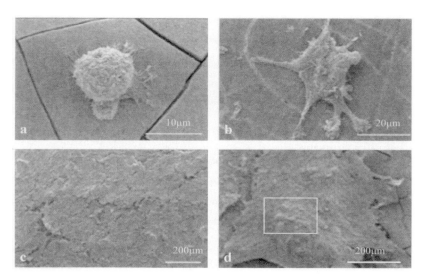

Fig. 1 Scanning electron micrograph showing human osteoblasts growing on Bioglass 45S5™ after (a) 2, (b) 6 and (c) 12 days of culture. (d) SEM image of a bone nodule on Bioglass after 12 days of culture. (Copyright 2000, Springer-Verlag New York Inc.)

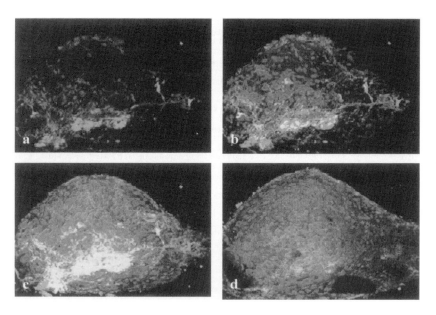

Fig. 2 Confocal scanning laser microscopy images (z-series) of a bone nodule (osteoblasts embedded in organic matrix) grown on Bioglass™ and stained for the extracellular matrix protein type I collagen (×250). Cell nuclei are counter-stained red with propidium iodide. (Copyright 2000, Springer-Verlag New York Inc.)

Although these structures have, as yet, only been grown in culture on a small scale, it is hoped that in the near future the use of bio-reactors will enable much larger pieces of tissue to be produced. Additionally, the capacity of bio-active glasses to form strong, stable bonds with living tissue when implanted *in vivo* means that many of the interface problems which dogged earlier types of implant may be overcome.

As well as populating materials with cells pre-implantation, correct physical and chemical properties can encourage invasion of cells into materials post-implantation. This has been demonstrated in animal studies of artificial blood vessels: an implanted acellular collagen tube will be invaded by endothelial and muscle cells within a few weeks, and remodelled into a morphologically and functionally normal vessel.[52] This approach avoids problems of rejection as the implant is populated by migratory cells from the host. If the host cells cannot support a sufficiently vigorous regenerative response, then re-population with non-host cells is an alternative provided that methods can be found for stopping rejection. In normal organ transplants, the need for long term post-operative use of immuno-suppressant drugs, often with associated side-effects, has a significant effect on long-term survival rates. The arrival of tissue engineering will therefore not end problems of bio-compatibility, but require input from both a materials science and biology viewpoint. What is hoped is that, in the future, tissue engineering can provide rapid access to biologically functional, regenerative implants which will remove the long and difficult wait for donor organs for transplant patients and reduce the need for revision or replacement of inert implants currently in use.

The Future Use of Materials in Medicine

The achievements being made today in developing new bio-active materials and combining them with living cells to create artificial tissues are remarkable and would probably astonish those early pioneers who had the vision to use materials to improve and save lives. The ability to control material properties on a much smaller scale, even to the degree of nanometer scale pores and tubes, has greatly enhanced our capacity to control the properties and effects of materials in a biological environment. By designing and fabricating materials whose properties match exactly the requirements of damaged tissue, implants with greater functionality and improved longevity will come about.

Acknowledgements

The authors gratefully acknowledge support from the Medical Research Council, The Wellcome Trust, The March of Dimes and US Biomaterials.

References

1. A. B. Wilson, *Limb Prosthetics*, 6th Edn, F. A. Demos Publications, New York, 1978.
2. S. Romm, Arms by Design: From Antiquity to the Renaissance, *Plast. Reconstr. Surg.*, 1989, **84**, 158.
3. M. Vitali, *Amputation & Prosthesis*, Cassell & Co. Ltd, London, 1978.
4. P. Klopsteg, *Human Limbs and their Substitutes*, McGaw-Hill, New York, 1945.
5. G. T. Sanders, *Amputation Prosthetics*, F. A. Davis Company, Philadelphia, 1986.
6. A. B. Wilson, History of Amputation Surgery and Prosthetics, in *Atlas of Limb Prosthetics: Surgical and Prosthetic Principles*, American Academy of Orthopaedic Surgeons, C. V. Mosby Company, St Louis, 1981.
7. P. Locci, L. Marinucci, C. Lilli, S. Belcastro, S. Staffolani, S. Bellochio, F. Damiani and E. Bechetti, Biocompatibility of alloys used in orthodontics evaluated by cell culture tests, *J. Biomed. Mater. Res.*, 2000, **51**, 561.
8. J. M. Anderson, Inflammation and the foreign body response, *Prob. Gen. Surg.*, 1994, **11**, 147.
9. K. Udipi, R. L. Ornberg, K. B. Thurmond, S. L. Settle, D. Forster and D. Riley, Modification of inflammatory response to implanted biomedical materials in vivo by surface bound superoxide dismutase mimics, *J. Biomed. Mater. Res.*, 2000, **51**, 549.
10. B. G. MacMillan, Present status of bioadherent materials, barrier dressings and biosynthetics as skin substitutes, in *Burn Wound Coverings*, Vol. 1, D. Wise (ed.), CRC Press Inc, Boca Raton, 1984.
11. S. Pellet, L. Menesi, J. Novak and A. Temesi, 'Freeze-dried irradiated porcine skin as a burn wound covering, in *Burn Wound Coverings*, vol. 1, CRC Press Inc., Boca Raton, 1984.
12. R. F. Oliver, R. A. Grant and C. M. Kent, The fate of cutaneously and subcutaneously implanted trypsin purified dermal collagen in the pig, *Br. J. Exp. Path.*, 1972, **53**, 540.
13. I. V. Yannas, Use of artificial skin in wound management, in *The Surgical Wound*, P. Dineen and G. Hildick-Smith (eds), Lea & Febiger, Philadelphia, 1981.
14. M. A. Karasek and M. E. Charlton, Growth of post-embryonic skin epithelial cells on collagen gels, *J. Invest. Dermatol.*, 1971, **56**, 205.
15. M. Eisinger, M. Monden, J. H. Raaf and J. G. Fortner, Wound coverage by a sheet of epidermal cells grown in vitro from dispersed single cell preparations, *Surgery*, 1980, **88**, 287.
16. I. V. Yannas, J. F. Burke, D. P. Orgill and E. M. Skrabut, Wound tissue can utilize a polymeric template to synthesize a functional extension of skin, *Science*, 1981, **215**, 174.
17. J. G. Ferl, K. L. Goldenthal and N. K. Mishra, FDA Regulation of prosthetic ligament devices, in *Prosthetic Ligament Reconstruction of the Knee*, M. J. Friedman and R. D. Ferkel (eds), W. B. Saunders Co., Philadelphia, 1988.
18. R. M. Rusch, E. F. Nelson and D. Noel, Integraft anterior cruciate ligament reconstruction: Arthroscopic technique, in *Prosthetic Ligament Reconstruction of the Knee*, M. J. Friedman and R. D. Ferkel (eds), W.B. Saunders Co., Philadelphia, 1988.

19. H. E. Cabaud, J. A. Feagin and W. G. Rodkey, Acute anterior cruciate ligament injury and repair reinforced with a biodegradable intraarticular ligament. Experimental studies, *Am. J. Sports Med.*, 1982, **7**, 18.
20. S.-J. Shieh, M. C. Zimmerman and J. R. Parsons, Preliminary characterization of bioresorbable and nonresorbable synthetic fibres for the repair of soft tissue injury, *J. Biomed. Mater. Res.*, 1990, **24**, 789.
21. J. Charnley, A biomechanical analysis of the use of cement to anchor the femoral head prosthesis, *J. Bone Jt. Surg.*, 1965, **47B**, 354.
22. R. Poss, G. W. Brick, R. J. Wright, D. W. Roberts and C. B. Sedge, The effects of modern cementing techniques on the longevity of total hip arthroplasty, *Orthop. Clin. N. Am.*, 1988, **19**, 591.
23. I. Oh and W. H. Harris, Proximal strain distribution in the loaded femur: An in-vitro comparison of the distributions in the intact femur and after insertion of different hip replacement femoral components, *J. Bone Jt. Surg.*, 1978, **60A**, 75.
24. H. G. Willert and M. Semlitsch, Reactions of the articular capsule to wear products of artificial joint prostheses, *J. Biomed. Mater. Res.*, 1977, **11**, 157.
25. J. K. Maguire, M. F. Coscia and M. H. Lynch, Foreign body reaction to polymeric debris following total hip arthroplasty, *Clin. Orthop. Rel. Res.*, 1987, **216**, 213.
26. W. R. Krause, J. Miller and P. Ng, The viscosity of acrylic bone cement, *J. Biomed. Mat. Res.*, 1982, **16**, 219.
27. D. W. Burke, E. I. Gates and W. H. Harris, Centrifugation as a method of improving tensile and fatigue properties of acrylic bone cement, *J. Bone Joint Surg.*, 1984, **66A**, 1265.
28. K. L. Ohashi and R. H. Dauskard, Effects of fatigue loading and PMMA pre-coating on adhesion and sub-critical debonding of prosthetic-PMMA interfaces, *J. Biomed. Mater. Res.*, 2000, **51**, 172.
29. J. Black, Requirements for successful total knee replacement, *Orthop. Clin. N. Am.*, 1989, **20**, 1.
30. F. H. Silver, *Biomaterials, Medical Devices and Tissue Engineering*, Chapman & Hall, London, 1994.
31. L. R. Sauvage, J. C. Smith, C. C. Davis, E. A. Rittenhouse, D. G. Hall and P. B. Mansfield, Dacron arterial grafts; comparative structures and basis for successful use of current prostheses, in *Vascular Graft Update: Safety and Performance*, H. E. Kambic, A. Kantrowitz and P. Sung (eds), American Society for Testing Materials, Philadelphia, 1986.
32. E. Vinard, R. Eloy, J. Descotes, J. R. Brudon, H. Guidicelli, P. Patra, R. Streichenberger and M. David, Human vascular graft failure and frequency of infection, *J. Biomed. Mat. Res.*, 1991, **25**, 499.
33. C. D. Campbell, D. Goldfarb, D. D. Detton, R. Roe, K. Goldsmith and E. B. Diethrich, Expanded polytetrafluoroethylene as a small artery substitute, *Trans. Am. Soc. Artif. Intern. Organs*, 1974, **20**, 86.
34. L. M. Graham, W. E. Burkel, J. W. Ford, D. W. Vinter, R. G. Kahn and J. C. Stanley, Expanded polytetrafluroethylene vascular prostheses seeded with enzymatically derived and cultured canine endothelial cells, *Surgery*, 1982, **91**, 550.

35. A. Starr, Total mitral valve replacement fixation and thrombosis, *Surg. Forum*, 1960, **11**, 258.
36. M. Thurbrikar, Replacement cardiac valves, in *The Aortic Valve*, M. Thurbrikar (ed.), CRC Press, Boca Raton, 1990.
37. L. H. Cohn, The long term results of aortic valve replacement, *Chest*, 1985, **85**, 387.
38. P. E. Oyer, E. B. Stinson, D. C. Miller, S. W. Jamieson, R. S. Mitchell and N. E. Shumway, Thromboembolitic risk and durability of the Hancock bioprosthetic cardiac valve, *Eur. Heart J.*, 1984, **5**, 81.
39. M. F. Refojo, Current status of biomaterials in ophthalmology, *Surv. Ophthalmol.*, 1982, **26**, 257.
40. J. I. Lippman, Contact lens materials: A critical review, *Contact Lens Association of Ophthalmologists Journal*, 1990, **16**, 287.
41. J. W. Moore, The Allergan Advent™ flexible fluoropolymer for daily or extended wear, *Contact Lens Forum*, August 27, 1989.
42. O. D. Schein, R. J. Glynn, E. C. Poggio, J. M. Seddon and K. R. Kenyon, The relative risk of ulcerative keratitis among users of daily-wear and extended-wear soft contact lenses: A case-control study, *N. Engl. J. Med.*, 1989, **321**, 773.
43. W. A. Franks, G. G. W. Adams, J. K. G. Dart and D. Minassian, The relative risks of different types of contact lenses, *Br. Med. J.*, 1988, **297**, 534.
44. J. M. Legeais, G. Renard, J. M. Parel, O. Serdarevic, M. Mei-Mui and Y. Pouliquen, Expanded fluorocarbon for keratoprosthesis: Cellular ingrowth and transparency, *Exp. Eye Res.*, 1994, **58**, 41.
45. F. H. Silver, J. Librizzi and D. Bernadetto, Use of viscoelastic solutions in ophthalmology: A review of physical properties and long term effects, *J. Long-Term Eff. Med. Implants*, 1994, **2**, 49.
46. K. R. Stone, W. G. Rodkey, R. Weber, L. McKinnley and J. R. Steadman, Meniscal regeneration with copolymeric collagen scaffolds – in vitro and in vivo studies evaluated clinically, histologically, and biochemically, *Am. J. Sports Med.*, 1992, **20**, 104.
47. M. D. Buschmann, Y. A. Gluzband, A. J. Grodzinsky, J. H. Kimura and E. B. Hunziker, Chondrocytes in agarose gel culture synthesize a mechanically functional extracellular matrix, *J. Orthop. Res.*, 1992, **10**, 745.
48. L. E. Freed, J. C. Marquis, A. Nohria, J. Emmanual, A. G. Mikos and R. Langer, Neocartilage formation in vitro and in vivo using cells cultured on synthetic biodegradable polymers, *J. Biomed. Mater. Res.*, 1993, **27**, 11.
49. H. Oonishi, S. Kushitani, E. Yasukawa, H. Iwaki, L. L. Hench, J. Wilson, E. I. Tsuji and T. Sugihara, Particulate bioglass compared with hydroxyapatite as a bone graft substitute, *Clin. Orth. Rel. Res.*, 1997, **334**, 316.
50. Y. Fujishiro, L. L. Hench and H. Oonishi, Quantitative rates of in vivo bone generation for Bioglass® and hydroxyapatite particles as bone graft substitute, *J. Mater. Sci. Mater. M.*, 1997, **8**, 649.
51. I. D. Xynos, M. V. J. Hukkanen, J. J. Batten, L. D. Buttery, L. L. Hench and J. M. Polak, Bioglass® 45S5 stimulates osteoblast turnover and enhances bone formation in vitro: Implications and applications for bone tissue engineering, *Calcif. Tiss. Int.*, 2000, **67**, 321.

52. T. Huynh, G. Abraham, J. Murray, K. Brockbank, P. O. Hagen and S. Sullivan, Remodelling of an acellular collagen graft into a physically responsive neovessel, *Nat. Biotechnol.*, 1999, **17**, 1083.

Research Council Priorities

RICHARD BROOK
Engineering and Physical Science Research Council, Swindon, UK

ABSTRACT
One risk of being recognised as a crucial encompassing and enabling discipline is that opportunities for focused strategy become less evident. Should we worry? A further risk is that the concept of being a specialist in that discipline becomes less distinct. Should we worry? The funding support of materials science is reviewed against such questions, with emphasis on the key issue of ensuring creativity and excellence in the resulting research.

Formula 1 Materials Engineering

G. SAVAGE
British American Racing

ABSTRACT
A significant proportion of a Formula 1 team's budget is spent on research and development, in what is sometimes cynically referred to as 'the search for the unfair advantage'. As a direct consequence the sport has become synonymous with high technology science and engineering. Through the media most people are by now familiar with the various aerodynamic aids employed by the designers, as well as electronic devices such as semi-automatic gearboxes and 'active' control systems. The use of advanced materials, although not hyped to anywhere near the same extent, has contributed enormously to the performance of Formula 1 cars in recent years. The cars are now faster and safer than ever before. All of the cars that make up the grid employ the complete range of materials, metals, ceramics, polymers, elastomers and composites in their construction.

Any engineering structure, irrespective of its intended purpose, must be made of one or more materials. More often than not it is the choice and behaviour of those materials that determine its mechanical performance. Materials science and engineering strives to optimise the operation of any device or construction by ensuring the correct choice of constituent materials and processing technology. To illustrate the influence of materials science on the cars' performance, three very different examples, carbon fibre reinforced composites, aluminium–beryllium alloys and polyurethane elastomers are discussed.

The Design of Formula 1 Racing Cars

In order to understand the design philosophy behind a Formula 1 racing car it is first necessary to appreciate the various subassemblies that make up the structure. The general arrangement of single seat racing cars has remained the same since the early 1960s. The central component, which accommodates the driver, fuel cell and front suspension assembly, is the chassis. This is a semi-monocoque shell structure known as the 'tub'. The tub of a Formula 1 car is more like a jet fighter aircraft cockpit, both in terms of shape and construction, than anything one would expect to find on the road. The engine, in addition to providing propulsion, has a structural function and is attached directly to the rear of this unit by high strength metal studs. The assembly is completed by the addition of the gearbox and rear suspension assembly (Fig. 1). The car's primary structure of chassis, engine and gearbox may be considered as a 'Torsion-beam' arrangement carrying the inertial loads to their reaction points at the four corners. The secondary structures (bodywork, undertray, wing configurations and cooler ducting etc.) are arranged around and attached to the primary structure at various points (Fig. 2).

The role of the chassis is of major importance to the operation of the vehicle. A racing car must be 'set-up' for each individual circuit. Changes are made to the aerodynamic devices and the suspension elements (springs, dampers, anti-roll bars and so on) in an attempt to modify and optimise handling to improve its lap time. Changes in the performance levels of the various sub-components should be manifest in the balance of the car. Clearly this will not occur if the structure transmitting the loads is not of adequate

Fig. 1 The car's primary structure of chassis, engine and gearbox.

Fig. 2 Complete car with 'secondary' structures added.

stiffness. In common with many other engineering disciplines, the designers of Formula 1 racing cars are required to comply with a stringent set of regulations. The rules are imposed by the FIA, the sport's governing body. Constraints are laid down on geometry, strength and weight. Strict limitations are placed on the overall dimensions of the cars and the sizing of the driver envelope within the cockpit. The regulation limiting the minimum weight of the car to 600 kg is of great significance. Building a car to the weight limit is a vital task if it is to be competitive.

It has been calculated that a mass of 20 kg above the weight limit equates to a loss of 0.4 seconds around a typical Grand Prix circuit. Less than half a second does not sound very much, but during a full race distance this amounts to half a lap or several grid positions during a qualifying session. With modern materials it is relatively easy to build a car which satisfies all of the statutory requirements whilst still being well under the minimum weight limit. As a result the majority of the cars are required to carry ballast (generally in the form of a heavy metal such as tungsten) in order to make up the deficit. One would consider that once a design had produced a structure that was close to the minimum weight limit, there would be little point in developing it to achieve further weight reduction for the same performance. On the contrary, studies of vehicle dynamics have shown the benefits in controlling the vehicle's mass distribution upon its handling. As a consequence every component on an Formula 1 car must be engineered to the absolute minimum weight. The more ballast that is needed to return the car to the legal minimum weight, the more scope is provided to achieve optimum performance by tuning its balance by appropriate positioning of said ballast. There is therefore an incentive to use weight efficient materials such as composites wherever possible.

Composite Materials

Composites are defined as materials in which two or more constituents have been brought together to produce a new material consisting of at least two chemically distinct components, with resultant properties significantly different to those of the individual constituents. A more complete description also demands that the constituents also be present in reasonable proportions. 5% by weight is arbitrarily considered to be the minimum. The material must furthermore be considered to be 'man made'. That is to say it must be produced deliberately by intimate mixing of the constituents. An alloy that forms a distinct two-phase microstructure as a consequence of solidification or heat treatment would not therefore be considered as a composite. If on the other hand, ceramic fibres or particles were to be mixed with a metal to produce a material consisting of a dispersion of the ceramic within the metal, this would be regarded as a composite.

On a microscopic scale composites have two or more chemically distinct phases separated by a distinct *interface*. As we shall see later this interface has a profound influence on the properties of the composite. The continuous phase is known as the *matrix*. Generally the properties of the matrix are considerably improved by incorporating another constituent to produce the composite. A composite may have a ceramic, metallic or polymeric matrix. The second phase is referred to as the *reinforcement* as it enhances the properties of the matrix. In most cases the reinforcement is harder, stronger and stiffer than the matrix.

The measured strengths of materials are several orders of magnitude less than those calculated theoretically. Furthermore the stress at which nominally identical specimens fail is subject to a marked variability. This is believed to be due to the presence of inherent flaws within the material. There is always a distribution in the size of the flaws and failure under load initiates at the largest of these. Griffith[1] derived an expression relating failure stress to flaw size, a.

$$\sigma_f = \frac{K_{IC}}{ya^{1/2}} \tag{1}$$

Where σ_f = failure stress, K_{IC} is the material's fracture toughness and y a geometrical constant.

As equation 1 shows the larger the flaw size, the lower will be the failure stress (Fig. 3). It follows therefore that eliminating or minimising such imperfections can enhance the strength of a material. Cracks lying perpendicular to

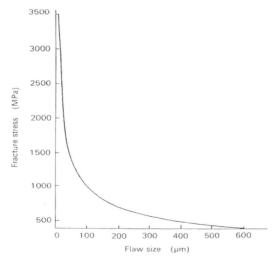

Figure 3 Effect of flaw size, a, on fracture stress, σ, of a brittle material.

the applied loads are the most detrimental to the strength. Fibrous or filamentary materials thus exhibit high strength and stiffness along their lengths because in this direction the large flaws present in the bulk material are minimised. Fibres will readily support tensile loads but offer almost no resistance and buckle under compression. In order to be directly usable in engineering applications they must be embedded in matrix materials to form fibrous composites. The matrix serves to bind the fibres together, transfer loads to the fibres and protect them against handling damage and environmental attack.

The driving force for the increasing substitution of metal alloys is demonstrated in Table 1. Contrary to many a widely held belief, composites are not 'wonder materials'. Indeed their mechanical properties are roughly of the same order as their metal competitors. Furthermore they exhibit lower extensions to failure then metallic alloys of comparable strength. What is important however is that they possess much lower densities. As a result fibre reinforced composites have vastly improved specific properties, strength and stiffness per unit weight for example. The higher specific properties enable the production of lower weight components. The weight savings obtained in practice are not as great as Table 1 implies because the fibres are extremely anisotropic, which must be accounted for in any design calculations. In addition specific modulus (E/ρ) and strength (σ/ρ) are only capable of specifying the performance under certain loading regimes. Specific modulus is useful when considering materials for components under tensile loading such as, for example, wing support pillars (Fig. 4). The lightest component that will carry a tensile load without exceeding a predetermined deflection is defined by the highest value of E/ρ. A compression member such as a suspension push rod on the other hand is limited by buckling such that the best material is

Table 1 Comparison of the mechanical properties of metal alloys and composites.

Material	Density (g cm^{-3})	Tensile strength, σ (MPa)	Tensile Modulus, E (GPa)	Specific strength (σ/ρ)	Specific Modulus (E/ρ)
Steel	7.8	1300	200	167	26
Aluminium	2.81	350	73	124	26
Titanium	4.42	900	108	204	25
Magnesium	1.8	270	45	150	25
E glass	2.10	1100	75	524	21.5
Aramid	1.32	1400	45	1060	57
IM Carbon	1.51	2500	151	1656	100
HM Carbon	1.54	1550	212	1006	138

that which exhibits the highest value of $E^{1/2}/\rho$ (Fig. 5). Similarly, a panel loaded in bending such as a rear wing main-plane, will produce minimum deflection by optimising $E^{1/3}/\rho$ (Fig. 6). Nevertheless, weight savings of between 30–50% are readily achieved over metal components.

Fig. 4 Front wing support pillars, loaded primarily in tension.

Fig. 5 Front push rod, compression member.

Fig. 6 Rear wing main plane, loaded in bending.

Composite Structures in Formula 1

Designers of weight sensitive, stiffness critical structures such as aircraft and racing cars, require materials that combine good mechanical properties with low weight. Aircraft originally employed wood and fabric in their construction, but since the late 1930s aluminium alloys have been the dominant materials. During the last two decades advanced composite materials have been increasingly employed for stressed members in aircraft. At this point it is appropriate to define what has become a 'buzz-word' in the composites industry, the term 'advanced'. An advanced composite is simply one in which the mechanical properties are dominated by the fibres. Structures are designed to have a precisely defined quantity of fibres in the correct location and orientation with a minimum of polymer to provide the support. The composites industry achieves this precision using 'prepreg' as an intermediate product. Prepreg is a broad tape of aligned or woven fibres, impregnated with polymer resin (Fig. 7). A composite structure is fabricated by stacking successive layers of prepreg and curing under temperature and pressure. Many components consist of 'sandwich construction'; Thin, high strength composite skins are separated by, and bonded to, thick, lightweight honeycomb cores. The thicker the core, the higher the stiffness and strength of the component, for minimal weight gain (Fig. 8).

A prepreg consists of a combination of a matrix (or resin) and fibre reinforcement. It is ready to use in the component manufacturing process.
It is available in :
- UNIDIRECTIONAL (UD) form (one direction of reinforcement)
- FABRIC form (several directions of reinforcement).

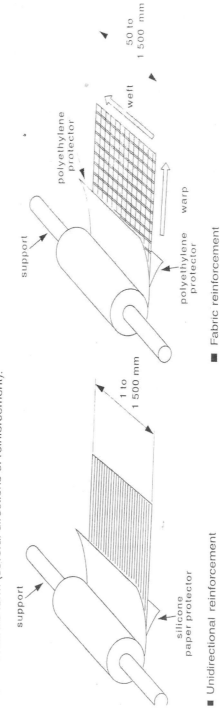

■ Unidirectional reinforcement   ■ Fabric reinforcement

Fig. 7 'Prepreg'.

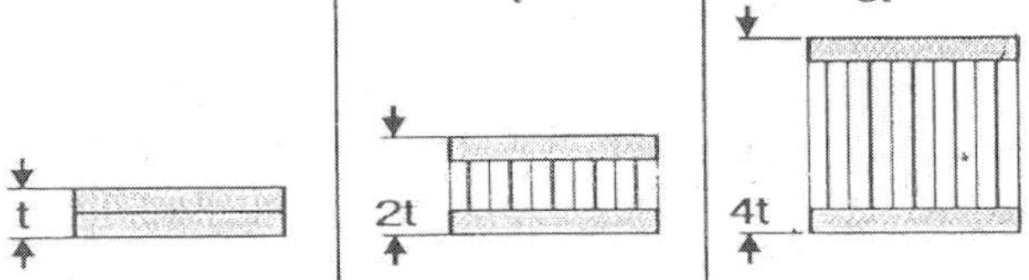

	Solid Material	Core Thickness t	Core Thickness 3t
Stiffness	1.0	7.0	37.0
Flexural Strength	1.0	3.5	9.2
Weight	1.0	1.03	1.06

Fig. 8 Improving structural efficiency using honeycomb structures.

In the mid-to-late 1970s the predominant method of Formula 1 composite chassis construction used aluminium skinned, aluminium honeycomb material fabricated using the 'cut and fold' method. The tubs were formed from pre-bonded sheeting, which was routed, folded and riveted, into the appropriate shape (Fig. 9). The various teams involved later pre-formed the skins prior to bonding to the core using an epoxy film adhesive. In 1980, McLaren introduced the first carbon fibre reinforced composite chassis. This pioneering work was to revolutionise the construction of racing cars. They were able to increase the torsional rigidity of the chassis by 66% whilst at the same time reducing its weight by 30% (Fig. 10). During the design of the MP4/1, McLaren used carbon composites wherever they offered advantages in mechanical properties or a reduction in complexity of design. Since that time there has been a continual process of metals replacement within the sport. In the early 1990s, Savage and Leaper from McLaren developed composite suspension members.[2] Composite suspension components are now used by the majority of teams (Fig. 11). The latest development was the introduction of a composite gearbox by the Arrows and Stewart teams in 1998 (Fig. 12). Figure 13 shows an exploded schematic of the composite constituents of a contemporary Formula 1 car. It is not intended to represent any particular vehicle but rather to show the range of components used. In addition to the chassis it shows composite bodywork, cooling ducts for the radiators and brakes, front, rear and side crash structures, suspension, gearbox and the steering wheel and column. Aside from the structural materials a number of 'speciality' composites are also used. These include carbon-carbon brakes and clutches, and ablatives in and around the exhaust ports.[3]

Fig. 9 Aluminium skinned honeycomb composite chassis.

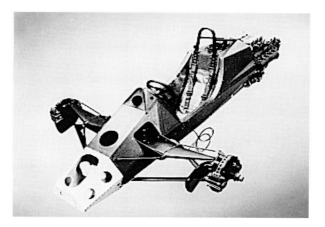

Fig. 10 McLaren MP4/1, the first carbon fibre composite chassis.

Fig. 11 Composite suspension members.

Fig. 12 Composite gearbox.

1. Chassis
2. Roll Hoop
3. Side impact structures
4. Bonnet panel
5. Air box
6. Nose box
7. Engine cover
8. Rear wing Assembly
9. Radiator Ducts
10. The Seat
11. Steering wheel
12. Steering Column
13. Brake Ducts
14. Front wing Assembly
15. Deflectors
17. Front suspension members
18. Rear suspension members
19. Heat shields
20. Gear box
21. Rear crash structure

Fig. 13 The range of composite components used on the latest generation Formula 1 cars.

Impact Performance

The changeover from aluminium alloy to carbon fibre chassis precipitated a degree of anxiety within Formula 1 with respect to the ability of such brittle materials to protect the driver in the event of a crash. The reality of the situation however was that composite racing cars afford vastly improved crashworthiness compared to their metallic predecessors. Much of the sport's improved safety record in recent years derives from the controlled fracture behaviour of composite materials.

166   Materials Science and Engineering: Its Nucleation and Growth

The large forces generated during major impacts of vehicle structures are sufficient to exceed the elastic limit of the materials from which they are made. Destruction of a metallic race car chassis may be illustrated by considering the axial collapse of a thin-walled metal tube under impact. Following an initial peak load, which initiates the process, energy will be absorbed as a consequence of the work done in forming 'plastic hinges'[4] which develop progressively along the tube. A load deflection curve typical of such an event is shown in Fig. 14, as is the plastic buckling typical of a metal energy-absorbing device. The fluctuations in the curve result in the area under the curve (i.e. the energy absorbed) being relatively low, thus rendering the device inefficient. By contrast, the failure of a composite chassis, comprising brittle carbon fibres in a brittle epoxy matrix, does not involve plastic deformation. The immense stiffness of a carbon fibre monocoque is such that its elastic limit will not be exceeded. This high stiffness serves to transmit the load from the point of impact further into the structure so that higher loads can be absorbed without permanent damage. Once the load in the locality of the impact has exceeded the absolute strength of the laminate, failure in that area is total as the laminate progressively tears itself to pieces.

Impact Response of Composite Materials and Structures

The energy absorbing capability of composite materials is a consequence of the 'work of fracture' arising from the mechanisms occurring during catastrophic fracture.[5] The inherent brittleness of composites ensures that they do not undergo the yield processes characteristic of ductile metals but on the application of load, deform elastically up to the point of fracture. A

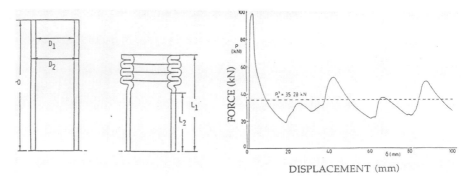

Fig. 14   Axial collapse by plastic buckling of a ductile metal tube with a typical load/deflection plot for the event.

number of modes of deformation are available to complex multiphase composite materials. The primary energy absorbing mechanisms in fibre reinforced plastics are:

- cracking and fracture of the fibres
- matrix fracture
- de-bonding (pull-out) of fibres from the matrix
- delamination of the layers making up the structure.

A composite body thus disintegrates both structurally and microscopically during impact. A typical load/deflection response for a composite tube is shown in Fig. 15. All of the deformation mechanisms can operate simultaneously. After the initial peak load, the curve is much flatter than the plastically deforming metal tube in Fig. 14. The area under the curve, is therefore, much greater. This combined with the lower density of the composite makes it far more efficient.

Safety and Survivability

The survivability of the pilot in an accident is achieved by a combination of the crash resistance of the car and its ability to absorb energy. This has been achieved by providing a survival cell (the chassis), which is extremely resistant to damage, around which energy absorbing devices are placed at strategic points on the vehicle. The energy absorbing devices operate to enable maximum deformation up to a specified limit. The devices used are designed

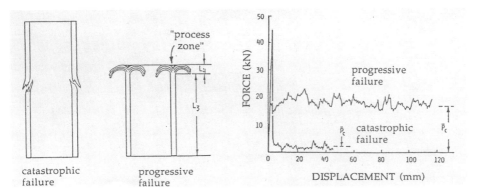

Fig. 15 Axial crushing of composite tubes.

168 Materials Science and Engineering: Its Nucleation and Growth

to dissipate energy irreversibly during the impact, thereby reducing the force and momentum transferred to the survival cell and hence the pilot. They are 'one-shot' items, being partially or totally destroyed so as to act as a load limiter. They are proportioned so as to possess a more or less rectangular force/displacement characteristic (Fig. 16).

Since the late 1980s the controlling body of Formula 1 (FIA) has introduced a series of regulations to ensure that the cars conform to stringent safety standards and build quality.[6] Each vehicle must satisfy a list of requirements, in the form of officially witnessed tests, before it is allowed to race. There are two groups of tests that must be passed. The first is a series of static loads applied to the chassis, which guarantees the strength and integrity of the survival cell. The second series defines the position, and effectiveness of the energy absorbing structures. Each year the number and severity of the tests increases in line with ongoing research and development into survivability, or in response to track 'incidents'. The regulations for the 2001 season are summarised in Tables 2 and 3 and examples are shown in Figs 17 and 18.

Table 2 FIA Static Test Requirements.

Structure	Applied Load (kN)
Primary (rear) roll over hoop	120.0
Secondary (dash board) roll over hoop	75.0
Cockpit side	30.0 (load must be held for 30 s)
Fuel tank floor	12.5 (load must be held for 30 s)
Cockpit rim	10.0 (load must be held for 30 s)
Nose box push off test	40.0 (load must be held for 30 s)

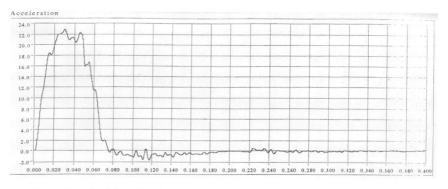

Fig. 16 Ideal load/deflection response of energy absorbing structure.

Formula 1 Materials Engineering 169

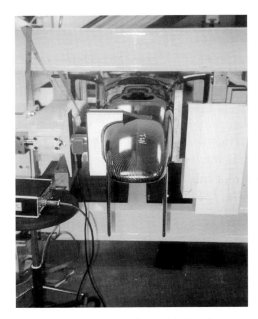

Fig. 17 FIA nose box side load test.

Fig. 18 FIA front impact test.

Table 3 FIA Impact Test Requirements.

Structure	Impact Mass (kg)	Velocity (ms^{-1})	Energy (kJ)	Peak Force Permitted	Maximum Mean Deceleration (g)
Nose box	780	14	76.44	60 g for 3 ms	40 g (must be <5 g for initial 150mm)
Side	780	10	39.0	80 kN for 3 ms	20
Rear	780	12	56.16	60 g for 3 ms	35
Steering column	8	7	0.196	80 g for 3 ms	na

Impact Structures

From the numerous experimental studies that have been carried out on composite energy absorbing devices, it is generally accepted that thin-walled tubes offer the most weight efficient solution.[6] The tubular devices have been shown to perform at their best when global Euler column buckling and local wall buckling, with corresponding bending collapse modes, has been precluded (Fig. 15). That is to say when geometric, material, and loading conditions are such that axial failure of the tubes is characterised by the progression of a destructive zone of constant size at the loaded end. This is called the crash 'process zone' or 'crash frond' (Fig. 19).

The challenge of design is to arrange the column of material such that the destructive zone can progress in a stable fashion. The energy absorption must be as high as possible by allowing the development of a sustained high level crushing force, with little fluctuation in amplitude as the process zone travels along the component's axis. Furthermore, the destruction should be initiated smoothly by avoiding large initial peak resisting forces that might cause global wall or column buckling rather than the beneficial load wall destruction mode.

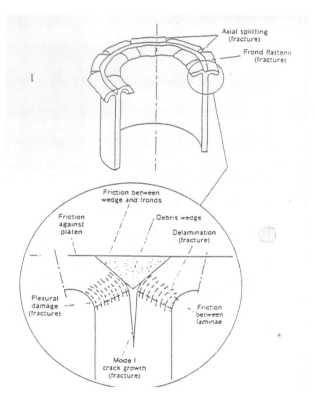

Fig. 19 Process zone development in axially crushed composite tube.[6]

For tubes with structures other than circular, i.e. 'real structures' crushing behaviour has been shown to be influenced favourably when the corners of polygonal thin-walled sections are rounded so as to represent segments of circular tubes.[7,8] For square sections, the greater the corner radius, the higher will be the efficiency of energy absorption.[9] Rounded corners prevent flat segments from failing by load plate buckling, with associated plate strip buckling and much lower specific energy absorption.

The successful axial crushing of composite structures hinges on the necessity to trigger the respective highly energy absorbing progressive failure mechanism. In practice, the 'trigger' is designed into the structure by tapering its geometry and lay-up within the confines of the envelope defined by the car's aerodynamic performance, the sport's technical regulations and the basic physics governing the event.[5]

The design of the component follows an iterative process. An initial test piece is produced by conservatively fitting the correct amount of material into the laminate's fracture zone. A 'remote' or practise test is then carried out, to FIA specification, on the component fixed to a solid metal plate (rather than a monocoque, for obvious reasons!) in order to test the theory. The design then follows a trial and error process, changing the lay-up and modifying the geometry in order to arrive at the most weight efficient solution. Wherever possible one aims to use a monolithic structure for the composite as this is the most efficient way of absorbing energy.

As one moves away from a simple tube to a more complex geometry, the energy absorbing efficiency is reduced. For example, a tube made from a woven carbon fabric will have an efficiency of $\approx 80 \, \text{J g}^{-1}$. In a more complex structure made from the same material, such as a side impact device, the efficiency drops to $\approx 60 \, \text{J g}^{-1}$. When the component has a high axial ratio, a nosebox for example, it is necessary to use a honeycomb-stabilised structure in order to increase the wall thickness at minimum weight penalty in order to prevent catastrophic failure. In a situation such as this, where the axis of the honeycomb cells is perpendicular to the impact, the energy efficiency is significantly reduced (to $\approx 35–40 \, \text{J g}^{-1}$) because of a tendency towards plate strip buckling. The effect of these considerations is to reduce the efficiency of the component. Nevertheless a nosebox weighing of the order of 4 kg, is capable of absorbing in excess of 76 kJ. There are moves to introduce specialist, finite element crash simulations into the design process.[10,11] At the time of writing however the quasi-numerical 'heuristic' approach is favoured. Once the remote test has been passed and the team is happy that it has produced the optimum design, an official test is carried out to homologate the car.

Fixing the component to a chassis tends to be a less harsh test than when attached to a rigid plate due to the increased compliance of the system. As a consequence the impact devices generally perform better in a 'real' rather than practice test. This does not however prevent the many nervous hours spent before it is all over and the car certified for the season!

Aluminium-Beryllium Alloys

With reference to Table 1, all of the common structural metals and alloys have roughly the same specific stiffness of 25 GPa/g cm^{-3}. One metal however, beryllium, has a vastly superior specific stiffness (Table 4). Beryllium is a dark grey, low-density metallic element, which can be machined to extremely close tolerances.[12] Historically, beryllium-based materials have been utilised in space and aerospace applications. Beryllium itself is extremely brittle and, although it is relatively safe in solid form, its dust is extremely toxic.[13] Furthermore, all machining operations, especially grinding, produce a damaged surface layer that must be removed by chemical etching a minimum of 0.05 mm per surface for stressed applications.[14]

Aluminium-beryllium alloys combine the high modulus and low-density characteristics of beryllium with the fabrication and mechanical property behaviour of aluminium. The effect of adding pure aluminium to beryllium is to produce a material with greatly improved ductility while retaining properties far superior to any other aluminium alloy. When compared with metal matrix and organic composites, aluminium-beryllium alloys are simpler to use and easier to fabricate. The alloys are weldable, and can be formed, machined, brazed and adhesively bonded in the same manner as more 'conventional' aluminium alloys. Aluminium alloys do not display sensitivity to machining damage and do not therefore require etching.[15]

The properties of aluminium-beryllium can be explained with reference to the Al–Be phase diagram (Fig. 20) and the typical microstructure shown in Fig. 21. The beryllium phase is dark and semi-continuous and is surrounded by the lighter continuous aluminium phase. The volume fraction of the individual phases depends on the alloy composition, the optimum being 62% by weight of Be,[16] while the beryllium size and morphology depends upon the processing history. The structure may be treated as a 2-phase

Table 4 Properties of Beryllium based materials.

Material	Density (g cm^{-3})	Tensile Modulus (GPa)	Tensile Strength (MPa)	E/ρ	σ/ρ
Beryllium	1.8	300	485	167	269
Al-62%Be	2.1	192	380	91	180

Formula 1 Materials Engineering 173

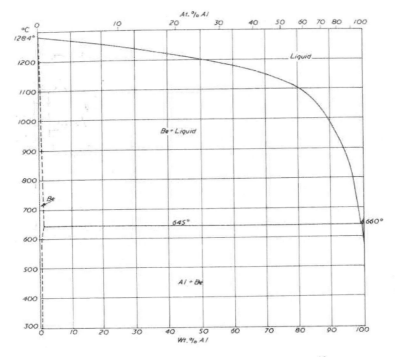

Fig. 20 Aluminium-beryllium phase diagram.[12]

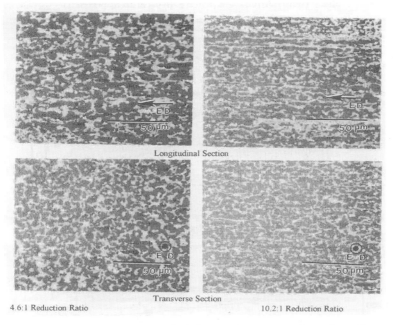

Longitudinal Section

Transverse Section

4.6:1 Reduction Ratio   10.2:1 Reduction Ratio

Fig. 21 Typical microstructures of Al-62%Be (Courtesy of Brush Wellman Ltd).

composite material in which the ductile aluminium matrix surrounds the beryllium reinforcement. It must be noted however that, although Al–Be behaves like a composite, the material is an alloy, to be designated a composite it must be regarded as 'man made' whereas in this case the microstructure formed is natural.[3]

The introduction of aluminium-beryllium produced a very significant improvement in brake calliper performance. Components were machined from solid forgings (Fig. 22) based on existing designs, simply replacing the aluminium/silicon carbide MMC with the new material. The effect of increased stiffness of the callipers was an immediate improvement in the braking performance.

Figure 23 shows results obtained from a bench test (Fig. 24) of the car's complete braking system. It was found that the aluminium-beryllium increased the rate of calliper pressurisation by up to 20%. It has been claimed that the new callipers produced a reduction in lap time of the order of 0.5 sec,[17] although 0.2 sec is probably a more realistic estimate.[18] Nevertheless, whatever value one accepts this was a significant improvement in performance achieved by simply changing the material. The introduction of Al–Be did however increase the cost of the callipers by a factor of 6, but the £24,000 (approx.) per car set represented a very good investment for the improved lap time. Extensive use was also made of aluminium-beryllium in Formula 1 engines.[19] Parts such as cylinder liners, con-rods, gudgeon pins and pistons etc. were all exploited to produce shorter, lighter and more powerful engines.

Fig. 22 Aluminium-beryllium brake calliper (Courtesy of Brush Wellman Ltd).

Formula 1 Materials Engineering 175

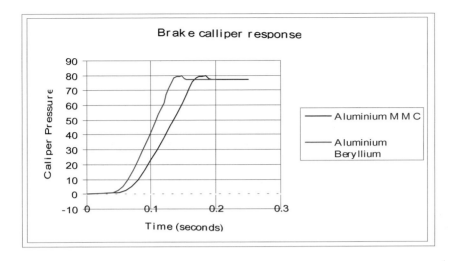

Fig. 23 The increased stiffness of an Al–Be calliper increases the braking efficiency by approximately 20%.

Fig. 24 Bench testing a Formula 1 braking system.

Al–Be alloys were banned in chassis from season 1998 and in engines from 2001. Prior to that they had contributed enormously to braking and the performance of race engines in the late 1990s, particularly those of Ferrari and Mercedes who were at the forefront of the technology.

Aerodynamics

Grand Prix racing cars have always had a 'sleek and smooth' appearance (Fig. 25) but it was not until the 1960s that aerodynamics became an integral part of the design criterion. Aircraft wings generate lift as a consequence of the pressure drop between the underside and top surface of the wing; Formula 1 cars produce negative lift, known as 'downforce' by turning the wing upside down. From the late 1960s onwards racecars began to spout wings and other aerodynamic devices (Fig. 26).[20] The aerodynamics of a modern racing car are important because the car gets its lateral grip from its tyres and this force is dependent on not only the tyre performance, but also the normal load on the tyre. This normal load is a combination of the static weight of the car, the load transfer due to driving manoeuvres and the aerodynamically induced downforce.[21] A typical Formula 1 car weighing 600 kg with pilot is capable of producing up to twice that load in downforce. Indeed for many races the lowest speed of the car on the circuit is capable of generating more than its own mass in downforce such that it could run the whole race upside down on an imaginary ceiling. The production of downforce is not the only part of the aerodynamic story. If downforce were the only consideration the cars would

Fig. 25 Racing cars have always sought aerodynamic efficiency.

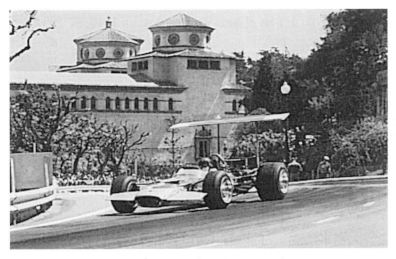

Fig. 26 Early aerodynamic aids.

always run with the maximum downforce possible at every circuit to produce the highest possible cornering speeds. The generation of downforce also leads to the production of drag which, except when braking, is unwanted. Hence the ratio of lift to drag of the vehicle is critical to the performance of the car. Along the straight, where no cornering force is needed, the only real effect of aerodynamics (aside from an increase in traction due to downforce on the rear wheels) is to slow the car down and limit the terminal velocity.

Formula 1 teams spend millions of pounds per annum on wind tunnel development. Wind tunnel testing techniques have progressed greatly in recent years. Testing is typically carried out using 40–50% scale models in a rolling road wind tunnel with rotating wheels. The rolling road is necessary to effectively simulate the full size ground conditions with a stationary model. This is particularly so for Formula 1 cars as they run so close to the ground and a large percentage of their downforce is produced from the undertray. The models themselves need to be extremely accurate and have every detail faithfully reproduced including scale suspension members and radiators. This huge amount of development time and financial expense may only gain around a few percentage points improvement in the aerodynamic performance of the car year on year. The days of a week's work in the wind tunnel gaining 15–20% improvement are long gone. These small improvements are critical in gaining an advantage over the opposition. Indeed, given that the teams in Formula 1 are supplied with engines from different manufacturers, and thus have no direct control over their performance, aerodynamics is a major area in which individual teams can obtain large gains in overall performance compared to the opposition.

Bump Stops and their Use on Formula 1 Cars as a Handling Tuning Aid

Unfortunately the aerodynamic centre of pressure of the car varies greatly with both the ride height and the rake angle of the car. As a result a car may be aerodynamically efficient, but if it is difficult to balance throughout the speed range and operating conditions, it will be very difficult to achieve a quick lap time. Typically a Formula 1 car is set up to be regressive, with the centre of pressure moving rearwards as the speed increases. Such a car will have more load on the front at low speed and thus have good turn-in. In high-speed corners however, it will tend towards understeer. The degree of 'regressiveness' will depend on the circuit and tyres and also the preferences of individual drivers. In order to predict and optimise this behaviour it is necessary to predict and control the ride height and rake angle of the car as a function of the variation in the speed.

Bump stops have been used by Formula 1 teams for more than 10 years to attempt to control the handling characteristics of the car with speed. The primary function of the bump stops (Fig. 27) is to avoid metal-to-metal contact

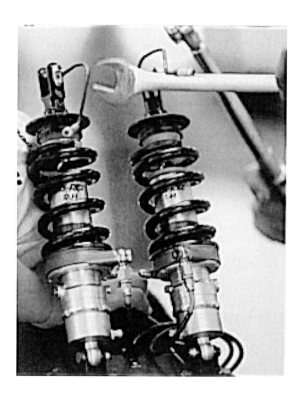

Fig. 27 'Bump stops'.

damaging the dampers, but as the aerodynamic forces increased more dramatically with varying ride heights, they took on a secondary function of trying to control the car's ride height. Control of the ride height affords the race engineers with a degree of control over the aerodynamics, providing a tuning device that may be used to reduce undesirable balance changes. These balance changes are largely due to the cars' aerodynamics being more sensitive to front ride height changes, requiring the car to be stiffened at the frond when the suspension has been partly compressed (i.e. at high speed and high load).

Essentially the bump stops act as a 'rising rate spring'. Initial use of bump stops for this purpose used standard, closed-cell, polyurethane foam[22] known as 'Koni' bump stops after one of the manufacturers (Fig. 27). Their foam construction meant that the bump stops supplied (in varying lengths with tapered ends) all had the same basic load/deflection characteristics once the air had been largely compressed out. There is an initial stiffness as the bump stop is loaded, a plateau as the air is compressed, followed by a harsh rising rate as the rubber is compressed (Fig. 28). The only affect of changing the length of this type of bump stop is to move the rising rate to a shorter or longer suspension deflection. Aside from their limited effectiveness, the pilots complained of being able to feel the bump stop being hit by the suspension. Other problems include the large hysteresis and strain rate dependency of the material, which meant that predictable performance could not be guaranteed. These difficulties precipitated the search for a material with a more 'gentle' rising rate and reproducible properties.

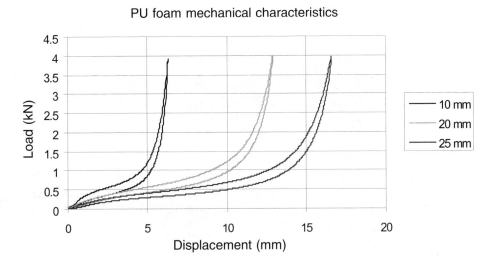

Fig. 28 Performance of standard 'Koni' bump stops.

The solution was to use a polyurethane rubber (Fig. 29) rather than foam, providing a uniform rising rate. Initially, cast rubber was used which, although possessing the desired load/deflection response, led to problems in variability and durability due to porosity from the processing route. Instead it was necessary to employ high quality 'aerospace specification' materials formed by pressing prior to curing of the rubber. Research and development showed that it was possible to further tune the mechanical response of the material by varying the degree of cross-linking as manifest in its Shore hardness,[22] these being denoted by colour coding the rubbers. Furthermore, by choosing a simple geometry, a whole range of load/deflection responses can be obtained by stacking multiple rubbers of differing hardness, allowing the suspension to be fine-tuned for individual circuits (Fig. 30).

It is difficult to quantify the effect of the bump stops on lap time since their influence on performance is generally subjective. Having said that, in a sport where the competitors are so closely matched, any performance enhancement must be taken. There have certainly been instances where the tuning of the bump stops has transformed the performance of the car in the opinion of the pilot and race engineer.[23]

Conclusion

In terms of materials technology, fibre reinforced composites have had the greatest influence upon the design of Formula 1 racing cars. All of the cars that make up the grid are now totally dependent upon composites in their construction. Not only have these materials provided levels of performance

Fig. 29 The use of polyurethane rubber bump stops.

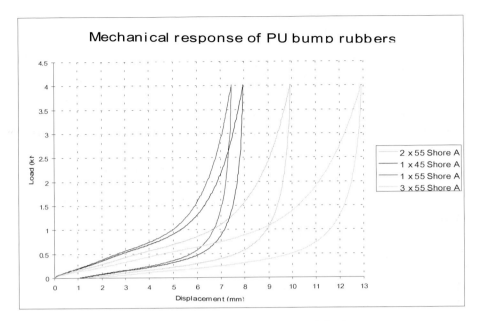

Fig. 30 'Tuning' the mechanical response of the bump stops.

that would otherwise be unattainable, they have acted as an 'enabling' technology facilitating advances in other areas such as aerodynamics. The inevitability in any Formula 1 race is that at some point there will be crashes, either between the competitors and involving cars and some part of the circuit. Over the years the teams and the governing body have worked continuously to enhance the safety of the sport. It must not be forgotten however that the present levels of pilot survivability would simply not be possible were it not for the impact physics of fibre reinforced composite materials.

Aluminium–beryllium alloys were only available to designers for a very short period of time, during which they precipitated very significant advances in braking performance and engine technology. Had they not been outlawed in the rules they would doubtless have found many other niche applications in the cars' construction. One area where their high specific stiffness, isotropy and processability could have been exploited would have been in the gearbox main case. It remains to be seen whether these alloys are reintroduced at a later date, in the same way that a number of electronic control systems have recently been legalised.

Finally, elastomeric bump stops demonstrate how a simple, cheap component can be employed to great effect to improve the handling and performance of the car. Formula 1 mirrors society, the correct choice of the correct material can reap enormous benefits.

Acknowledgements

The author is indebted to Pat Murphy of Brush Wellman Ltd. for the supply of information and figures on beryllium based materials. Thanks are also due to Ian Wright and Steve May of BAR for their help with figures and sections of the text.

References

1. A. A. Griffith, *Phil. Trans. Roy. Soc. London*, 1920, **221A**.
2. G. M. Savage, *Race Tech.*, 1995, **1**, 1.
3. G. M. Savage, *J. STA.*, 2000, **140**, 18.
4. G. M. Savage, *Met. Mater.*, 1992, **8**(3), 147.
5. G. M. Savage, *Proc. Anales De Mecanica de la Fractura*, Baiona (Vigo), Spain, (Mar. 2001), **18**, 274.
6. A. G. Mamalis, D. E. Manolakos, G. A. Demosthenous and M. B. Ioannidis, *Crashworthiness of Composite Thin-walled Structural Components*, Technomic Pub. Co. Inc., 1998.
7. P. H. Thornton and R. A. Jeryan, *Int. J. Impact Eng.*, 1988, **7**(2), 167.
8. D. W. Schmueser, L. E. Wickliffe and G. T. Mase, 'Front impact evaluation of primary structural components of a composite spaceframe', Engineering Mechanics Department, General Motors Research Labs., paper nb 880890.
9. J. N. Price and D. Hull, 'How to apply advanced composites technology', ASM Int., 1988.
10. E. Haug, O. Fort and A. Tramecon, SAE Technical Paper Series, 910152, 245, 1991.
11. E. Haug and A. de Rouvray, *Proc 3rd Int. Symp. On Structural Crashworthiness and Failure*, Univ. Liverpool, 1993.
12. E. A. Brandes, (ed.), *Smithells Metals Reference Book* (6th edn), 1983, Butterworths.
13. *Health Effects Related to Beryllium Exposure*, Brush Wellman Inc. Cleveland, Ohio, USA.
14. *Metals Handbook*, ASM, Metals Park, Ohio, USA.
15. F. C. Grensing and D. Hashiguchi, *Proc Int. Conf. On Powder Met. & Particulate Materials*, May 1995, Seattle, USA.
16. C. Pokross and A. Carr, *Proc Int. Conf. On Powder Met. & Particulate Materials*, May 1995, Seattle, USA..
17. J. M. Marder, *Adv. Mats. & Processes*, Oct. 1997.
18. I. Wright, Senior Vehicle Dynamicist, BAR (Private communication).
19. A. Wood, *Race Tech.*, Feb. 1998.
20. J. Katz, *Race Car Aerodynamics*, Robert Bentley Publishers, 1995, Cambridge, MA, USA.
21. W. F. Milliken and D. L. Milliken, *Race Car Vehicle Dynamics*, SAE International, 1995, Warrendale, PA, USA.

22. J. Brandrup, E. H. Immergut and E. A. Crulke (eds), *Polymer Handbook* (4th edn), 1999, John Wiley and Sons.
23. J. Robinson, Chief Race Engineer, BAR (Private communication).

Some Features of Iron- and Steelmaking in Integrated Steel Plants

AMIT CHATTERJEE
General Manager (Tech), Tata Steel, Jamshedpur-831001, India

ABSTRACT
In daily life, steel is a material used extensively by mankind, be it in the form of razor blades, utensils, vehicles for transport, storage aids, etc. As a result, society and steel are intimately related to progress, the consumption of steel has increased. At the same time, competition from alternative materials like aluminium, plastics and ceramics has come to stay and steel has to fight to stay ahead.

In this paper, developments that have taken place in integrated steel plants to maintain the competitive edge have been highlighted, with particular reference to process changes introduced in Tata Steel – India's premier private sector steel plant making 3.5 million tonnes of steel annually. Changes made in the areas of ironmaking, steelmaking and processing as well as hot and cold rolling are dealt with.

It is clear that steel will continue to have a bright future, particularly in developing regions like India, if appropriate technological changes are introduced to lower the cost of production.

Introduction

Is steel a sunset industry? If the answer were yes, it would not be opportune to talk about developments, which have taken place in this industry over the last few years. On the other hand, if the answer is in the negative, modern developments have to be taken on board. It is in this context that mention must be made of the fact that steel is not really one metal – in a way, it is a galaxy of metals containing various amounts of iron, carbon, several alloying elements as well as the so-called 'impurities'. It is, therefore, not surprising that steel is the most widely used metal by mankind. Consequently, over the last 30–40 years, several revolutionary changes have been made in the steel industry such as: the introduction of oxygen steelmaking, continuous casting, etc. to retain steel's premier position in the league of traditional as well as emerging materials. What has been the effect of these revolutionary changes and have they influenced the relative importance of steel in any way? Consider these facts: the annual world production of steel is second only to that of cement; at 820 million tonnes per annum, steel production is several times the production of all other metals taken together; the fact that the entire amount of steel used is recyclable and that many integrated steel plants reuse over 90% of the solid wastes they generate; and finally, the astounding statistic that by 2005, steel consumption is expected to go up to nearly 850–880 million tonnes annually! In this scenario, concerns regarding the decline of the steel industry appear to be misplaced. For those of us from the steel industry, this is *manna from heaven*. Nevertheless, complacency has to be avoided and we in the steel business have luckily woken up, sooner rather than later, following several wake-up calls given by aluminium, composites, man-made materials, etc.

Efforts to ensure survival by remaining competitive, have resulted in the structure of the steel industry itself undergoing a major transformation – electric arc furnace steelmaking, which earlier was restricted to special, localised markets, today accounts for just over 30% of the world steel production. Indeed, the day is not very far off when EAF steelmaking and the basic oxygen process will have equal shares in world steel production. Subsequent to this shift, there are today, basically two types of steel plants – integrated steel plants which use blast furnaces fed with by iron ore followed by steelmaking in basic oxygen furnaces (BOFs), and mini steel plants based on EAF steelmaking which process re-circulated steel scrap (plus direct reduced iron and in many cases, hot metal). Mini mills have the advantages of flexibility, lower capital cost, ability to cater to niche markets, etc.; the integrated plants are larger, process virgin iron ore, are often located far from their market, etc. This makes them often produce steel at relatively high cost for the customer. The efforts made by integrated steel plants to remain competitive, with special reference to the changes introduced in Tata Steel will now be discussed.

Tata Steel – the first integrated iron and steel plant in India – was set up very early in the twentieth century and over the years, has grown into one of the most modern steel plants in the world. This has been possible because Tata Steel has kept pace with technology and, in doing so, the production facilities have been updated during various phases of expansion and modernisation. The state-of-the-art 1.2 Mtpa Cold Rolling Mill Complex, which was commissioned in April 2000, is perhaps the best example of Tata Steel's mission to keep pace with (and in many cases, be ahead of) changing times. With the progress of the country, the Indian automobile industry is witnessing a radical transformation. In anticipation of the changing steel demand, Tata Steel installed this CRM for essentially producing automobile grade strips. Some other important technological innovations introduced in the areas of ironmaking, steelmaking and processing over the last two decades have enabled Tata Steel to retain its competitive advantage in the global market. In the subsequent sections, some of these major innovations as well as process changes, which have been introduced in Tata Steel to produce world-class goods and services, will be discussed.

Developments in Ironmaking

Amongst all the ironmaking processes, the blast furnace, which is the oldest, still holds the dominant position. Many of the more recent ironmaking technologies which have been developed can at best complement the blast furnace in the years ahead, but, it is believed, that even after these technologies have achieved maturity, the blast furnace will continue to be the principal route of ironmaking. This sustained dominance of blast furnaces can be ascribed to the developments that have taken place over the years in blast furnace process technology as well as in the improvements made in the design and engineering aspects of the process. Hot metal production rates of up to 10 000 tpd, fuel rates of around 450 kg per tonne hot metal (320/350 kg/thm coke plus 100/80 kg/thm coal), and furnace availability ranging between 95–98% are the results of improved process control, better understanding of the process as well as stability of operations. Low silicon (<0.20%) and low sulphur (<0.030%) hot metal is produced consistently nowadays. A campaign life of 15 years and more between two major relining is today quite common.

Innovations in measuring techniques, like the introduction of profilometres, over- and under-burden probes, top gas analysers, etc. along with the use of mathematical models have radically improved the ability to monitor blast furnace operations. The importance of the cohesive zone, its position, size and shape have been fully recognised. Raceway and gas distribution measurements as well as room temperature/high temperature model studies have helped in establishing the flow dynamics in the furnace, particularly in an effort to quantify the effects of burden distribution. The use of the latest charging systems and control devices: central working philosophy, improved cast house and drainage practices, use of tuyère injectants, introduction of high top pressure, etc. have greatly enhanced blast furnace performance. Preventive maintenance techniques and computer assisted procedures have drastically reduced the down time. Use of better quality refractories, optimal cooling as well as gas flow control have extended the campaign life. Lower energy consumption and higher productivity achieved with increased oxygen enrichment and massive coal injection are today the norm. Expert systems – 'intelligent computer programmes using existing knowledge and inference rules to solve complicated problems by simulation' – have been developed to reduce the dependence of blast furnace operation and control on the empirical knowledge of the operators. These developments, over the years, have ensured that the blast furnace has remained the most efficient method for tonnage production of liquid iron.

At Tata Steel, the philosophy of phase-wise modernisation on a continuing basis was adopted since the early 1980s with great advantage in the earlier two million tonne steelworks at Jamshedpur. Four phases of modernisation were successfully completed in 2000 to increase the rated capacity to the current level of 3.5 million tonnes per annum of liquid steel. The new facilities which were added during modernisation in the area of ironmaking included:

- A raw material bedding and blending yard to ensure consistent chemistry of the feed material for sintermaking.
- A new 1.37 Mtpa sinter plant (SP 2) equipped with deep bed sintering facilities to increase the proportion of sinter in the blast furnace burden to 60–65% (the total annual production of sinter at Tata Steel is 2.5 Mtpa from SP 1 and SP 2).
- Adoption of stamp charging technology for cokemaking – the five stamp charged batteries now available can produce 2 Mtpa of blast furnace coke.
- A new 1 Mtpa 'G' blast furnace for augmenting the production of hot metal to 3.5 Mtpa.
- External desulphurisation of hot metal for quality steelmaking using BOFs and 100% continuous casting.

Some of the innovative measures adopted for increasing the productivity of the blast furnaces to match global standards included:

- Improved burden distribution using the movable throat armour (MTA) method in the older furnaces ('A', 'C', 'D' and 'F') and the bell-less Paul Wurth (PW) distribution system in the new 'G' furnace.
- Control of alkali input to less than 3.0 kg/thm in the burden.
- High blast humidity (60 to 90 gm Nm^{-3}) operation ensuring 8–9% hydrogen in the bosh gas, which is beneficial for better indirect reduction, increased efficiency of gas utilisation, scavenging of coke fines and increased flushing of alkalis through the slag.
- Use of 30-40% blue dust (very pure iron ore in fine form) in sinter-making, decrease in the alumina content of sinter to less than 2.6%, production of low basicity-sinter using dunite (resulting in improved reduction degradation index and superior softening-melting character-istics).

- Use of 20–30% low ash imported coal (and even semi-soft coals) in the blend for cokemaking using stamp charging – this coke has superior CSR (coke strength after reaction) and CRI (coke reactivity index) values compared with conventional top charged coke.
- Pulverised coal injection (60–130 kg/thm with replacement ratios of 0.8–1.0) in 'D', 'F' and 'G' blast furnaces.
- Tar injection (40–45 kg/thm with a replacement ratio of around 1.5) in 'A', 'B', 'C' and 'E' furnaces.
- Increase in hot blast temperature to 1000°C in the older furnaces and 1160°C in 'G' blast furnace.
- Increased oxygen enrichment (4–6%) and high RAFT (2130°C) particularly in 'G' furnace.
- Operation with low slag volumes (less than even 300 kg/thm in 'G' blast furnace).

Some of these techniques which have helped in making blast furnace ironmaking at Tata Steel come up to international standards are now described in greater detail.

Deep Bed Sintering

With the changes in the characteristics of the raw materials as well as increasingly stringent requirements on sinter quality following the use of higher percentages of sinter in the blast furnace burden, the operating philosophy of the sinter plants had to be modified to achieve higher productivity levels. Deep bed sintering was identified as one of the potent means of achieving higher sinter plant productivity levels.

Success of deep bed sintering is governed by the size distribution of the sinter mix, prevention of segregation of the mix on the strand, choice of the appropriate size of coke breeze, extensive lime addition, use of the optimum suction pressure, moisture control, etc. A proper balance has to be maintained between the bed depth, suction pressure and the rate of sintering to produce acceptable quality sinter at high productivity. In order to exploit the advantage of high suction pressure in SP 2 fully, the bed depth was increased to a maximum of 600 mm. With the addition of around 30% blue dust, the ultra fines (i.e. −100 mesh) in the green mix increased from about 5% to 11–12%. The adverse effect of higher bed depth and increased ultra fines content of the mix on the permeability of the bed was nullified by the introduction of as much as 3% lime fines to the green mix (30–35 kg lime per tonne of sinter). Additionally, steps were taken to improve the granulation behaviour in the mixing drum by optimising the amount of moisture added and by increasing the residence time of the green mix in the drum. This improved the bed permeability and facilitated higher productivity along with acceptable sinter quality.

Figure 1 shows the effect of bed depth on sinter productivity and quality at Tata Steel. Another parameter, which has an influence, is the size of the coke breeze, which determines the efficiency of combustion during sintering. Excessively crushed coke breeze undergoes rapid and incomplete combustion resulting in the formation of larger amounts of carbon monoxide than carbon dioxide. The heat thus evolved is inadequate for the surrounding mix to attain the required sintering temperature. On the other hand, very coarse coke particles cause localised overheating, thereby prolonging the combustion cycle and broadening the sintering zone. This results in excessive fuel consumption and deterioration in sinter quality as well as productivity. The size of coke breeze used in Tata Steel is 100% below 6 mm with 85% of −3.15 mm; the −1 mm fraction is restricted to a maximum of 55–60%.

Higher suctions at comparatively low bed depths make the sinter relatively weak, although very high productivities can be achieved. The importance of bed depth on sinter productivity and quality is, therefore, obvious. The gradual decrease in the alumina content of sinter over the years and the consequent improvement in sinter RDI (reduction degradation index) are shown in Figs 2 and 3 respectively. Figure 4 depicts the increase in the productivity of SP 2 over the last few years. Very recently, a major alteration

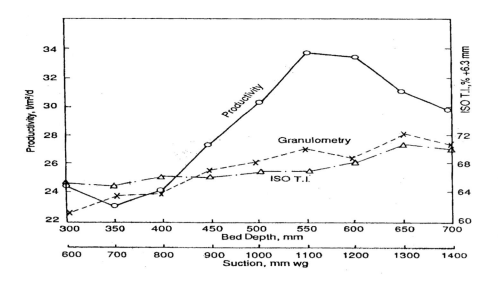

Fig. 1 Effect of bed depth on sinter productivity and quality.

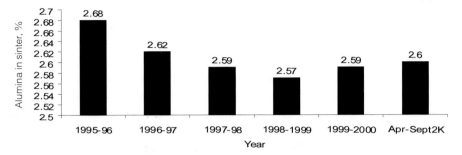

Fig. 2 Decreasing trend in alumina content of sinter.

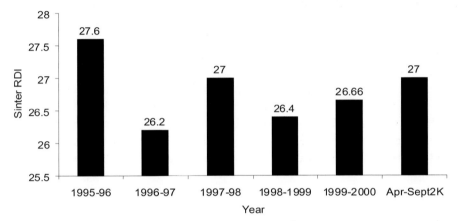

Fig. 3 Variation in RDI of sinter.

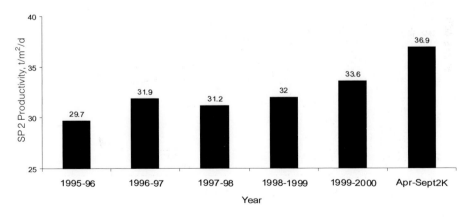

Fig. 4 Improvement in SP 2 productivity.

in the bed depth – machine speed combination has been made (Fig. 5) which has resulted in further improvement in productivity (+38 t/m²/d) without compromising quality, which today is at least as good as, if not better than, the best international benchmark levels.

Stamp Charging

The overall economics of producing iron and steel in any integrated steel plant is extremely sensitive to the performance of blast furnaces in terms of the fuel rate. Out of the many variables affecting the fuel rate of blast furnaces, the quality of coke is perhaps the most paramount. Therefore, it was only natural that improvement in the quality of coke received particular attention during the modernisation programmes at Tata Steel.

The quality of coke can be improved by adopting one, or both, of the following two approaches:

- Use of a better coal blend for cokemaking.
- Adoption of a suitable pre-carbonisation technology during cokemaking.

Although India has large reserves of coal, the availability and quality of coal for producing blast furnace grade coke fall far short of expectations. The limited quantity of coking coals available are rich in 'inerts' (both organic and inorganic), and deficient in coking properties, owing to their low vitrinite contents and low reflectance values. Incorporation of appropriate prime coking coals in the coal blend, through imports, is now practised in all the integrated plants in India. However, the scope is narrow and, beyond a certain limit, has adverse economic implications. It was, therefore, decided to adopt a pre-carbonisation technology which would improve the quality of blast

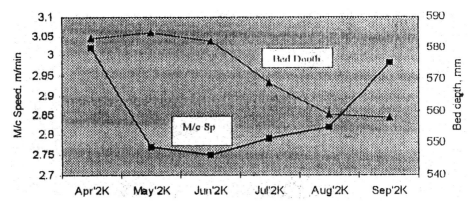

Fig. 5 Correlation between machine speed and bed depth of high productivity.

furnace coke from a given coal blend. Following intensive in-house R&D efforts, stamp charging was found to be the most suitable technology for the conditions prevailing at Tata Steel. The first stamp charged battery in the country, consisting of 54 ovens, was commissioned in 1988–1989 at Jamshedpur and over the next decade, Tata Steel rebuilt four more batteries incorporating stamp charging so that, at present, 2 Mtpa of blast furnace grade coke is produced using this pre-carbonisation technique.

Stamp charging was chosen since the improvement in coke quality was the highest compared with all the other methods e.g. briquette blending, pre-heating, etc. Studies on the role of coke quality on blast furnace productivity indicated that a 5% increase in productivity could be expected by lowering the Micum 10 value of coke (a measure of the strength of coke at room temperature) by two points – this was easily achieved in stamp charging. What was more meaningful was that the quantum change achieved in coke quality following stamp charging was found to be virtually independent of the blend. This gave rise to new opportunities in terms of using inferior coals, thereby reducing the cost of coke, which normally accounts for over 50% of the cost of hot metal.

At the time when the first stamp charged battery was commissioned, Tata Steel was using a coal blend containing 25–30% of low-ash, high-rank, prime coking coal from Australia along with 70–75% captive coals in the top charged batteries. A similar blend was also used in the stamp charged battery. Once it was established that even with distinctly inferior and less expensive blends, stamp charging could produce coke with similar (if not better) room temperature and high temperature properties than top charged coke, a switch was made to around 20% semi-soft Australian coal in the blend in order to arrive at a reasonable coke strength after reaction at high temperature (CSR). The replacement of expensive prime coking coals by relatively inexpensive semi-soft grades, without affecting the coke quality, resulted in a 32% decrease in the cost of coke. This reduction in coke cost was complemented by a higher yield of sized coke. The revised coal blend for stamp charging shown in Table 1 gave the same high CSR coke as the earlier top charged coke and,

Table 1 Coal Blends for Top Charged and Stamp Charged Ovens and Their Relative Costs.

Source of coals for blending	Top charged, %	Stamp charged, %
Australian	30	20
Jamadoba	45	0
Imported	25	80
Relative cost	100	68

in many cases, the coke properties, in fact, improved. It will be clearly noticed that superior grade coke can be made using stamp charging with a lower cost coal blend.

CRI and CSR are important for improving blast furnace performance. The low strength after reaction of coke made conventionally in top charged ovens is often the principal factor limiting blast furnace productivity. Stamp charged coke made from only 20% prime coking coal, showed CSR values 3–4 points higher than top charged coke made from 75% prime coal. Furthermore, the CSR of stamp charged coke was also found to be insensitive to changes in the coal blend, within limits. The significant improvement in coke quality achieved in Tata Steel, as indicated by the CSR values can be seen in Fig. 6. It can be unhesitatingly stated that stamp charging is the most appropriate pre-carbonisation technology for cokemaking in countries like India with large reserves of low rank coals. Successful use of non-coking coals in the stamp charging blend has opened a new horizon, and is a breakthrough in widening the coal base for cokemaking and, to that extent, reducing the dependence on high grade coals, which are at least US$8–10 and often even US$15 per tonne more expensive.

Pulverised Coal Injection in Blast Furnaces

Pulverised coal injection has the maximum potential in reducing the coke rate in blast furnaces, thereby conserving scarce metallurgical coking coal. All over the world, efforts are being made to maximise the rate of injection and achieve a high coke to coal replacement ratio, without any compromise on the productivity and hot metal quality. The complete combustion of coal in the raceway is the single most important parameter that limits the rate of injection.

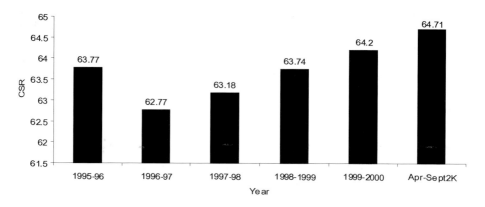

Fig. 6 Improvement in CSR of stamp charged coke.

Incomplete combustion in the raceway results in deposition of unburnt carbon particles on the walls of the deadman and the cohesive zone, thereby affecting the burden permeability. Deterioration in permeability affects the wind acceptance and hence, the furnace productivity becomes lower. Injection rates of up to 200 kg/thm with replacement ratios of 0.80–0.90 have been achieved in several blast furnaces – present research aspires to raise the rate to 250–300 kg/thm.

At Tata Steel, three out of the seven blast furnaces are operating successfully with coal injection. The extent of injection in 'D' and 'F' furnaces is limited to a maximum of 100 kg/thm, while in 'G' blast furnace, which is equipped with modern instrumentation and control facilities, the rate of injection ranges between 120–150 kg/thm. It must be mentioned that the operational philosophy of blast furnaces with high pulverised coal injection rates is quite different from that of all-coke operations. High PCI rates (130–175 kg/thm) demand:

- Superior quality coke (CSR >65).
- Lower RDI of sinter (RDI <26).
- Revised burden distribution (ensuring predominant central working conditions within the furnace).
- Softer blowing (tuyère velocity around 210 ± 10 m/sec).

In order to keep the lower part of the furnace (cohesive zone and raceway) permeable on a sustained basis, the use of relatively large sized coke (40–100 mm) and operation with proportionately more coke in the central part of the furnace (5% above the coke base in the feed) are better. This also decreases the average depth of the molten iron pool at the end of a cast, thereby protecting the hearth refractories. If raw materials of the desired quality are not available and the furnace conditions are unstable (with regard to wind acceptance and furnace movement), then, the burden distribution and blowing conditions (wind volume, hot blast pressure, top gas pressure) have to be suitably adjusted. Furnace operations with a flatter burden profile, reduced rate of PCI, and lower RAFT are beneficial. In all cases, attempts have to be made to operate the furnace with the minimum possible pressure drop.

Hearth management also plays a decisive role in successful blast furnace operation with PCI. Hearth drainage has a marked bearing on the furnace performance in general and hearth refractory temperatures and build-ups in particular. When coke in the deadman becomes impermeable, it promotes peripheral flow of liquids, particularly in the area opposite the tap hole. The contact time between the liquid and the hearth refractory lining also increases, to some extent. Predominant peripheral flow of liquids and increased contact time result in high heat loads on the hearth walls giving rise to erosion of the hearth refractories and/or build-ups (the so called 'elephant foot'). While this is true for blast furnace operation in general, it has to be borne in mind that the situation becomes worse at high PCI rates. Poor hearth drainage results in the liquid pool interfering with the raceway, forcing excessive peripheral gas flow in the bosh region – this also increases the heat load on the refractory wall. An impermeable/dense deadman and high slag volume promote peripheral drainage of liquids under the deadman, rather than through the deadman. Virtually the entire liquid flows along the side walls or under the deadman instead of through the deadman which would minimise the heat load on the hearth refractories. The maximum velocity of liquid flow in the coke-free layer is attained when the height of the coke free zone is equal to 0.015 times the hearth radius – an increase in the height of the coke free layer decreases the velocity.

A strong correlation exists between the average depth of the molten iron pool at the end of any cast and the thickness of hearth build-ups/hearth erosion. The lower this depth the better. Central coke charging is beneficial for improving the permeability of the cohesive zone and the deadman coke. Important technical issues associated with high PCI rates include:

- Change in the raceway gas composition profile.
- Alteration in the burden distribution.
- Accumulation of unburnt coal particles leading to the formation of a bird's nest (Fig. 7).
- Impeding the reaction of coke with CO_2.
- Disintegration of the limited amount of coke available.

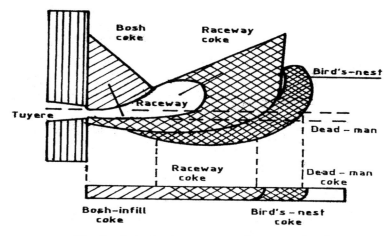

Fig. 7 Coke bed structure at the tuyere level.

- Increased wall heat losses.
- Slower dissolution of the solid charge.
- Raceway collapse cycle.

At Tata Steel, furnace operation with superior quality raw materials (especially in terms of sinter RDI and coke CSR), a permeable deadman, and a softer blowing regime have resulted in dramatic improvements in 'G' furnace productivity and fuel rate (Figs 8 and 9) following coal injection. Considering the enormous economic benefits associated with PCI, Tata Steel is in the throes of increasing the extent of coal injection to around 175 kg/thm the world average injection rate at present.

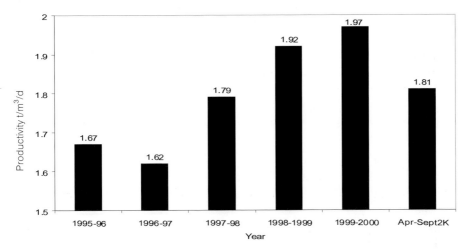

Fig. 8 Improvement in productivity (based on working volume) of 'G' blast furnace.

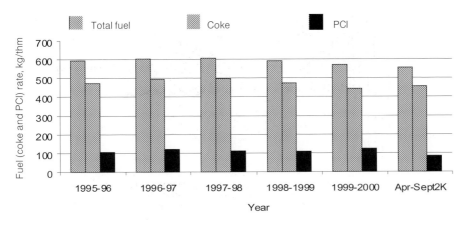

Fig. 9 Fuel rate in 'G' blast furnace.

Developments in the Area of Primary Steelmaking

Not very long ago, there were two equally popular steelmaking processes – open hearth (OH) steelmaking and electric arc furnace (EAF) steelmaking. Around the 1950s, when Elvis Presley started making a splash in the field of music, another type of 'wave' (literally) was generated by a new steelmaking process. Started in Linz and Donawitz in Austria, this process (LD) stirred-up the steelmaking world, like nothing before, by vertically injecting high purity oxygen on to the molten iron bath at Mach Numbers greater than 2! The heat making time which used to be a sluggish eight hours in the OH process, was suddenly compressed to forty-five minutes and what was more meaningful, no external fuel was necessary. Today, the LD process is more popularly known as the BOF process.

The increased efficiency of the BOF process became a boon for all integrated steel plants. It reduced steel costs and improved productivity – it was, therefore, not surprising that it steadily became the dominant steelmaking process all over the world accounting for over 60% of the world steel production today. The EAF is the other player – the obituary of the OH process has already been engraved in several tombstones scattered all over the world.

Tata Steel began the gradual shift from OH to BOF steelmaking in 1982 and by 1993, the entire steel production was through this process. However, as the demand on steel quality increased, in particular the need for low phosphorus steel, the BOF process itself needed modifications. This involved changes in the design of the lance used for injecting oxygen vertically from the top and the simultaneous introduction of inert gas through tuyères/canned refractory elements from the bottom. Canned elements were at first used in Tata Steel, but later, the German steel producer Thyssen, with whom Tata Steel has a technical tie-up, helped in introducing bottom tuyères (TBM process). All the five 130 t converters in Tata Steel (2 in LD 1 and 3 in LD 2) are now equipped with the TBM system. This has enabled the steel company to reduce the cost of liquid steel (higher yield, lower consumption of ferro-alloys, etc.) as well as to enrich the product mix by producing more exotic steel grades (e.g. steels required by the automobile industry).

Production of Low Phosphorus Steel

After the upgrading of the Hot Strip Mill at Tata Steel to 2 Mtpa in 1999, over 1.2 Mtpa of the steels processed through the BOF-Slab Caster–Hot Strip Mill route needed 0.020% max. phosphorus in the product. In some cases, the acceptable phosphorus level was even lower – around 0.015% max. Production of low phosphorus steels in Tata Steel is a big challenge because the phosphorus content of Tata Steel's hot metal (which constitutes around 90% of the BOF charge) is as high as 0.25–0.28%. It is not easy to lower the phosphorus content of hot metal since captive iron ores and coals (which have high phosphorus but are available at relatively low cost) are used. These raw materials contribute almost equally to the phosphorus input to hot metal.

A two-pronged strategy involving modification in the way oxygen is blown from the top along with optimised inert-gas stirring from the bottom was adopted to take care of this situation. A six-hole lance (SHL) was installed in Tata Steel's LD 1 in January 1993 to replace the three-hole lance (THL) which had been used since the inception of the shop in early 1982. The lance operating pressure was increased by about 60%, resulting in an increase in the Mach Number and jet exit velocity to 2.255 and $524\,\mathrm{m\,s^{-1}}$ respectively compared with 2.01 and $488\,\mathrm{m\,s^{-1}}$ for the THL. With the increase in the number of nozzle openings, it became possible to increase the rate of oxygen supply without increasing the metal loss in the droplets formed because of excessive turbulence. The risk of burning the converter bottom because of too deep a penetration of the oxygen jet on account of excessive gas flow was also precluded. These advantages accrued from the total energy of the jets getting distributed over a larger surface area of the bath. As a result, the slag–metal interaction was better without impeding the decarburisation rate. For example, because of the effective 'harder' blowing with the THL, the slag formed was 'drier' than in the case of the SHL with the result that dephosphorisation was often hindered, particularly if rapid decarburisation was also desired for which relatively low lance heights were used.

With the introduction of the optimum amount of inert gas through the bottom tuyères using the TBM process, the control of phosphorus to fairly low levels became possible. This has given rise to the following advantages:

- Improved bath mixing, mass transfer, heat transfer and a quieter blow.
- Chemical reactions closer to equilibrium conditions.
- Higher phosphorous and sulphur partitions between the metal and the slag.
- Higher manganese recovery and lower dissolved oxygen contents in the steels, resulting in a higher recovery of ladle additions.
- Possibility of producing low carbon steels without over oxidation of the bath.
- Lower nitrogen contents and better process reproducibility.
- Concurrent process improvements such as less slopping, better gas recovery and increased lining life.

Tata Steel today is making 0.012–0.015% phosphorus steels using the combined blown BOFs followed by extensive secondary steelmaking which includes ladle furnace treatment, vacuum arc degassing/RH degassing.

Developments in the Area of Secondary Steelmaking

With the revolutionary changes in automotive styles, steel grades with guaranteed formability and excellent elongation are in great demand. Such grades of steels require stringent controls in terms of C, N, O and H contents so much so, that the steel compositions virtually approach that of pure iron. This demand has led to the evolution of a large number of secondary steelmaking processes for treating liquid steel after BOF steelmaking. For decarburisation to a very low level, say below 25 ppm, the thermodynamics of the carbon–oxygen reaction at reduced pressure plays a key role. RH degassing is one technological application of decarburisation, which is widely used. Originally, the RH degasser was developed for hydrogen removal. In the RH process, liquid steel is sucked into a vacuum chamber, which encourages the removal of dissolved gases like hydrogen (and oxygen).

The C–O reaction equilibrium at reduced pressure is shown in Fig. 10. Thermodynamically, it appears to be easy to achieve low levels of carbon in liquid steel, but the kinetics of the reaction is slow. Conventional RH degassing suffers from large temperature drops during the liquid steel treatment. This has led to different modifications of the RH process – RH–OB, RB–KTB and RH–PB. In the RH–OB (Oxygen Blown) process,

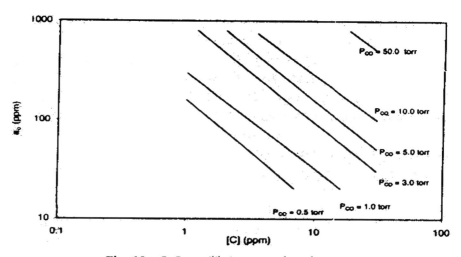

Fig. 10 C–O equilibrium at reduced pressure.

oxygen is introduced into the lower part of the RH vessel, whereas, in the KTB (Kawasaki Top Blowing) process, oxygen is blown from the top of the vessel through a lance. In the RH–PB (Powder Blowing) process, a flux in powder form is blown on the liquid steel surface during vacuum treatment. For powder blowing, several systems are available RH–PB (IJ) (powder flux injected into the ladle through a lance positioned beneath the up snorkel), RH–PB (OB) (powder blown into the liquid steel through the oxygen blowing nozzles located in the lower part of the vacuum chamber) and RH–PTB (powder blown through a top lance at the top of the vacuum chamber). Low sulphur and nitrogen levels in liquid steel, of the order of 5 and 14 ppm respectively, can be achieved using any of these modified RH systems. The loss in temperature during vacuum treatment can also be tackled by aluminium addition – the resultant exothermic reaction leads to a temperature rise of as much as $6°K\,min^{-1}$.

A comparison of various operating parameters of RH and RH–OB is given below:

	RH	RH–OB
Decarburisation level, ppm	20	20
Decarburisation rate	Satisfactory	Highest
Decarburisation period, min	12–15	10–15
Total treatment time, min	30–40	30
Hydrogen removal	Satisfactory	Satisfactory
Cleanliness	Reasonable	Cleanliness deteriorate, (7–10 min post treatment required)
Exothermic/chemical heating	No	Yes
Relative capital cost RH–OB = 1.0	0.7–0.8	1.0
Maintenance cost		15–20% higher owing to higher refractory consumption

Tata Steel has a 130 t RH degasser in LD 2 where all the steel grades for flat products are made. To enable less than 0.012% phosphorus steels to be produced in this shop, investments to modify the RH degasser to reduce temperature drop during the treatment of liquid steel are under consideration.

Developments in the Area of Continuous Casting

In the 1960s there was excitement in the air. The steel industry had adopted a technology that could easily and efficiently convert molten steel into billets. The process eliminated the roughing operation normally needed for an ingot, leading to a tremendous increase in yield and a large saving in energy. This new revolutionary technology was capable of cutting production cost. It was, of course, continuous casting – bread and butter stuff today for all steel producers, big and small. The next thirty years saw developments in continuous casting in the areas of machine design, longer sequence casting, faster casting rates, and so on. This fledgling technology of the not too distant past now processes more than 60% of the total liquid steel produced in the world.

The start of the 1990s once again saw tremendous excitement in the air. Towards the end of the last decade, July 1989 to be exact, a new technology was tried in Nucor Steel, Crawfordsville, USA which, by directly linking the casting process to rolling was able to bring about further improvements in yield, saving in energy, and a reduction in product cost compared with classical continuous casting. This epoch-making technology became known as thin slab casting (TSC).

The continuous casting machines of the 1960s and the 1970s produced slabs that were 200–250 mm thick; the slabs of today are on a slimming diet! They turn out a trim 50–60 mm thick (thin?)! Thus, if in the 1960s it was fashionable to compare conventional ingot casting with the then new continuous casting process, it is the 'in-thing' now to do the same for the new thin slab casters and by now conventional, continuous casting. Whatever benefits were attributable then to the conventional casters are now applicable to thin slab casters – only they stand several times multiplied! Concerns had been raised in the 1960s whether the quality of the material from a cast product which underwent only limited reduction could ever compare with that of a product made from an ingot, which was extensively and severely 'worked'. Questions are similar now – only the comparison, this time, is in terms of strips from a thin slab caster *vis-à-vis* from a conventional caster.

This is one option that Tata Steel has not yet incorporated in its plant, but is actively assessing its implications. In the meantime, TSC has found worldwide favour (Table 2). Two popular players in this technology are:

- Schloemann Siemag (SMS) – Compact Strip Process (CSP) – Fig. 11.
- Mannesmann Demag Huttentechnik (MDH) – In-line Strip Process (ISP) – Fig. 12.

Important Issues in Thin Slab Casting

It has been estimated that the investment cost per tonne for a two-strand 1.5 Mtpa thin slab caster is only 65% of that for a conventional caster of 3.6 Mtpa, and the operating cost is expected to be lower by US$12–15 per tonne. However, excessive turbulence while feeding liquid steel into the

Table 2 Steel Plants with Thin Slab Casting Machines (Installed/Planned).

Names of plant	Rated thin slab capacity, Mtpa	Start up	Caster machine
Nucor Steel, Crawfordsville, Indiana, USA	1.8	1989 and 1994	CSP
Nucor Steel, Hickman, Arkansas, USA	2.0	1996 and 1994	CSP also ITSC
Nucor Steel, Berkeley, South Carolina, USA	1.5	1996 (Oct.)	CSP
Steel Dynamics, Butler, Indiana, USA	1.2	1996 (Jan.)	CSP
AST, Terni, Italy	–	1992	CSP
Hylsa, S.A., Monterrey, Mexico	0.9	1995	CSP
Hanbo Steel, Asan Bay, Korea	1.0	1995	CSP
Gallatin Steel, Warsaw, Poland	1.0	1995	CSP
Kentucky, USA	1.0	1996 (Oct.)	CSP
Acme Steel, Riversale, Illinois, USA	1.0	1996 (Oct.)	CSP
Nippon Denro, Dolvi, India	1.2	1996 (Aug.)	CSP
Aceria Compacta Bizkaia, S.A., Bilbao, Spain	0.9	1996 (July)	CSP
NSM, Chonburi, Thailand	1.2	1997 (Aug.)	CSP
ASM, Malaysia	2.0	1998 (May)	CSP
Zhujiang Steel, China	0.8	1998 (Nov.)	CSP
Baotou Steel, China	2.0	2000	CSP
Handan Iron, China	1.2	2000	CSP
POSCO, Kwangyan, Korea	1.8	1996 (Sept.)	ISP
Arvedi, Cremona, Italy	0.5	1992	ISP
Saldanha Steel, Rivoria, South Africa	1.4	1997 (Dec.)	ISP
Algoma Steel, Sault Ste, Marie, Ontario, Canada	2.0	1997 (Aug.)	fTSC
Avesta-Sheffield, AB, Sweden	0.6	1998	CONROLL
Armco-Mansfield, Ohio, USA	1.1	1993	CONROLL
North Star BHP Steel, Ohio, USA	1.35	1996 (July)	Sumitomo
Trico Steel, Alabama, USA	2.2	1996 (Dec.)	Sumitomo

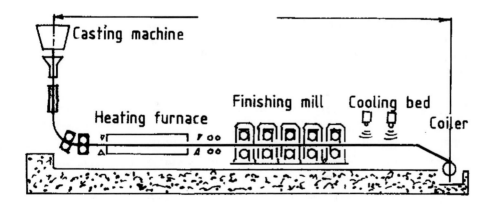

Fig. 11 Layout of a CSP machine.

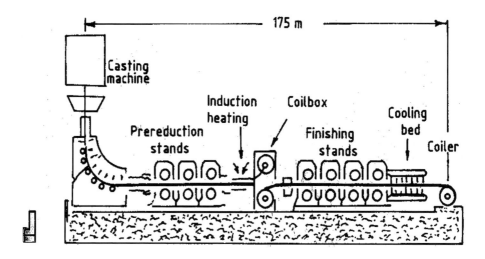

Fig. 12 Layout of an ISP machine.

mould of a thin slab caster to achieve reasonable production rates, leads to a slab surface that produces a strip unsuitable for use in exposed auto body parts (particularly in the absence of slab scarfing). Thus, the product mix of a conventional caster can potentially be richer than that from a TSC. Additionally, casting of peritectic and crack sensitive steel grades requires reduction in the speed to one-third of the normal casting speeds to prevent breakouts. While this is also true in conventional slab casting, the lower speeds needed for the peritectics are too low in a thin slab caster leading to premature freezing of the liquid in the restricted space available in the mould. Thus, these grades have to be altogether avoided in a TSC product mix, which is not always the case in conventional casters.

Product Quality

The cooling in TSCs is rapid and, therefore, there is little chance of centre-line segregation. Faster cooling makes the strip almost free from segregation and prevents the precipitation of alloying elements. Some of the important features of TSC cast and directly rolled products are:

- Low C–Mn steels give superior HAZ (heat affected zone) toughness owing to delayed MnS precipitation and finer precipitate size. It also influences the precipitation of iron carbide. The efficiency of over-ageing is expected to be more effective.

- Higher yield strengths of around 485 MPa compared with 435 MPa of C–Mn steel sheets. Owing to faster cooling, the size of titanium nitride precipitates decreases from 100 nm to a more desirable value of 20 nm.
- Centre-line segregation decreases after reduction and the solidified structure changes to a fine equiaxed structure in a small zone at the centre of the slabs – no 'V' or inverse 'V'-shaped segregation exists.

The production per strand of a TSC for different widths and thicknesses as a function of the casting speed is illustrated in Figs 13 and 14 respectively. It is possible to use this data in assessing the potential of tonnage production from thin slab casters. On the flip side, the BF–BOF–TSC route offers a few challenges. On account of lower availability of a TSC compared with that of a blast furnace, arrangements have to be made to dispose off the hot metal produced during the down time of the caster (e.g. by using pig casting machines). Operational experience at Algoma Steel and also at Thyssen has shown that this linkage is possible and may not be as critical an issue as was apprehended earlier.

Ferrite Hot Rolling of Low Carbon Steel

Rolling is energy intensive phenomena, which is very critical because of its direct impact on the finished product quality. In conventional hot rolling, the finishing operation is performed above the Ar_3 temperature so as to get a

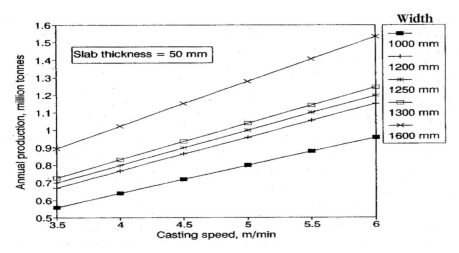

Fig. 13 Production per strand of a thin slab caster arrangement for different slab widths.

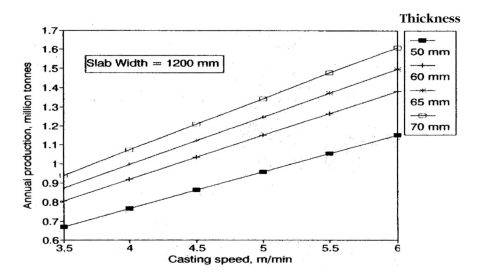

Fig. 14 Production per strand of a thin slab caster arrangement for different slab thicknesses.

uniform transformed ferrite structure after cooling. Lowering of the finishing temperature slightly below the Ar_3 temperature, results in microstructural inhomogeneity in terms of varying ferrite grain size and texture. Transformation of retained austenite to ferrite in the finishing stand may cause cobbles, especially in the case of IF (interstitial free) steels. If finish rolling is conducted entirely below the Ar_1 temperature, popularly known as ferritic rolling, the steel strip microstructure is predominantly ferrite during rolling. This is an excellent practical application of the low carbon end of the Fe–C diagram to rolling technology. Time–temperature diagrams for both conventional rolling and ferrite rolling are schematically illustrated in Fig. 15. Following recrystallisation of the deformed ferrite, a uniform structure is obtained in the as-rolled product. The goal is to produce low cost and ductile hot rolled sheets from low C ($\sim 0.03\%$ C) or ultra low C steel slabs. The metallurgical advantages of ferritic rolling are:

- Uniform ferrite structure which results in good formability in the as-rolled product.
- The size of the recrystallised ferrite grains of low C steels is typically larger than in conventional rolling. These larger grains promote lower yield strength and higher elongation, which are beneficial for formability.

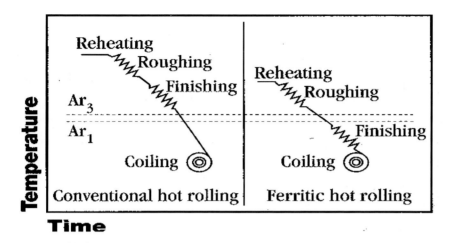

Fig. 15 Schematic representation of conventional and ferritic rolling.

Ferritic rolling also gives rise to cost advantages on account of the following:

- Lower operating temperature, which decreases fuel consumption.
- Lower reheat temperatures are associated with slower oxidation rates, thereby increasing the average yields of slabs.
- The wear of the work rolls decreases because of lower temperature in the finishing stand.
- Since the ferrite grains are fully recrystallised in the hot rolled strips, higher reduction is possible during cold rolling.
- For the production of fully recrystallised hot strips, a high coiling temperature of around 680°C is used. This minimises the need for extensive water spraying on the run-out table, thereby reducing the water consumption.
- Thinner gauge sheet production is facilitated because of lower rolling temperatures.
- Ferritic hot rolling can be used to improve the edge condition of IF steels by work hardening of the ferrite – it gives a clean break during trimming.
- Low solubility of AlN as well as the relatively high mobility of N and Al in BCC iron permits rapid precipitation of AlN during processing. The kinetics of AlN precipitation is known to be much faster in ferrite than in austenite.

As mentioned earlier, Tata Steel has a 2 Mtpa HSM capable of producing HR coils from 1.2 to 12 mm in thickness. Ferritic rolling has been conducted successfully corroborating some of the advantages listed above.

Cold Rolling

To add value to HR coils, world over, at least 60% of HR coils are cold rolled. Tata Steel has just set up a Cold Rolling Mill at 'global speed and cost' to serve the customers' needs of today and tomorrow. The makers of passenger cars, two-wheelers, refrigerators, domestic appliances, computers, drums, barrels, containers and a host of others use cold rolled sheets. They look for:

- Cold rolled sheets with a flawless surface, consistent mechanical properties and close dimensional tolerance to ensure top class performance of the end products they make.
- A service quality, which would enable them to meet just-in-time (JIT) conditions in spite of variations in demand of their end products.

Tata Steel's Cold Rolling Mill, which began operations in April 2000 satisfies these requirements, and even more, because it incorporates the most advanced facility of its kind in the world. The important facilities installed in the CRM Complex are:

- PL–TCM (pickling line connected with the tandem cold rolling mill) of 1.2 Mtpa capacity. Pickling is carried out with hydrochloric acid and for rolling, a 5-stand 6-high tandem cold rolling mill is used.
- 100% hydrogen batch annealing for 0.8 Mtpa of strips.
- Single-stand 4-high skin pass mill with a capacity of 1 Mtpa.
- Continuous galvanising line (CGL) 1 of 0.1 Mtpa to make zinc coated strips for roofing and engineering applications.
- CGL 2 of 0.3 Mtpa for catering to the automobile and white goods sector demands for zinc coated strips.
- Recoiling lines with a total capacity 0.75 Mtpa to aid 100% inspection.
- Coil packaging line of 1 Mtpa.

Tata Steel's CRM can process 1.27 Mtpa of hot rolled coils received from the HSM. The products are characterised by:

- Low carbon and ULC/IF steels.
- 0.8 Mtpa annealed coils (width: 800–1560 mm, thickness: 0.25–3.2 mm).
- 0.4 Mtpa of zinc coated coils (width: 800–1560 mm, thickness: 0.2–2.0 mm).
- Coating thickness (gm/m^2): galvanised (60–600 both sides together) and galvannealed (60–200 both sides together).
- Coil weight: 28 tonnes (max.)
- Coil OD: 1000–1900 mm.
- Coil ID: 508–610 mm.

Competitiveness of Tata Steel in the Global Steel Market

The vision of Tata Steel is to be one of the world's lowest cost producers of hot rolled coils. The Company has gradually progressed along this path and is now in the lowest position on the cost league of global players including some mini mills (Table 3). Much of this competitive advantage arises out of the availability of low cost captive iron ore and coal which constitute over 81% of the total cost of raw materials, Fig. 16(a) and (b).

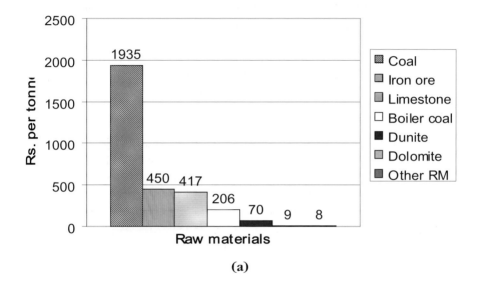

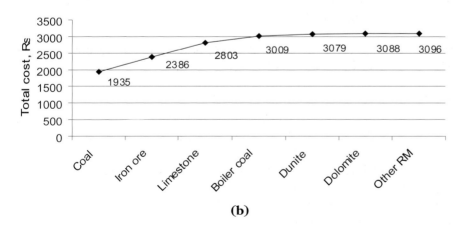

Fig 16 (a) and (b): Raw materials cost at Tata Steel.

Table 3 Cost of HR Coils (Tata Steel and Other Global Players) as Reported by CRU 2001.

	Tata Steel India	CST Brazil	Corus Netherlands	Magnitogorsk Russia	Sidor V'zuela	Aceria Spain	Aceralia Spain	CSN Brazil	Shanghai Baosteel China	Sollac France	Iscor S.Africa	Posco K'yang	Nucor USA
	ISP	ISP	ISP	ISP	Mini-mill	Mini-mill	ISP	ISP	ISP	ISP	ISP	ISP	Mini-mill
Hot metal													
$/thm	97	93	107	102	DRI	DRI	106	107	107	111	120	113	DRI
Rank	2	1	6	4			5	7	8	11	22	13	
Slab caster													
$/tonne slab	148	136	151	156	155	167	153	153	154	156	161	155	179
Rank	2	1	3	10	8	19	5	4	6	11	14	9	35
HR coil													
$/tonne coil	177		178	179	180	190	181	182	183	185	185	186	202
Rank	1		2	3	4	12	5	6	7	8	9	11	26

The international prices of HR coils in March 2000 were around US$300–320 per tonne compared with April 1999 prices of US$200–220 per tonne. The prices of HR coils had been predicted to increase by 20–30% in the year 2000; it is obvious that the actual increase was more. Nonetheless, the volatility of HR coil prices is quite clear and Tata Steel's goal is to cut costs so that returns are guaranteed even under the most trying market conditions. This is particularly important since the product mix of Tata Steel in the year 2000–2001 contains 27% HR coils, 34% cold rolled coils, 29% long products and 8% semis which the company is constrained to sell for want of adequate finishing facilities – an aspect which is receiving constant attention.

Conclusions

As a result of innovation and technology upgrading, Tata Steel has become an efficient, cost competitive and ultra modern integrated steel plant – one of the best in the world. The cost of hot rolled coils has reached US$175–180 per tonne, which is in tune with the international benchmark. Adoption and absorption of the latest technologies, fundamental changes in operating philosophy and setting of stretched performance targets have brought about this remarkable transformation. Adoption of thin slab casting technology is essential for remaining competitive in the world market and Tata Steel is looking for an opportunity to adopt this technology. This would be the 'icing on the cake' since low cost of raw materials such as iron ore and low cost of labour, etc. make the BF–BOF–TSC route extremely attractive in India. In the Indian context, EAFs continue to be at a disadvantage on account of high power costs.

So where is the question of sunset in the steel industry? The sun continues to shine and it is going to be a long hot day(s) ahead. A typical Indian summer is in store. The sun may have set for some (British Empire?) but, not on the steel industry, and certainly not in the Indian steel horizon.

Interaction of Materials Science with Extractive Metallurgy

DEREK FRAY

Department of Materials Science and Metallurgy, University of Cambridge, UK

ABSTRACT

The basic thermodynamics and kinetics of extractive processes were generated several decades ago and, since that time, the subject has been in relative decline as an academic subject. In contrast to this situation, materials science has flourished over the same time span during which our understanding of existing materials and new materials has dramatically increased. Many of these developments can be used to great effect to improve existing extraction processes that, in turn, can offer new ways of synthesising novel materials. Examples will be given from the titanium, aluminium and ferroalloy industries where the structure of oxides, electrochemical properties and intercalation behaviour can lead to new processes.

Introduction

The understanding of extractive metallurgy has been extremely successful in transforming inefficient polluting processes into the modern streamlined energy efficient processes that we see today. This has been achieved by a combination of academic studies, coupled with imaginative management teams who are able to apply this technology and create new processes. From the academic point of view, the development essentially falls into three distinct developments, understanding of the thermodynamics by Chipman,[1] Elliott[2] in the United States and Richardson[3] in the Nuffield Group at Imperial College, studies of heat and mass transfer, which enabled the speed of metallurgical processes to be greatly enhanced, by Richardson and Elliott and, more recently, the computer modelling of metallurgical processes.[4-8] During the period of the latter two developments, Metallurgy departments have gradually evolved into Materials Science Departments incorporating other materials, polymers, ceramics, composites, electronic materials and, more recently, nano-materials. These developments have led to the expansion of the Departments in terms of total staff and a diminution of staff in the area of extractive metallurgy. In the UK, in common with the rest of the world, the discipline has essentially been banished to Departments of Mining and Mineral Engineering, coupled with a vastly reduced number of scientists and engineers devoted to the subject. Is this a wise move and what has been the impact on Extractive Metallurgy and Materials Science? This presentation will demonstrate that the natural place for the extractive metallurgy or chemical metallurgy is in a materials department and there are considerable benefits to be gained from the interaction of the chemical side of the subject and the knowledge base of materials science. In this short presentation, I will cover just three examples, physical metallurgy, solid state chemistry and electrochemistry.

Physical Metallurgy

Nitriding of iron alloys has been studied in physical metallurgy for decades and the mechanism and kinetics are well defined.[9] Nitrides have some interesting properties, usually they are not mutually soluble, resistant to attack by acids and, at high temperatures, become unstable.[10–11] This leads to some interesting possibilities in the separation of valuable elements from ferroalloys as these alloys are generally much easier to prepare than the individual elements. For example, ferrochromium and ferroniobium can simply be prepared by reducing the concentrate that contains both iron and the valuable element to the alloy and, in some cases, there is an order of magnitude difference in cost between the pure metal and the ferroalloy. Conventionally, the pure metals are obtained by purifying the ore and then reducing the oxide. Provided the ferroalloy is in the form of fine particles it can simply be nitrided to the nitride of the valuable element and the iron, either remains unreacted or forms an unstable nitride. Due to the lack of mutual solubility, the alloy will have separated into two phases with an excellent separation of the two elements. The incorporation of nitrogen into the structure causes the material to dramatically increase in volume which induces cracking in the particle and this allows easy ingress of the acid into the centre of the particle and complete access to the iron phase.[13–15] Figure 1 shows an example of nitrided ferrochromium.

The iron phase can, therefore, simply be removed leaving the pure nitride phase behind. Furthermore, by careful control of the electrolytic conditions, it is possible to recover the iron and regenerate the acid.[16] As the nitrides are unstable at elevated temperatures, it is possible to obtain the pure metal by simply heating the nitrides at elevated temperatures under a moderate vacuum.[15] This approach has been shown to be feasible, in the laboratory, to obtain chromium from ferrochromium and molybdenum from ferromolybdenum.[13–15,17] On an industrial scale, this approach is now used to obtain niobium nitride from ferroniobium by nitriding ferroniobium by Wah Chang in Oregon.[18,19] After leaching of the iron, the niobium nitride is oxidised to niobium oxide, that is subsequently reduced by aluminium.

Fig. 1 SEM photograph of ferrochromium nitriding in ammonia argon (6:4) at 973 K; 4 hours; 251×.

Prior to the use of nitriding, the ferroniobium was chlorinated, the chlorides of niobium and iron separated by fractional distillation and the chlorides hydrolysed to the oxides that can lead to an excess of polluting chloride solutions. If the nitrides of chromium can be prepared by this method, what other nitrides can be synthesised? Boron nitride and silicon nitride are usually prepared by reacting the elements together. Using the above approach, boron nitride can simply be prepared from ferroboron by nitriding to form boron nitride and then leaching away the iron phase.[20] Silicon nitride is even easier to form as the silicon nitride can simply be synthesised by passing ammonia over ferrosilicon and the ferrosilicon forms as a mat of fibres on top of the alloy.[21] Aluminides and some carbides have also be prepared by this technique.[22] This is an example where starting with a technique used in physical metallurgy it is possible to devise a new extractive route for some valuable elements. Exactly the same technique can be applied to prepare nitrides and other intermetallic compounds, used in materials science, from cheap sources. The overall sequence is that the understanding of physical metallurgy can lead to new extractive processes which, in turn, lead to novel methods for the preparation of materials.

Electrodes

Anodes

One of the most pressing needs in fused salt extractive metallurgy is to develop an inert anode for the Hall-Heroult cell for the electrolysis of alumina, dissolved in cryolite, to produce aluminium and oxygen. At the present time, the only suitable anode material is carbon and this reacts with the oxygen when it is discharged at the anode to form carbon dioxide with a small amount of carbon monoxide. As a result, the anode is continually being consumed and needs to be replaced periodically. The carbon anodes are expensive to prepare and design of the cell could drastically be improved if an inert anode were used. A further consideration is that it is far better to release oxygen into the atmosphere than carbon dioxide.

This is a very demanding problem in that the material has to last several years and to survive the attack of oxygen as well as the fluoride electrolyte. In addition, if the alumina content of the electrolyte falls below a critical amount, fluorine is liberated which is even more corrosive than oxygen. There have been several approaches to overcome this problem and these include the use of oxides, cermets, coatings and oxidation resistant alloys. All have disadvantages as well as advantages. The essential requirements are physical stability, resistance to attack from gases and the electrolyte, an electronic conductor and good thermal shock resistance.

Oxides do not suffer from the problems of oxidation but the materials may dissolve in the electrolyte especially as most fluoride melts dissolve oxides, especially cryolite which is selected for its ability to dissolve alumina. Although many oxides are conducting, their electronic conductivity is unlikely to match that of carbon and this will introduce an additional resistance to the cell. The cell voltage will increase anyway by about 1 V as the fuel cell effect given by the reaction of the oxygen with the carbon

$$2O^{2-} + C = CO_2 + 4e$$

will be replaced by

$$2O^{2-} = O_2 + 4e$$

However, this voltage increase is more than offset by the advantages that will accrue.

In order to overcome the problems of the low conductivity various cermets have been investigated where it is assumed that the metal component will contribute to the electronic conduction and toughness and the oxide the resistance to attack. Copper–nickel alloys and ferrites have been investigated. Thonstad has examined nickel ferrite–nickel oxide–copper cermets and found that, at low alumina concentrations, catastrophic corrosion of the anodes occurred and high current densities and high as well as low NaF/AlF_3 molar ratio were also detrimental.[23,24] Post-electrolysis examination of the anodes showed a reaction layer of approximately 50 μm formed. Another solution is to use a coating on a metal substrate but, given that the anode might be expected to last years, coatings have never been entirely satisfactory as once failure of the coating occurs, the metal substrate rapidly corrodes.

The ultimate solution is to use a corrosion resistant alloy in which the alloy is highly conducting, does not suffer from thermal shock, and the oxide film prevents the metal being attacked.[25] There is also the further advantage that the oxide film should be self-healing if damaged. However, the development of this alloy will require considerable effort from both materials scientists to perfect the alloy and process metallurgists for its implementation. This is a long term goal and is far from being implemented.

Cathodes

In the Hall-Heroult cell there are also problems with the cathode in that aluminium does not wet carbon and as the electrode potential for the deposition of sodium on carbon is less cathodic than the deposition of aluminium, due to the fact that the sodium can intercalate into the carbon at a reduced activity. This causes the carbon lining to expand and to decrepitate. In order to prevent this, operators always ensure that there is a substantial layer of liquid aluminium at the bottom of the cell. However, due to the intense magnetic fields, this layer is curved and the surface is highly unstable leading to waves and turbulence on the surface of the molten aluminium. This, in turn, dictates that the anode to cathode distance is substantial leading to increased resistive losses in the cell. The ideal material would be one which has the following properties – wetted by molten aluminum, good electrical conductivity, low reactivity with aluminium and cryolite, reliable attachment to the cathode substrate, low cost, easy fabrication into any desired shape, high resistance to wear, impact and thermal shock, resistance to oxidation and corrosion by reactive gases and, lastly, low thermal conductivity. Titanium diboride has many of these properties and is readily wetted by molten aluminium and there have been many attempts over about 40 years to introduce titanium boride into the Hall-Heroult cell. However, the thermal shock resistance of the material is very poor so that only relatively small tiles could be used initially. Fixing the tiles on the bottom of the cell caused considerable problems as no suitable adhesive existed which would survive the high temperatures and the fluoride electrolyte. Overall, early work, using hot pressed titanium diboride tiles was hindered by mechanical breakage, thermal shock, bath attack along oxide-rich grain boundaries and attachment problems. These difficulties were finally overcome by using a composite of carbon, carbon fibre, pitch and titanium diboride.[26] This composite material managed to combine all the best properties of titanium diboride and carbon and could simply be plastered on to the bottom of the cell and fired. There have been several long term plant trials. The great advantages of using a wetted cathode is that the anode to cathode distance can be dramatically reduced and coupled with inert anodes, it is possible to devise a bi-polar cell, rather like the Alcoa cell for the electrolysis of aluminium chloride. This approach would greatly increase the output of metal from a given cell and eliminate carbon dioxide mixtures. However, with all step changes in technology years of trials are needed before they become industrially accepted. It is no use having a revolutionary technology that only works sporadically.

Electrochemistry

A recent paper highlights the possibility of electrochemically reducing metal oxides directly to metals without dissolving the metal oxides in a solvent.[27] In aqueous electrochemistry, the oxides such as copper, nickel and zinc oxides are dissolved in dilute sulphuric acid and as the potential for deposition of the metal is less cathodic than that for the evolution of hydrogen, the metal is deposited. In the case of zinc, hydrogen evolution is less cathodic but due to the very high overpotential of hydrogen on zinc, the metal is preferentially deposited. Alumina is reduced by dissolution in cryolite (Na_3AlF_3), followed by electrolysis and, in the electrowinning of magnesium, magnesium chloride is dissolved in alkali chlorides. In the 1960s Ward and Hoar reduced cuprous sulphide by making it the cathode in a molten bath of barium chloride.[28] The cathodic reaction was assumed to be

$$Cu_2S + 2e = 2Cu + S^{2-}$$

and the anodic reaction

$$4Cl^- = 2Cl_2 + 4e$$

More recently, Chen et al. have performed experiments using insulating oxides, such as titanium dioxide, and have found that it is possible to reduce these oxides directly to the metal.[27] Figure 2 shows a schematic diagram of the process[29] and Fig. 3 shows the microstructure of the reduced titanium.

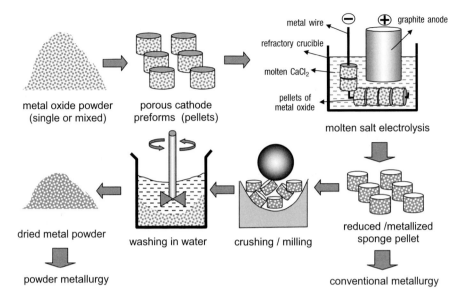

Fig. 2 A schematic diagram representing the new electrolytic process for the direct extraction of metals from their metal oxide.

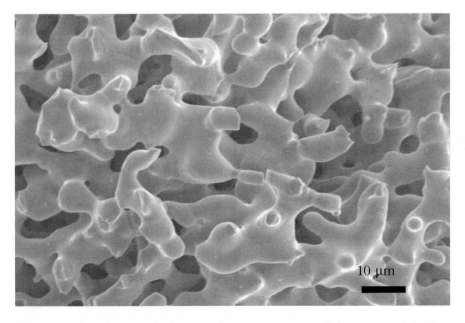

Fig. 3 A SEM photograph showing the microstructure of the extracted titanium by the new electrolytic process.

From Table 1, it does not matter whether the oxygen is present as a gas, dissolved in a metal or as an oxide, provided the driving force for the ionisation of oxygen to oxygen ions, dissolved in the salt, is less cathodic than the deposition of calcium, ionisation of the oxygen will occur. One particularly interesting aspect of this research is that pure titanium dioxide is an insulator and, initially, it is difficult to see how it can become sufficiently conducting to allow the electrons to travel the oxide/electrolyte interface. Examination of the Ti–TiO$_2$ phase diagram shows that any removal of oxygen results in the formation of Magnelli phases that are known to have high electronic conductivity.[30] It is surmised that this the mechanism whereby the material becomes sufficiently conducting for the process to work. The voltage that is applied is below the decomposition potential of the calcium chloride and the initial currents while the titanium dioxide is being reduced to titanium are of the order of 10,000 Am^{-2}. The slow step is the removal of oxygen dissolved in titanium which is a relatively slow process as it is controlled by the diffusion of oxygen in solid titanium. In spite of this the process appears to be much faster, less polluting, less labour and capital intensive than the Kroll process. It is hoped that this new approach will allow the cost of titanium production to be dramatically reduced so that the remarkable properties of titanium (lightness, strength and corrosion resistance) can be more widely applied. In the laboratory the experiments have graduated from the gram scale to the kilogram quantities of material are

Table I. Electrode Potentials in Fused Chlorides, Calculated from Thermodynamic Data $[E_{Na} = 0]$ at 973 K.

$O_2 + 4e^- = 2O^{2-}$	$E^0 = 2.713$ V
$2PbO + 4e^- = 2O^{2-} + 2Pb$	$E^0 = 2.082$ V
$SnO_2 + 4e^- = 2O^{2-} + Sn$	$E^0 = 1.734$ V
$MoO_2 + 4e^- = 2O^{2-} + Mo$	$E^0 = 1.650$ V
$\frac{2}{5}Nb_2O_5 + 4e^- = 2O^{2-} + \frac{4}{5}Nb$	$E^0 = 1.209$ V
$\frac{2}{3}Cr_2O_3 + 4e^- = 2O^{2-} + \frac{4}{3}Cr$	$E^0 = 1.189$ V
$\frac{2}{5}Ta_2O_5 + 4e^- = 2O^{2-} + \frac{4}{5}Ta$	$E^0 = 1.038$ V
$TiO_2 + 4e^- = 2O^{2-} + Ti$	$E^0 = 0.750$ V
$Zr + 4e^- = 2O^{2-} + Zr$	$E^0 = 0.349$ V
$\frac{2}{3}Al_2O_3 + 4e^- = 2O^{2-} + \frac{4}{3}Al$	$E^0 = 0.348$ V
$2TiO + 4e^- = 2O^{2-} + Ti$	$E^0 = 0.338$ V
$UO_2 + 4e^- = 2O^{2-} + U$	$E^0 = 0.337$ V
$HfO_2 + 4e^- = 2O^{2-} + Hf$	$E^0 = 0.211$ V
$2MgO + 4e^- = 2O^{2-} + 2Mg$	$E^0 = 0.143$ V
$2Ca^2 + 4e^- = 2Ca$	$E^0 = -0.06$ V

being prepared. At the end of the reduction process, the pellets are simply withdrawn and the salt washed away. Calcium chloride has some attractive properties; it is a waste product form the chemical industry so it is pure and inexpensive, it has a very high solubility for oxygen ions, unlike many other metal chlorides and, lastly, its toxicity is the same as sodium chloride so that disposal is not a problem. This approach can also be applied to many other oxides and a large number of oxides have been reduced in the laboratory. Some examples are shown in Table 1. It should be noted that the oxides need to be free of other cations, if a pure metal is required, as it is only the anionic species that are removed. The process appears to be particularly successful at removing oxygen, sulphur and nitrogen.

One can take advantage of the fact that all the cationic species in the oxide are reduced to produce alloys. The oxides are simply mixed together and reduced to give a highly homogeneous alloy. It is felt that this approach may offer considerable advantages when the alloy constituents have high melting points or have different densities that make homogenisation difficult. Intermetallic compounds can also be easily synthesised from oxide precursors. The other considerable advantage is that the oxides are frequently very much cheaper than the metals so that as well as giving better properties, the end result may also be cheaper. In many cases the product is in the form of a homogeneous powder, virtually oxygen free, which will make it an ideal product for powder metallurgical techniques and near net shape forming. The savings in downstream processing costs may exceed the cost reduction in the primary extraction. Again, this is an example where the combination of materials science and extractive metallurgy can lead to dramatic improvements both in metal extraction and materials processing.

Conclusions

Extractive metallurgy has been extremely successful over the past few decades in improving performance and making metallurgical processes far more efficient and less polluting. This has been brought about by a combination of thermodynamics, kinetics, heat and mass transfer and modelling. Most of the research is now done outside conventional materials science departments but this paper has shown that the combination of materials science and extractive metallurgy can lead to significant benefits to both disciplines and the examples quoted here are just a small fraction of the opportunities that can be explored. Much of the intensive effort in fuel cells, batteries, both at high and low temperatures, can simply be applied to transform many metallurgical operations. Overall, the future looks very exciting.

References

1. J. Chipman, in *The Chipman Conference*, J. F. Elliott and T. R. Meadowcroft (eds), MIT Press, Cambridge, Mass. 1965, xvii–xxi.
2. J. F. Elliott, in *Proceedings of the Elliott Symposium on Chemical Process Metallurgy*, P. J. Koros and G. R. St Pierre (eds), The Iron and Steel Society, Warrendale, PA. 1991, 613–620.
3. F. D. Richardson, *Physical Chemistry of Melts in Metallurgy*, Volumes 1 and 2, Academic Press, London and New York, 1974.
4. P. V. Barr, E. J. Osinski, J. K. Brimacombe, M. A. Khan and P. J. Readyhough, *Ironmaking and Steelmaking*, 1994, **21**(1), 44–55.
5. J. S. Woo, J. Szekely, A. H. Castillejos and J. K. Brimacombe, *Metall. Trans. B*, 1990, **21B**, 269–277.
6. J. Szekely and G. Trapaga, *Modelling and Simulation in Materials Science and Engineering*, 1994, **2**(4), 809–828.
7. G. Trapaga and J. Szekely, *Metall. Trans. B*, 1991, **22B**, 901–914.
8. D. Xu, D. Jones and J. W. Evans, *Appl. Math. Model.*, 1998, **22**(11), 883–893.
9. B. Mortimer, P. Grieveson and K. H. Jack, *Scand. J. Metals*, 1972, **1**, 203–209.
10. T. Mills, *J. Less Common Metals*, 1970, **22**, 373–381.
11. L. Duparc, P. Wenger and W. Schssele, *Helv. Chim. Acta*, 1930, **13**, 917–929.
12. P. C. Van Wiggen, H. C. F. Rozendaal and E. J. Mittemeijer, *J. Mat. Sci.*, 1985, **20**, 4561–4582.
13. A. W. Kirby and D. J. Fray, *Trans. IMM*, 1988, **98**, C33–34.
14. A. W. Kirby and D. J. Fray, *Trans. IMM*, 1988, **98**, C89–95.
15. A. W. Kirby and D. J. Fray, *Trans. IMM*, 1988, **98**, C96–98.
16. C. Liu and D. J. Fray, *Mat. Sci and Tech.*, in press.
17. A. W. Kirby and D. J. Fray, *J. Mat. Sci. Let.*, 1993, **12**, 633–636.
18. A. K Suri, K. Singh and C. K. Gupta, *Trans. Met. Soc.*, 1992, **23B**, 437–442.
19. J. A. Sommers, J. R. Pearson, T. R. McQueary and M. A. Rossback, Materials from ferroalloys: I Nitride processing of ferroniobium to niobium metal, in *Light Metals 1992*, E. R. Cutshall (ed.), (The Minerals, Metals and Materials Society, Warrendale PA 1992), 1303–1304.
20. T. E. Warner and D. J. Fray, *J. Mat. Sci.*, 2000, **35**, 5341–5345.
21. T. E. Warner and D. J. Fray, *J. Mat. Sci. Let.*, 2000, **19**, 733–734.
22. J. A. Sommers, L. J. Fenwick and V. Q. Perkins, Materials from ferroalloys: II Processing of ferroniobium to niobium nitride, carbide, oxide and aluminide, in *Light Metals 1992*, E. R. Cutshall (ed.), (The Minerals, Metals and Materials Society, Warrendale PA 1992, 1305–1307.
23. H. Xiao, R. Hovland, S. Rolseth and J. Thonstad, *Met Trans B*, 1996, **27**, 185–193.
24. E. Olsen and J. Thonstad, *J Appl. Electrochem.*, 1999, **29**, 301–311.
25. D. R. Sadoway, *JOM*, 2001, **53**(5), 34–35.
26. L. G. Boxall and A. V. Cooke, Successful application of carbon/TiB_2 composite refractories to conventional aluminium reduction cells, in *Extractive Metallurgy '85 IMM*, IMM, London 1985, 473–496.
27. G. Z. Chen, D. J. Fray and T. W. Farthing, *Nature*, 2000, **407**, 361–364.

28. T. P Hoar and R. G. Ward, *Trans IMM*, 1957–58, **67**, 393–401.
29. D. J. Fray, *Met. Trans B*, 2000, **31B**, 1153–1162.
30. H. Gruber and E. Krautz, *Phys. Status Solidi A*, 1992, **69**, 287–295.

Single Crystal Superalloys

KEN HARRIS
Cannon-Muskegon Corp. [SPS Technologies]

ABSTRACT
A review is made covering the development and application success of single crystal nickel-base superalloys for turbine engine applications over the last 20 years. These include both flight engine (AGT) and industrial gas turbine (IGT) blade, vane and seal components. Reference will be made to single crystal castability issues, grain defect intolerance, mechanical properties, alloy phase stability, oxidation and thermal barrier coating performance.

The development and application potential for a new grain boundary strengthened single crystal superalloy designated CMSX®-496 is also discussed, along with an improved version of CMSX-4® containing residual ppms of lanthanum and yttrium.

Materials Processing: The Enabling Discipline

L. CHRISTODOULOU
Department of Materials, Imperial College of Science, Technology and Medicine (on leave of absence at DARPA)

J. A. CHRISTODOULOU
US Navy, NSWC Carderock Division, West Bethesda, MD 20817-5700, USA

ABSTRACT

Materials processing is a vast but uniquely 'Materials' discipline, for the role of the materials scientist is not simply to define a formula or a material's composition, but to identify and exploit its optimum properties in a specific application. Thus, while materials can be synthesised by chemists and physicists, it is the materials scientists who have the unique facility and responsibility to *process chemicals into useful materials with enabling properties.* Optimised mechanical, magnetic, electrical or optical properties are functions not only of chemistry but are, within a class of materials or compositions, determined primarily by microstructure.

Traditionally, control of microstructure is achieved primarily through thermo-mechanical processing: the use of force and temperature. From the first blacksmith's furnace, anvil and hammer to the highly instrumented rolling mills, isothermal forges and hot isostatic presses of modern industry, the exploitation of materials properties has come through the art and science of manipulating microstructures. Over the past 150 years, the RSM has made significant contributions to understanding how processing of materials defines their attributes and capabilities. In recent years, this understanding has begun to be captured in computer simulation and prediction capabilities that complement, guide and greatly accelerate the pace of empirical research on which much of our understanding of processing/property relationships is founded.

As we look to the 21st century, we see that materials science is on the verge of a revolution. Focusing on identifying ever more efficient means of achieving the optimum properties of materials, our repertoire is rapidly expanding to include more sophisticated physically faithful and quantitative processes simulation, self-assembly techniques and bio-inspired, low temperature processing approaches. These more environmentally friendly processes will enable more rapid, affordable and assured introduction and exploitation of new materials.

This paper discusses some immediate challenges faced by the materials processing community and indicate some of the opportunities for the 21st century.

The Materials Science of Cement and Concrete: An Industrial Perspective

KAREN L. SCRIVENER
Department Materials, École Polytechnique Fédérale de Lausanne, Switzerland

ABSTRACT
Cement and concrete are the most widely used materials on earth (1 tonne/person per year), without them the infrastructure of the developed world could not exist. However as materials their complexity and heterogeneity has favoured empirical developments, rather than the establishment of a microstructure-based understanding of properties. Nevertheless the concepts of materials science have found successful application – as illustrated by three examples drawn from the author's experience in industry.

1. Study of the microstructural and microchemical changes occurring in concretes and mortars after curing at elevated temperatures, has enabled the mechanism by which expansion may sometimes occur to be identified.
2. A transmission electron microscope study of calcium aluminate cement clinker enabled anomalous X-ray diffraction results to be explained, allowing XRD to be used with confidence to determine the relative amounts of the different phases present.
3. Finally the combination of particle packing optimisation, fibre reinforcement and heat curing has enabled ultra high strength, ductile concrete to be developed, offering the possibility of lighter and more durable structures.

Introduction

Concrete is the most widely used material in the world today. The annual production is more than 10 billion tonnes – that is more than one tonne for every person on the planet, and more than ten times the amount of steel used. Due to these enormous volumes no other material can replace it. So it is important that we are able to make good concrete and use it in the best way possible.

The enormous success of concrete as a material is due to several factors:

- Cement, can be made from locally available raw materials (limestone, clay, shale, etc) in almost all countries of the world.

- Cement, supplied as an easily transportable powder, can be transformed with the simple addition of rocks and water into a solid mass of almost any shape, at room temperature with minimal volume change.
- Concrete, correctly made, is a highly durable material, which will last for centuries.

And all this at a bargain price!

Nevertheless concrete involves:

- complex chemical reactions;
- producing a physically complex porous solid;
- destined to be used by engineers;

and is therefore an ideal case for the interdisciplinary approach of Material Science.

How Does it Work?

For the uninitiated it is useful to explain the basics of how cement reacts with water to give concrete – most importantly the hardening of concrete is *not* a matter of drying!

Cement clinker is produced in large rotary kilns (~80 m long by ~4 m diameter), and is then ground with a small addition of calcium sulphate to produce cement powder. The phases present in cement are thermodynamically unstable relative to their hydrates and so dissolve in water to give a solution supersaturated with respect to these hydrates, which then precipitate. This *hydration* reaction takes place 'through solution', whereby the ions present in the *anhydrous* (unreacted) cement compounds dissociate in solution and precipitate hydrates of different stoichiometry. For example the reaction of Portland cement is dominated by the hydration of tricalcium silicate, Ca_3SiO_5 or C_3S,* which reacts to give calcium silicate hydrate, C–S–H, a poorly crystalline phase with a Ca to Si ratio of around 1.7 and calcium hydroxide, $Ca(OH)_2$ or CH:

$$C_3S + 5.3H \Longrightarrow C_{1.7}SH_4 + 1.3CH \qquad (1)$$

The hydrates occupy a larger volume than the anhydrous phases, so the volume of solid increases during hydration. This solid fills in the space originally occupied by water to bridge and connect the solid particles into a

---

*Cement chemists notation: $C = CaO$; $S = SiO_2$; $A = Al_2O_3$; $F = Fe_2O_3$; $\$ = SO_3$.

rigid network. However, the total volume of the anhydrous phases *plus water* before hydration is about 10% larger than the volume of hydrates. A small amount of this overall decrease in volume is manifested as shrinkage, but most is accommodated by the entry of air into the pores. Thus even a fully hydrated cement paste will still contain significant porosity, typically of the order of 10 to 20%.

After an initial induction period in which the concrete remains fluid and can be put in place, the chemical reactions proceed rapidly, until the deposition of the hydration products around the hydrating cement grains leads to a decreasing rate of reaction after the first few hours. At one day the degree of reaction is around 30–50%, but further reaction leads to 2- to 4-fold increase strength over the next month and slow continuing increase thereafter, if moisture is available to sustain the chemical reactions.

It is well known that porosity is a limiting factor in the strength of a material. The ultimate strength of concrete is thus very strongly dependent on the amount of water added at the mixing stage, as this determines the initial porosity to be filled by the hydration products. The amount of water added is usually expressed as a 'water to cement' (weight) ratio, or W/C. Figure 1 shows schematically the relation of this to typical mature strengths for concretes, with indications of the typical ranges for good quality 'standard' concrete and high strength concrete. Unfortunately, the addition of too much water is the most frequent cause of bad, poor quality concrete.

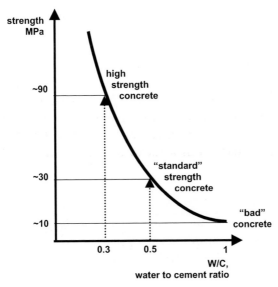

Fig. 1 Schematic illustration of the relationship between the amount of water added (W/C ratio) and mature strength of concrete.

A Materials Science Approach

The basis of materials science is understanding the relationships between processing, microstructure and properties on a mechanistic basis. However, the complexity of cementitious materials has made the establishment of such relationships difficult and favoured an approach based on empirical relations between process variables and performance. This has often resulted in a very conservative approach to the use of these materials and failure to realise their potential. As behaviour cannot be predicted from understanding, but only from previous references, the introduction of new materials is impeded by the need to establish their durability over time scales commensurate with the life of structures (decades/centuries) and the difficulty of predicting long term performance from short term laboratory testing (particularly within the constraints of the typical three year PhD study!).

The critical element in moving towards a Materials Science approach is the capacity to characterise and preferably quantify microstructure (the development of a quantitative approach as one of the corner stones of the development of materials science for metals has been discussed in another contribution to this conference). A typical microstructure of a mature concrete is shown in Fig. 2.

It can be seen that the microstructure is extremely heterogeneous. This is mainly due to the use of starting ingredients of widely varying sizes – from centimetres for the aggregates, through the cement grains ($\sim$1–100 µm) and sometimes including fine additives such as silica fume which have a typical size of 0.1 µm (a size range of $1:10^6$). The wide range of sizes is the most

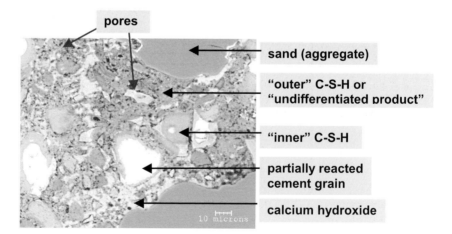

Fig. 2 Microstructure of mature cement paste. Polished section, backscattered electron image in the SEM.

efficient from the point of view of packing the particles, but means that a multi-scale approach is needed for microstructural characterisation. Also the extreme heterogeneity makes it difficult to derive 'average' data for a sample. As a consequence, microstructure property relationships are much less well established than for materials such as metals and ceramics. This is therefore a potentially fruitful field for further research, with the information technology revolution opening up new possibilities for data extraction and interpretation by simulation models.

Despite the difficulties posed by the heterogeneity and complexity of the microstructure, the rest of this paper deals with case studies where a materials science approach has been successfully applied within the context of an industrial research laboratory.

Example 1: Understanding Durability Problems in Heat Cured Concrete

In the 1970s in Germany, it was noticed that concrete railway sleepers which had been precast and heat cured to accelerate strength development, started to show significant cracking and damage after only 4 or 5 years in service. Extensive research was launched in Germany and then elsewhere. The research showed that some concretes subject to temperatures in excess of 70°C (typically 80–90°C) and subsequently stored in wet conditions, expanded but that not all concretes expanded even when subject to these severe treatments. Many research programmes tried to relate the occurrence or non-occurrence of expansion and damage to chemical and physical characteristics of the cement used. Despite the testing of hundreds of different cements it has not been possible to establish a general relation between the propensity for expansion and the cement characteristics.

This failure of empirical studies to provide a simple answer, gave impetus to the search for the mechanism by which expansion occurred. In expanded samples, gaps are observed around aggregate particles, usually filled with crystals of the phase ettringite ($C_3A.3C\$.32H$). This phase is a normal product of cement hydration, formed from aluminate ions from the anhydrous phase tricalcium aluminate, C_3A (and to some extent by the slowly reacting ferrite phase, $C_2(A,F)$) and sulphate from calcium sulphate (gypsum, plaster or anhydrite) added to the clinker during grinding (reaction (2)). At normal temperatures the calcium sulphate controls the reaction of the aluminate phase and consequently the setting of the cement, allowing adequate time for mixing and placing.

$$C_3A + 3C\$ + 32H \implies C_3A.3C\$.H_{32} \qquad (2)$$
normal ettringite formation at ambient temperatures

However, when the concrete is exposed to elevated temperatures during curing, the solubility of ettringite increases: Sulphate ions are absorbed onto the calcium silicate hydrate (C–S–H) phase, and aluminate ions are partially incorporated into the C–S–H structure and partially combined with lower amounts of calcium sulphate as calcium aluminate monosulphate, $C_3A.C\$.12H$ or AFm phase (reaction (3)).

$$C_3A + C\$ + 12H \Longrightarrow C_3A.C\$.H_{12} \qquad (3)$$

monosulphate formation at higher temperatures combines less sulphate

When the concrete returns to ambient temperatures, and moisture is available to facilitate ion transport, ettringite forms within the mature concrete (reaction (4)). This has led to the use of the term 'delayed ettringite formation' or DEF, for this type of degradation of heat cured concrete.

$$C_3A.C\$.H_{12} + 2\$_{(from\,C-S-H)} + 2C_{(from\,C-S-H\,or\,calcium\,hydroxide)} + 20H \Longrightarrow C_3A.3C\$.H_{32} \qquad (4)$$

delayed ettringite formation on wet exposure at ambient temperature after initial heat exposure

However, it has been clearly demonstrated that the amount of expansion is not related to the amount of ettringite which forms at later times and indeed, substantial quantities of this phase may be precipitated at late ages without any expansion or damage occurring. The formation of more or less even gaps around the dimensionally stable aggregate particles indicates that expansion is occurring throughout the cement paste component of the concrete, which expands away from the aggregate particles.

Thermodynamics indicate that for crystallisation to exert a pressure that can produce expansion, firstly it must occur from a supersaturated solution. However as a crystal will always prefer to grow into open space rather than exerting pressure, it must also be confined within a pore network. The level of supersaturation determines the maximum curvature of the crystal and therefore the minimum size of the pores into which the crystal can grow. Thus the pressure which can be exerted is inversely related to the size of the pores in which the crystal is growing. To exert a crystallisation pressure of 2–3 MPa (the typical tensile strength of concrete, a crystal would need to form in a pore network through which the size of the percolating pores was about 50–100 nm in size.

These considerations led to our study of the chemical changes occurring in the C–S–H product around the larger cement grains.[1,2] Samples were studied by energy dispersive analysis of X-rays in the SEM. The information from this technique concerns an interaction volume some 1–2 µm in size. In cementitious materials the individual phases are often intermixed on a submicron scale and so cannot be analysed separately. However, such mixtures can often be interpreted by looking for trend in atom ratios – analyses of binary phase mixtures will lie on a line joining the composition of the two phases present. Despite this difficulty of interpretation, the advantage of the technique lies in being able to study large cross-sectional areas and make many analyses in similar areas. This is not possible with TEM samples (where better spatial resolution of analyses is possible), as it is extremely difficult to prepare sizeable electron transparent areas.

Figure 3 shows the microchemical changes (in terms of ratios of sulphur to calcium and aluminium to calcium) occurring in the 'inner' C–S–H product of samples immediately after heat curing and after 200 days storage in water, for a mortar that did not expand and for one that did. The changes in atom ratios occurring in this inner product are similar for both mortars. Immediately after heat curing the ratios of sulphur to calcium are around 0.05 to 0.08 and the ratios of aluminium to calcium around 0.05 to 0.06. After 200 days' immersion in water the S:Ca ratios have dropped dramatically to around 0.01, while those for Al:Ca have changed relatively little to around 0.04–0.05. ^{29}Si and ^{27}Al NMR studies confirmed that the alumina ions were incorporated into the C–S–H structure, which explains why they do not leave the C–S–H during the water storage after heat treatment. The state of the sulphate is not well defined, but it seems to be only loosely bound and thus available to react and form ettringite at later ages.

After the water storage the formation of ettringite could be observed by X-ray diffraction and by ^{27}Al NMR. Ettringite reformed both in mortars that expanded and in those that did not – there was no relationship between the amount of ettringite that formed and the degree of expansion. Following the thermodynamic considerations it was suspected that the difference might lie in

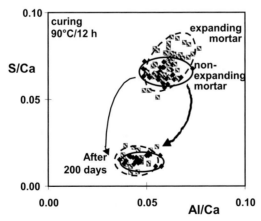

Fig. 3 Microanalyses made in the 'inner' product C–S–H immediately after heat curing and after 200 days' storage in water for a mortar which expanded and for one which did not expand.

the size of the pores in which the ettringite crystallised. The 'inner' C–S–H immediately surrounding the reacting grains is dense. This C–S–H is known to contain nanopores about 2 nm in size, but no evidence was found for ettringite forming in such small pores. Immediately surrounding the 'inner' C–S–H the 'outer' C–S–H also contains some small (~10–100 nm), so called capillary pores, which are nevertheless likely to be relatively isolated from the bulk pore solution. In addition, this outer C–S–H is in close contact with the inner C–S–H that provides a reservoir of sulphate ions after the heat curing. Microanalyses of these regions revealed very interesting differences between mortars that expanded and those that did not (Fig. 4).

Immediately after the heat curing the pattern of analyses is fairly similar and can be interpreted as C–S–H (with a similar amount of absorbed sulphate and combined aluminium observed in the inner C–S–H) finely mixed with calcium monosulpho-aluminate ($C_3A.C\$.H_{12}$), as indicated by the analyses lying along the dashed line extending to the upper right. In the case of the mortar that did not expand (Fig. 4 (left)), after 200 days' storage in water, the C–S–H loses its sulphate during curing, while the content of aluminate remains the same. After storage there is some intermixing with phases of low sulphate content. On the other hand, in the expanding mortar (Fig. 4 (right)), whilst the changes in the C–S–H cluster are similar after storage, there is a distinct population of points (circled) which extend away from the C–S–H composition towards the composition of ettringite. These are interpreted as mixture of small amounts of ettringite within the C–S–H. Consequently it is deduced that the expansion arises from the formation of ettringite within small pores (~50 nm) in the C–S–H from calcium aluminate monosulphate and sulphate (and calcium) released by the C–S–H (reaction (4)).

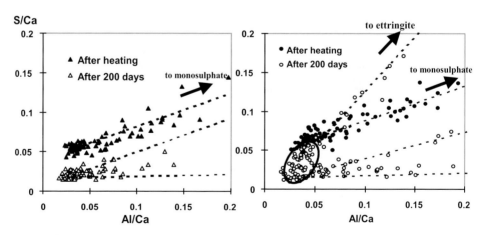

Fig. 4 Microanalyses of 'outer' C-S-H product for non-expanding (left) and expanding (right) mortars.

The Materials Science of Cement and Concrete: An Industrial Perspective 237

The studies of the mechanism of expansion have made a major contribution to the debate about how to avoid degradation of this type. It is now clear that many different chemical effects are involved and interconnected, and it is not possible to prescribe simple chemical limits for the cement composition. The most important and surest parameter to control is the initial curing temperature.

Example 2: Mineralogical Analysis of Calcium Aluminate Cement Clinker

Calcium aluminate cements (CACs) are used for specialist applications; in refractory concretes for metallurgical, glass, petrochemical and similar industries; as part of premixed mortars for the building trade; and in concretes with high resistance to chemical aggression and abrasion. Due to the complex nature of such applications it is very important to control the mineralogical regularity of these cements. X-ray diffraction (XRD) is the classic method used to determine cement mineralogy. However, due to the similarity of lattice parameters between the different phases and to the presence of solid solution in the phases, accurate quantitative phase analysis is difficult. In recent years this has led to increasing interest in 'Rietveld' type methods to deconvolute diffraction patterns. Such techniques have been developed successfully to characterise the higher alumina content CACs.[3] However studies of the lower alumina content CACs, which contain significant quantities of iron oxides and silica in addition to calcium and aluminium oxides, revealed a gap in our knowledge of their mineralogy.

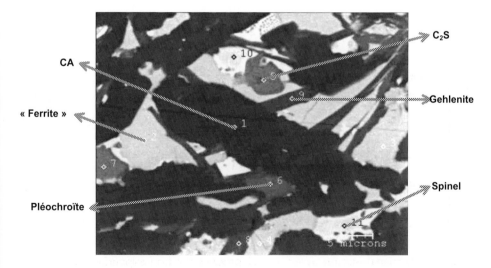

Fig. 5 BSE image of microstructure of calcium aluminate cement clinker.

A backscattered electron image of a typical clinker is shown in Fig. 5. In this microstructure 6 phases can be identified. However, when these known phases were used for the Rietveld analysis an important discrepancy was noted related to the peaks arising from the 'ferrite' phases. From the X-ray study the existence of a new phase was proposed related to merwinite ($Ca_3MgSi_2O_8$).[4] The magnesia content of the clinker was too low for this phase to have its stoichiometric composition and no evidence of such a phase could be found by SEM. To resolve this issue a PhD study was launched to study the microstructure by TEM with EDS and EELS analysis.[5]

This study found that the so called ferrite phase, which was considered to be a solid solution based on brownmillerite $Ca_2(Al,Fe)_2O_5$, was in fact a mixture of two closely related phases, one having the brownmillerite structure and the other having a perovskite structure. In many cases these phases were finely interlayered on a nanometer scale (Fig. 6) indicating a probable

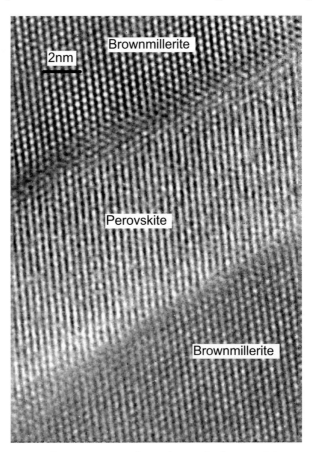

Fig. 6 Atomic resolution image of interlayered phases of Brownmillerite and perovskite structure in the 'ferrite' phase of calcium aluminate cement.

The Materials Science of Cement and Concrete: An Industrial Perspective 239

spinodal decomposition from a single phase, stable at high temperatures. This TEM study enabled the crystal structure, chemical composition and valency of the iron to be determined[6] and hence to complete the Rietveld analysis to determine the relative amounts of the different phases.

Example 3: Development of Ultra High Performance Concrete (UHPC)

The final example is the development of an ultra high performance concrete, now commercialised under the trade name Ductal®.[†] This development brings together several different technologies:

- Optimisation of particle packing in part through use of components of progressively smaller size, such that the smaller particles tend to fill the holes left by larger particles. This allows the amount of water, which needs to be added during mixing to be reduced, and so higher strengths to be obtained (as explained earlier, Figure 1).
- Incorporation of fibres to increase ductility.
- Use of superplasticisers – additives that increase the fluidity of the fresh mix – to allow mixtures to be placed at such low water to cement ratios.
- Heat curing to improve the strength of the hydrate assembly.

The effect of these can be seen in Fig. 7, where the flexural strength of this ultra high strength ductile concrete is compared to other concretes. A traditional concrete (C25), with a compressive strength of about 25 MPa breaks in bending at about 2–3 MPa, in a brittle fashion, with virtually no load bearing capacity post peak. A similar concrete with fibre reinforcement (FRC 30) has a similar peak load, but a more ductile fracture, with significant load carrying capacity post peak. A typical high strength, fibre reinforced concrete (FRC 80) has considerably improved performance, with a higher 'yield' stress, strain hardening and considerable capacity for deflection post peak. Nevertheless the improvements seen in the FRC80 concrete are well surpassed by those of the 'Ductal®' concrete which reaches a flexural stress of over 45 MPa.

The significance of such developments can be seen in Figure 8, which shows the cross section of beams of equivalent load bearing capacity.

[†] A range of ultra high strength concrete developed by Bouygues, Lafarge and Rhodia.

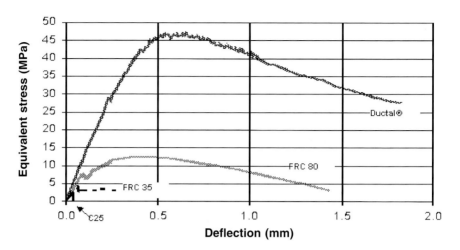

Fig. 7 Flexural strengths of conventional concrete (C25), Fibre reinforced concrete (FRC30); High Strength Fibre reinforced concrete (FRC80) and ultra high strength fibre reinforced concrete (Ductal®).

Traditional reinforced concrete and pre-stressed concrete beams are considerably larger and heavier than the equivalent steel beam. Use of the higher strength and ductility of the UHPC enables the beam weight to be considerably reduced, with reinforcing steel only needed in the tensile face. This illustrates how UHPC can lead to new lighter designs, coupled with improved durability due to the virtually impermeable nature of the concrete.

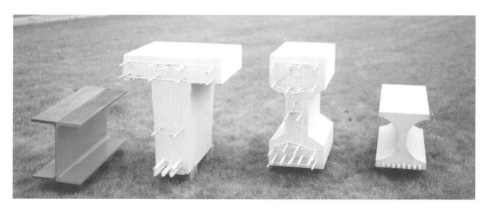

Fig. 8 Beams of equivalent load bearing capacity with weight per m in brackets: from left, steel (117 kg); reinforced concrete (530 kg); pre-stressed concrete (467 kg); ultra high strength fibre reinforced concrete (Ductal®) (140 kg).

Concluding Remarks

Concrete is a highly successful material. In the present form, concrete has been available for just over 100 years and most development to date has been on an empirical basis. The heterogeneity and complexity of this material poses many challenges for microstructural study. Nevertheless the application of the concepts of materials science offers considerable benefits within an industrial context, as illustrated by the examples presented here. As the development of ductile ultra high strength concrete shows, improvements in concrete technology, through the application of materials science, offer the possibility of lighter structures, which consume less raw materials and last longer; providing economic and environmental benefits.

Acknowledgements

The author would like to thank Lafarge who supported all the work presented here.

References

1. C. Famy, 'Expansion of Heat-Cured Mortars', PhD Thesis, University of London 1999.
2. C. Famy, K. L. Scrivener, A. R. Brough, A. Atkinson and E. Lachowski, 'Characterisation of C–S–H products in expansive and non-expansive heat-cured mortars: an electron microscopy study' in *Proc. 5th CANMET/ACI International Conference of Concrete Durability*, Barcelona, Spain, Vol. 1, pp. 385–402.
3. T. Füllmann, G. Walenta, T. Beir, B. Espinosa and K. L. Scrivener, 'Quantitative Reitveld Analysis of Calcium Aluminate Cement', *World Cement*, 1999, **30**(6).
4. T. Füllmann, 'Quantitative Rietveld Phaseanalyse von Tonerdezementen', PhD thesis University Erlangen-Nürnberg, 1997.
5. A. Gloter, 'Etude de l'état chimique du fer et de la structure à l'échelle nanométrique d'oxydes cimentiers par spectroscopie des pertes d'énergie d'électrons et microscopie électronique en transmission', PhD thesis, University of Paris-Sud, Centre d'Orsay, 2000.
6. A. Gloter, J. Ingrin, D. Bouchet, K. Scrivener, C. Colliex, 'TEM evidence of perovskite-brownmillerite coexistence in the $Ca(ALxFe_{1-x})O_{2.5}$ system with minor amounts of titanium and silicon', *Phys. Chem. Minerals*, 2000, **27**, 504–513.

Materials for Ceramic Ion Conducting Membrane Devices*

J. A. KILNER
Department of Materials, Imperial College of Science Technology and Medicine, London, SW7 2BP, UK

ABSTRACT
Ion conducting ceramic membrane materials offer the possibility of novel electrochemical devices with distinct economic and environmental advantages over their conventional counterparts. Several such ceramic membrane devices can be defined. These include oxygen sensors, solid oxide fuel cells and oxygen separation devices. This article focuses on one aspect of this broad spectrum: the development of materials for oxygen separation membranes for (i) the bulk industrial separation of oxygen, and (ii) Ceramic Membrane Reactors (CMRs) for the conversion of natural gas to syngas, as part of a gas-to-liquid fuel conversion plant. The mass transport properties required for candidate materials for membrane devices are discussed including the key parameter of the surface oxygen exchange coefficient.

Introduction

The aim of this article is to examine the current state of the art in a field that was started 100 years ago by Nernst, with the discovery of oxygen ion conductivity in zirconia. The development of materials such as these, with high ionic conductivity, has been one of the central themes of research in the Department of Materials through the efforts of such groups as the Nuffield Research group, the Wolfson Unit for Solid State Ionics (WUSSI) and more recently the Centre for Ion Conducting Membranes (CICM). The materials to be discussed below display mixed conductivity and are some of the most recent materials to be investigated. Potentially they have a very large number of technological applications, and hence are the subject of much current academic and industrial interest.

This current interest in mixed conductors was instigated by Japanese workers,[1,2] who, starting in 1985, investigated dense oxide membranes, which were able to permeate substantial fluxes of oxygen at temperatures

---

*This article is based on an earlier conference presentation given at the CIMTEC conference (June 2000) where a modified version of the text can be found.

of 800–1000ºC, with 100% selectivity. Since that time, there has been a rapid growth of both scientific and industrial studies aimed at the use of these membranes for the separation of oxygen. The industrial interest has been driven by the possibility of providing compact oxygen separation plants for applications including aerospace, medical, petrochemical and manufacturing industries.

Devices

The different devices made possible with these ion conducting membranes, split into two main groups; those that are pressure driven where the membranes are mixed conductors, and those that are electrically driven in which the major material is an ionic conductor. The electrically driven membranes are of interest for the production of small quantities of high purity oxygen in applications such as medical, aerospace and for portable gas generation. They are being actively pursued by a number of industrial gas and aerospace companies.[3,4,5] These devices will not be the main topic of this article, instead we shall focus on the pressure driven devices, which are more demanding in terms of the materials requirements, however the potential for application of such devices is much greater.

Pressure Driven Devices

Pressure driven devices are perhaps the simplest form of membrane device. The membrane consists of a dense, gas-tight mixed ionic-electronic conductor (MIEC) that allows the transport of both oxygen ions and electronic species. The driving force for oxygen transport is an oxygen potential gradient applied across the membrane. This can be achieved in two ways:

(i) by applying a higher partial pressure of oxygen on the membrane feed side than on the permeate side. The whole device may be operated well in excess of atmospheric pressure to achieve a pressurised permeate stream.
(ii) by promoting a chemical reaction at the permeate side in a catalytic membrane reactor (CMR), which keeps the net oxygen activity very low and provides a much larger driving force than (i). An example of CMR use is in the partial oxidation of methane to synthesis gas (CO and H_2).

Dyer et al.[5] coined the acronym (ITM) inorganic transport membrane for the two cases noted above and called the two processes ITM Oxygen and ITM Syngas. ITM Syngas is an important step along the way to the conversion of methane to synthetic liquid fuels, such as methanol, and to other liquids by the Fischer Tropsch process. There are vast reserves of natural gas in remote gas fields, such as Alaska's North Slope and in deep offshore locations that are uneconomic to exploit at present. If the gas could be converted into a liquid fuel the economics would change dramatically, and effectively boost the world's oil reserves by an equivalent to 30 years consumption.[6] CMRs offer a number of significant advantages over commercially available syngas technology including the major advantage of replacing the need for an air separation unit (ASU) and thereby significantly decrease the capital costs.[5,7]

The two ITM systems mentioned above have quite different applications however, the underlying principle of operation remains the same. No matter how the partial pressure difference is introduced to the membrane, the flux of oxygen through the membrane ($J_O / cm^3 \, cm^{-2} \, s^{-1}$) is given by a consideration of Wagner's tarnishing equation (neglecting surface reactions). At any temperature T and under a partial pressure gradient $P'_{O_2} > P''_{O_2}$,

$$J_O = \frac{RTV}{16F^2L} \int_{P''_{O_2}}^{P'_{O_2}} t_{ion} t_{elec} \sigma_{total} \, d \ln P_{O_2}$$

which can be approximated, under certain conditions, to:

$$J_O = \frac{RTV_m}{16F^2L} t_{ion} t_{elec} \sigma_{total} \ln \left\{ \frac{P'_{O_2}}{P''_{O_2}} \right\}$$

where, σ_{total} is the total conductivity of the material, and t_{ion} and t_{elec} are the ionic and electronic transport numbers. V_m is the molar volume of oxygen in the solid, and L is the thickness of the membrane. The main purpose of introducing these equations at this point is to show that the oxygen flux is dependent upon the pressure gradient and inversely proportional to the thickness of the membrane. Note that for a membrane where $\sigma_{elec} \ggg \sigma_{ion}$, as is the case for the majority of mixed conducting materials discussed here, the product $t_{ion} t_{elec} \sigma_{total}$ reduces to σ_{ion}. Thus for these membranes the ionic conductivity (and hence tracer diffusion coefficient for oxygen) is a major determining factor for the attainable oxygen flux.

Oxygen Transport

Of great interest for membrane materials is the way in which they exchange oxygen with the surrounding ambient gas, and the way in which that oxygen is transported in the bulk of the material. These properties are defined by two parameters, the surface exchange coefficient k and the oxygen tracer diffusion coefficient D, both of which can be measured by a variety of isotopic and electrochemical methods. The tracer diffusion coefficient is well understood, however much less is known about the surface exchange coefficient.

The Oxygen Tracer Diffusion and Surface Exchange Coefficients

The oxygen tracer diffusion coefficient, D, can be derived in terms of atomistic parameters from random walk theory. From this it is then easy to show that the oxygen tracer diffusion coefficient can be rewritten as $D = [V_o^{\cdot\cdot}]D_v$. Where D_v is the vacancy diffusion coefficient and $[V_o^{\cdot\cdot}]$ is the vacancy concentration expressed as a site fraction. To obtain materials with high diffusivities thus requires materials with highly mobile vacancies (high D_v) and a large vacancy population.

The surface exchange coefficient is a measure of the neutral oxygen exchange flux crossing the surface of the oxide, at equilibrium, as described by the following reaction.[8]

$$\frac{1}{2}O_2 + V_o^{\cdot\cdot} + e' \leftrightarrow O_o^x$$

This flux will be dependent upon the surface vacancy concentration, the surface electron concentration, and the dissociation rate of the di-oxygen molecule, however, at present, the rate-limiting step has yet to be identified. Kilner et al.[9] have derived a simple relationship for the surface exchange coefficient in terms of the bulk and surface vacancy concentrations. Adler et al.[10] have also arrived at a similar relationship for k, by consideration of the a.c. electrode behaviour of symmetrical cells with a 'double' cathode structure. As mentioned above, the exact mechanisms of oxygen surface exchange remain elusive, however the vacancy concentration is clearly a very important parameter.

On a more practical level, the influence of the surface exchange on membrane performance has been recently emphasised by both Adler et al.[10] and Steele.[11] They have both stressed the importance of a characteristic length L_c defined as the ratio of the two parameters D/k. This length is the changeover point at which the permeation flux of oxygen through a thin mixed conducting membrane changes from being limited by diffusion in the membrane, as implied earlier, to being limited by the surface exchange process. Figure 1 shows how the flux of oxygen varies with membrane thickness for a membrane in a set P_{O_2} gradient. This D/k relationship has

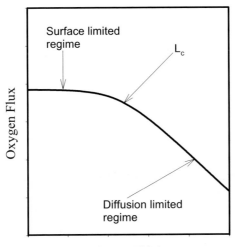

Fig. 1 A diagrammatic representation of the oxygen flux through a mixed conducting oxygen separation membrane as a function of thickness showing the changeover from bulk to surface control.

an important bearing in defining the limiting mechanism operative in any given membrane.

Lane et al.[12] have shown that the situation is somewhat more complicated and that, because the surface exchange coefficient is dependent upon partial pressure of oxygen (as is the conductivity), the flux of oxygen through such a membrane is determined by an equation which contains $L_{c(P_{O_2})}$ corresponding to the oxygen partial pressures at *both* the feed and permeate sides of the membrane. J_O in this case becomes:

$$J_O = \frac{RTV_m}{16F^2(L + L_{C(P'_{O_2})} + L_{C(P''_{O_2})})} \int_{P''_{O_2}}^{P'_{O_2}} \sigma_{total} t_{ion} t_{elec} \, d\ln P_{O_2}$$

Clearly to obtain high fluxes of oxygen through a membrane it must be made of a material with a high ionic conductivity (and consequently a high oxygen diffusion coefficient), plus a fast surface as embodied in a low L_c value (high value of k). In most cases the P_{O_2} dependence of k is positive,[12] it is probable that the low oxygen activity side of the membrane is the one with the largest L_c value, and is thus the most limiting for the flux of oxygen in thin devices. These parameters described above set the *electrochemical* criteria for a good working membrane; however these are not enough, and several further thermomechanical and thermodynamic criteria must be met in order for the material to be a good candidate for use in a membrane device.

Materials

Perovskite Materials

Following Teraoka's initial work, many investigators have studied materials with the perovskite structure with the general formula $La_{1-x}A_x(Tm_{1-y}Tm'_y)O_{3-\delta}$, where Tm represents the first row transition metal element, Cu, Fe, Cr, Mn and Co, and A an alkaline earth. A large number of possible materials exist with this general formula, and a correspondingly large number of materials have been synthesised. It is outside the scope of this article to attempt to review this data, instead the reader is directed to two authoritative articles on membrane technology by Bouwmeester and Burggraaf.[13] A comprehensive review of the work carried out in the former Soviet Union on the development of perovskite materials for use in membrane devices has also recently been given by Kharton et al.[14]

Of the materials investigated, the series $La_{1-x}Sr_xCoO_{3-\delta}$ has given some of the highest oxygen permeabilities and oxygen diffusion coefficients measured,[15] particularly at high strontium contents. Unfortunately these materials suffer from the mechanical instability mentioned below and are thus usually co-doped on the B site with iron, giving the cobalt-ferrate materials $La_{1-x}Sr_xFe_{1-y}Co_yO_{3-\delta}$. This iron substitution has the effect of increasing the stability of the materials but it does reduce the oxygen diffusivity.

Work at Imperial College has been undertaken to measure the rates of oxygen diffusion in this material at temperatures of interest for operation as either a ceramic separation membrane, or as a cathode in a SOFC. Benson and co-workers[16] have measured the oxygen diffusivity by the IEDP–SIMS method over a range of temperatures at oxygen pressures close to 1 atm. These data are shown below in Fig. 2. It is clear that the oxygen diffusivity is high at temperatures greater than 700°C, and that the same is true for the surface exchange coefficient. What is also apparent is that the ratio of the two changes as the temperature changes and that L_c becomes larger at higher temperatures.

From data such as these, oxygen permeation fluxes can be calculated and compared with experimental results obtained from permeation measurements[12], with considerable success, however some discrepancies still occur mainly due to the uncertainties in the surface exchange coefficient. From the transport data shown in Fig. 2 above, it is also clear that the high temperature value of the diffusion and surface exchange coefficient give a value for L_c that is close to 100 micrometers. This value is quite interesting and is one that is found to be the limiting case for a number of these high temperature perovskite oxides.[17] This result demonstrates that there is no great advantage

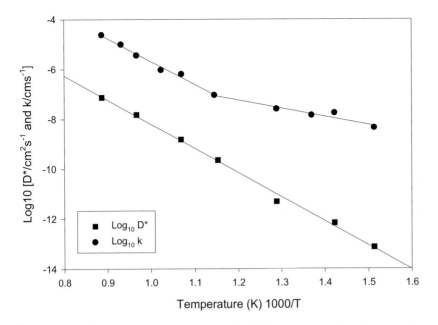

Fig. 2 Arrhenius plot of the oxygen self diffusion, D and surface exchange coefficient k for $La_{0.6}Sr_{0.4}Fe_{0.8}Co_{0.2}O_{3-\delta}$.

to be gained from going to very thin membranes to increase the oxygen flux, as the flux will be surface limited.

This group of LSCF materials has also been recently investigated by Lin[18,19,20] and co-workers, for CMR applications. They found that the material is able to sustain moderate fluxes of oxygen when tested in a permeation experiment at 850°C with helium as the sweep gas on the permeate side (~0.3 cm^3 cm^{-2} min^{-1} for a 2 mm thick disk. NB a flux of 3 cm^3 cm^{-2} min^{-1} is equivalent to a current density of 1A cm^{-2}). They also found that the material could be used in a syngas reactor but, as had been noticed by others, the material degrades with time and ultimately fails due to decomposition of the material.[20]

Degradation of these materials is an issue that has also been investigated by Benson et al.[21] Materials similar to the cobalt ferrates are also useful for processes such as the ITM Oxygen process, where the membrane is downstream of a combustor in a combined membrane/gas turbine system. They found that upon exposure of the materials to a simulated exhaust gas mixture containing wet oxygen and CO_2 at 5 atm and at 750°C, surface degradation took place. In particular many chromate species were formed due to the transport of Cr species from the inconel furnace tube used to contain the pressurised atmosphere. $SrCO_3$ was also found to be one of the reaction products and it is thought that this could be a problem in causing inhibition of the oxygen exchange process in CO_2 bearing atmospheres. This can be avoided by working at temperatures above the decomposition temperature of the carbonate.

Finally when discussing these materials it must be stated that the mechanical properties must be taken into account, and little work has been done in this area. For instance, stress under working and thermal cycling conditions is a problem with the perovskite membranes. The materials undergo a 'chemical' expansion, due to loss of oxygen from the crystal lattice. Oxygen loss occurs with increasing temperature, at a constant oxygen partial pressure, and with decreasing oxygen partial pressure at a constant temperature. This has severe implications for the heating and cooling of a membrane and for the establishment of a partial pressure gradient across the membrane during operation. Both can lead to the failure of the membrane. Atkinson and Ramos[22] have calculated the stresses in flat plate and tubular membranes of ceria-gadolina and doped lanthanum chromites when subjected to chemical gradients corresponding to a fuel cell environment similar to the conditions to be found in a syngas CMR. They find that at high temperatures both materials are at risk of failure due to the chemically induced stresses, particularly under transient conditions.

Perovskite Related and Other Novel Materials

The perovskite materials exhibit some high fluxes for oxygen, however their stability is questionable, particularly when used in aggressive environments such as the CMR applications for syngas production. Partly because of these limitations, the search for possible materials has been spread to perovskite-related materials. Balachandran and co-workers are perhaps the group that has had the most success in this area. They have developed the perovskite related material $SrFeCo_{0.5}O_{3.25\pm\delta}$ that has been reported to sustain very high fluxes of oxygen when used in a syngas type reactor (oxygen permeation rates as high as $4\,cm^3\,cm^{-2}\,min^{-1}$).[23,24] Balachandran et al. report that this material is stable[25] in this environment, unlike the $La_{1-x}Sr_xFe_{1-y}Co_yO_{3-\delta}$ materials mentioned earlier, and has a high efficiency for methane conversion.

Some controversy has surrounded these claims since the first publication by Balachandran. In a recent article, Armstrong et al.[26] report that the compositions described by Balachandran are in fact three-phase mixtures of an intergrowth phase $Sr_4Fe_{6-x}Co_xO_{13+\delta}$, the cubic perovskite $SrFe_{1-x}Co_xO_{3-\delta}$

and the cubic phase $Co_{1-x}Fe_xO$. They further show that the perovskite material with $x = 0.25$ has the highest permeability of oxygen and that this phase is the origin of the high fluxes originally reported.

Sammells and co-workers[27] are currently investigating materials with the brownmillerite structure of the general composition $A_2B_2O_5$ for use in CMRs. These materials are substituted on both the A and B sites. Very high fluxes of oxygen are reported for these materials (10–12 cm^3 cm^{-2} min^{-1}) at 900°C under syngas reactor conditions. It should be noted that these high fluxes are partly a result of the extremely large driving force present in a syngas reactor.

The final material to be described in this section is a material adopting the K_2NiF_4 (A_2BO_4) structure $La_2NiO_{4+\delta}$. The crystal structure of the A_2BO_4 group of materials is shown in Fig. 3. This material is unusual because it is hyperstoichiometric with δ being as high as 0.25.

The excess oxygen is accommodated by oxygen *interstitials*, which appear to be highly mobile. Measurements of the oxygen diffusivity[28] and of the permeability[29] have confirmed the high mobility of the oxygen in this structure. This material is significant for a number of reasons. (i) The material

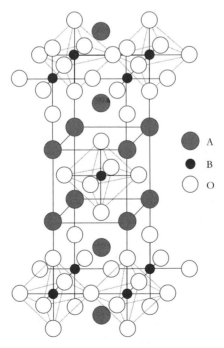

Fig. 3 The A_2BO_4 crystal structure.

is not substituted in any way, most of the other perovskite and perovskite-related materials are substituted with an alkaline earth such as Sr, which raises questions of the stability of those materials in CO_2-containing environments. (ii) They have low expansion coefficients ($\sim$12 ppm K^{-1})[29] and do not seem to be as sensitive to chemical expansion effects as the perovskite materials. The only drawback with these materials is that they are not stable in very low oxygen partial pressures making them unsuitable for some CMR applications.

More recently, work at Imperial has centred upon the effect of substitution of the Ni in these materials with Co. Shaw and Kilner[30] have shown the effect of the substitution of cobalt on the transport properties of these oxygen excess perovskite related samples.

They showed that whilst the diffusion coefficients remained as high as in the undoped materials, the surface exchange coefficient is markedly affected by the addition of Co. This is most clearly seen in the changes in activation enthalpies of the two processes as a function of the cobalt content, as shown in Fig. 4. The activation enthalpy for the diffusion process is hardly changed by the substitution of cobalt whilst that for the surface exchange drops dramatically from a modest value of $\sim$120 kJ mol^{-1} to the very low value of 20 kJ mol^{-1}. Clearly the Co addition is having a marked effect on the surface of these materials and this interesting effect needs further investigation.

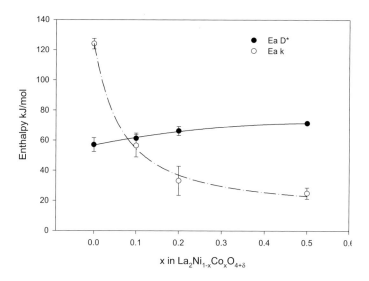

Fig. 4 The activation enthalpy for oxygen self diffusion and surface exchange for the materials $La_2Ni_{1-x}Co_xO_{4+\delta}$.

Oxygen Fluxes

Having shown the principle of the devices and having discussed the materials involved, it is of interest to look at the 'state of the art' in terms of the fluxes of oxygen that can be achieved. Figure 5 shows a range of fluxes measured from samples of single and dual phase materials, operated in a pressure driven mode, plotted as a function of inverse temperature.

This figure is intended to give some appreciation of the fluxes that are attainable; however the fluxes are not normalised to a given partial pressure gradient or thickness of membrane, and thus the fluxes are not directly comparable. Taking a value of between 1 and $10\,cm^3\,cm^{-2}\,min^{-1}$, as the level of oxygen flux needed for practical applications, it can be seen that the cobalt-containing single phase materials give appreciable oxygen fluxes above about 900°C. It is interesting to note on this graph that the dual phase material, fabricated from $(Bi_2O_3)_{0.75}(Y_2O_3)_{0.25}Ag(35\,vol.\%)$, approaches the lower bound of the practical fluxes at temperatures of 800°C.

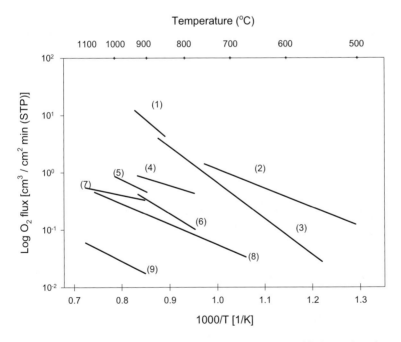

Fig. 5 Oxygen permeation fluxes for (1) $SrFeCo_{0.5}O_{3.25-\delta}$, (2) $(Bi_2O_3)_{0.75}(Y_2O_3)_{0.25}$-Ag(35 vol.%), (3) $SrCo_{0.8}Fe_{0.2}O_{3-\delta}$, (4) $La_{0.2}Sr_{0.8}Co_{0.8}Fe_{0.2}O_{3-\delta}$, (5) $La_2NiO_{4+\delta}$, (6) $La_{0.6}Sr_{0.4}Co_{0.2}Fe_{0.8}O_{3-\delta}$, (7) $La_{0.3}Sr_{0.7}CoO_{3-\delta}$, (8) $La_{0.5}Sr_{0.5}CoO_{3-\delta}$, (9) YSZ-Pd (40 vol.%). Figure modified from Ref. 6.

Conclusions

The study of ceramic membranes for the separation of oxygen is one that is receiving a great deal of attention. The potential for applications, such as the CMRs for syngas production, is enormous and this is the main driving force behind the large effort being expended worldwide. It is widely acknowledged that the materials aspects are the key to this technology and provide an enormous challenge for materials scientists working in this field. The newer material described above are going some away to meeting this challenge, however there remains the possibility for significant advances to be made.

References

1. Y. Teraoka, H.-M. Zhang, S. Furukawa and N. Yamazoe, *Chem. Lett.*, 1985, 1743.
2. Y. Teraoka, H.-M. Zhang, K. Okamato and N. Yamazoe, *Mater. Res. Bull.*, 1988, **23**, 51.
3. J. McAleese, M. D. Vasquez-Navarro, F. P. F. van Berkel and J. A. Kilner, in *Proc. Sixth International Symposium on Solid Oxide Fuel Cells* (SOFC VI), S. C. Singhal and M. Dokiya eds, *Electrochemical Society Proceedings*, 1999, Vol. 99–19, 1185.
4. K. R. Sridhar, *ibid.* 1163.
5. P. N. Dyer, R. E. Richards, S. L. Russek and D. M. Taylor. *Solid State Ionics*, 2000, **134**, 21.
6. J. A. Kilner, S. Benson, J. Lane and D. Waller, *Chemistry and Industry*, 17 Nov. 1997, 907.
7. H. W. G. Saracco, J. P. Neomagus, G. F. Versteeg and W. P. M. van Swaaij, *Chemical Engineering Science*, 1999, **54**, 1997.
8. J. A. Kilner, in *Proc. of 2nd Int. Symp. on Ionic and Mixed Conducting Ceramics*, T. Ramanarayanan, W. L. Worrell and H. L. Tuller eds, Electrochemical Soc., 1994, 174–190.
9. J. A. Kilner, R. A. De Souza and I. C. Fullarton, *Solid State Ionics*, 1996, **86–88**, 703.
10. S. B. Adler, J. A. Lane and B. C. H. Steele, *J. Electrochem. Soc.*, 1996, **143**, 3554.
11. B. C. H. Steele, *Solid State Ionics*, 1997, **94**, 239.
12. J. A. Lane, S. J. Benson, D. Waller and J. A. Kilner, *Solid State Ionics*, 1999, **121**, 201.
13. H. J. M. Bouwmeester and A. J. Burggraaf, Chapter 7 of *The CRC Handbook of Solid State Electrochemistry*, P. J. Gellings and H. J. M. Bouwmeester eds, 481–553, CRC Press 1997; H. J. M. Bouwmeester and A. J. Burggraaf, Chapter 10 of *Fundamentals of Inorganic Membrane Science and Technology*, A. J. Burggraaf and L. Cot eds, Elsevier, 1996, 435–528.
14. V. V. Kharton, A. A. Yaremchenko and E. N. Naumovitch, *J. Solid State Electrochem.*, 1999, **3**, 303.
15. R. H. E van Doorn, I. C. Fullarton, R. A. DeSouza, J. A. Kilner, H. J. M. Bouwmeester and A. J. Burggraaf, *Solid State Ionics*, 1997, **96**, 1.

16. S. J. Benson, R. J. Chater and J. A. Kilner, in *Ion and Mixed Conducting Ceramics III, Electrochemical So. Proceedings*, Vol. 97–24, 1998, 596.
17. R. A. De Souza and J. A. Kilner, *Solid State Ionics*, 1998, **106**, 175.
18. Y. Zheng, Y. S. Lin and S. L. Swartz, *J. Membr. Science*, 1998, **150**, 87.
19. S. Li, W. Jin, N. Xu and J. Shi, *Solid State Ionics*, 1999, **124**, 161.
20. W. Jin, S. Li, P. Huang, N. Xu, J. Shi and Y. S. Lin, *J. Membr. Science*, 2000, **166**, 13.
21. S. Benson, D. Waller and J. A. Kilner, *J. Electrochem Soc.*, 1999, **146**, 1305.
22. A. Atkinson and T. M. G. M. Ramos, *Solid State Ionics*, 2000, **129**, 259.
23. U. Balachandran, J. T. Dusek, R. L. Mieville, R. B. Poeppel, M. S. Kleefisch, S. Pei, T. P. Kobylinski, C. A. Udovitch and A. C. Bose, *Applied Catalysis A*, 1995, **133**, 19.
24. U. Balachandran, J. T. Dusek, P. S. Maiya, B. Ma, R. L. Mieville, M. S. Kleefisch and C. A. Udovitch, *Catalysis Today*, 1997, **36**, 265.
25. B. Ma and U. Balachandran, *Mat. Res. Bull.*, 1988, **33**, 223.
26. T. Armstrong, F. Prado, Y. Xia and A. Manthiram, *J. Electrochem. Soc.*, 2000, **147**, 435.
27. A. F. Sammells, T. F. Barton, D. R. Peterson, S. T. Harford, R. Mackay, P. M. van Calcar, M. V. Mundschau and J. B. Schutz, AICHE Meeting, Atlanta, Georgia, March 2000.
28. S. J. Skinner and J. A. Kilner, *Solid State Ionics*, 2000, **135**, 709.
29. V. V. Kharton, A. P. Viskup, E. N. Naumovich and F. B. Marques, *J. Mater. Chem.*, 1999, **9**, 2623.
30. J. A. Kilner and C. K. M. Shaw, to be published in *Solid State Ionics*.

Grain Boundaries in High T_c Superconductors: The Key to Applications

DAVID LARBALESTIER

L. V. Shubnikov Professor and David Grainger Professor, Director, Applied Superconductivity Center, Departments of Materials Science and Engineering/Physics University of Wisconsin

ABSTRACT

The key to all large-scale applications of superconductivity is the ability to carry very high current densities (10^5–10^6 A cm^{-2}) in polycrystalline wires, which are kilometres long. This condition is easily fulfilled for the metallic Nb-based superconductors. As metals they have high carrier densities and supercurrents pass through their grain boundaries without difficulty, even though the decay length of the superconducting condensation energy (2–5 nm) is not very much larger than the effective width of grain boundaries (0.5–2 nm). All of this changes for the high temperature copper oxide superconductors. Decay lengths are comparable or only slightly smaller but carrier densities are 2 to 3 orders of magnitude smaller and the compounds are strongly anisotropic. Progress in understanding and in overcoming these obstacles from a materials aspect will be summarised and reviewed.

Lightweight Materials and Structures

A. G. EVANS
Princeton Materials Institute, Princeton University, Princeton, NJ 08540, US

ABSTRACT
New categories of ultra-light load bearing structures are described. They are derived from topology optimisation and fabricated from advanced structural alloys. Maximum realisable load capacities are attained by designing panels with truss core architectures that exclude bending modes. Metrics are presented that permit these structures to be compared in a direct manner with other concepts. The comparison enables weight benefits to be specified for flat and curved panels subject to bending and compressive loads. Models that relate the performance to the core architecture and to the properties of the constituent alloys are described, as well as experimental measurements used to validate the models. The relative merits of different methods for fabricating the desired topologies are discussed, as needed to establish the basis for a cost/performance trade-off.

Introduction

Many technologies require that the materials being used support appreciable loads and, moreover, that the components be as lightweight and compact as possible. This requirement is especially true in the transportation arena, where lightness equates with fuel efficiency and range. Concept implementation is dictated by the trade-off between performance and manufacturing cost, manifest in a value factor.[1-4] The attainment of the minimum weight has a long history.[5-14] Four technical factors are involved: (i) materials selection,[1,2] (ii) utilisation of shape,[2,15-17] (iii) topology optimisation[2,18-21] and (iv) multi-functionality.[2,22] The new developments to be addressed here arise in topic (iii). Topic (iv) has been elaborated elsewhere.[2,22]

The ability to sustain loads has as much to do with component shape as the material of construction.[2,15-17] Shortcomings in material density can be circumvented by designing shaped components that optimise load capacity. Well-known examples include I-beams, box beams, and hat-stiffened panels. Other topologies can be used to achieve load capacity at yet lower weight, exemplified by truss structures and honeycomb panels.[18-21] The truss topology has the further benefit that the open spaces can be used to impart functionalities in addition to load bearing, such as cooling:[2,22] whereupon, the extra weight of an additional component normally needed to imbue that extra function can be saved.

Novel categories of topologically-designed metallic alloys are addressed in this paper,[18–21,23–25] along with the performance attributes that underlie their implementation. The metals are configured in truss-like arrangements with open domains that occupy most (>90%) of the volume. The choice of the topology is crucial. It must be designed such that, when sheared, the trusses are in either tension or compression, *with no bending*.

A procedure has been established for selecting the materials most applicable to aerospace structures, which are limited by their ability to support bending and compressive loads.[1,2] The associated metrics are encapsulated as diagonals superposed on maps of Young's modulus, E, as a function of density, ρ (Fig. 1a) and the ultimate tensile strength, σ_Y, against ρ (Fig. 1b). One metric is for beams ($\sqrt{E}/\rho, \sigma_Y^{2/3}/\rho$) and another for panels ($E^{1/3}/\rho, \sqrt{\sigma_Y}/\rho$).[2] The best materials are those that reside at the top left of the diagram, furthest from the diagonals. Evidently, for beams, the preferred materials sequence is:

$$\text{Composites} \rightarrow \text{Mg alloys} \rightarrow \text{Al alloys} \rightarrow \text{Ti alloys} \rightarrow \text{Steels}$$

For panels, the sequence is essentially the same, except that certain structural polymers can have performance superior to steels. These are the materials invoked in the subsequent assessments.

In the aerospace sector, the value factor attributed to lightweight characteristics is substantial:[1–4] whereupon alloys of Al, Ti and Mg, as well as composites, are preferred. In the automotive sector, the value attributed to lightness is lower, such that steel and polymer components are commonly used.

Minimum Weight Structures

Performance indices are needed to ascertain minimum weight configurations and to compare designs. The indices are based on overall structural weight, W, load, P, stiffness, S, and yield strain, ε_Y^s. When the faces and the core are made from the same alloy, the weight index is:[1,6,10,18,22]

$$\Psi = W/L^2 B\rho \tag{1}$$

where L is the length of the panel/beam and B the width.

For designs based on *load capacity*, explicit comparisons can be made using one of the following load indices. For *axial compression*, a commonly used index is:[1,10,18]

$$\Pi_e = P/E_s LB \tag{2a}$$

Lightweight Materials and Structures 261

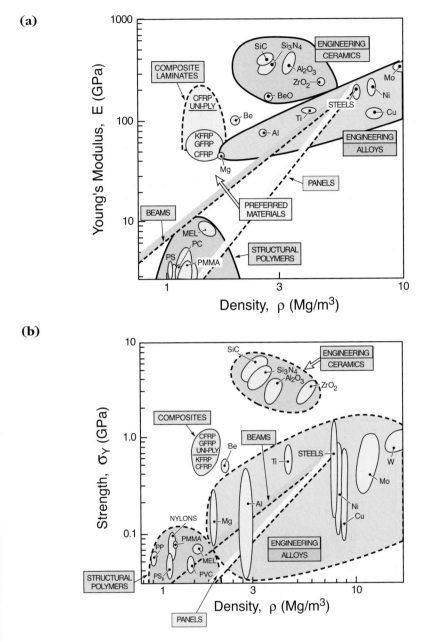

Fig. 1 Property diagrams used for materials selection: (a) stiffness and (b) strength. The diagonal lines indicate equivalent performance for lightweight beams and panels subject to bending, using mild steel as the reference material. Materials that reside furthest away from these lines, in a vector toward to top left of the diagrams, are preferred. The ceramics are discounted because of their brittle failure susceptibility.

An alternative index, Π_p, most suitable when the design involves face yielding, replaces E_s with σ_Y^s ($\Pi_e \equiv \Pi_p \varepsilon_Y^s$), such that for panels:

$$\Pi_p = P/\sigma_Y^s LB \tag{2b}$$

and for beams:

$$\Pi_p = P/\sigma_Y^s L^2 \tag{2c}$$

For *bending*, the corresponding indices are:[15]

$$\Pi_e^b = M/E_s/L^3 \tag{2d}$$

$$\Pi_p^b \equiv M/\sigma_Y^s L^3 \tag{2e}$$

where M is the moment ($M \equiv PL$).

Stiffness governs the weight at smaller loads, Π_p instead of load capacity[1] (Fig. 2). Comparative indices for *stiffness-limited* designs are not as all-encompassing. For beams, an index referred to as the shape factor, ϕ, has

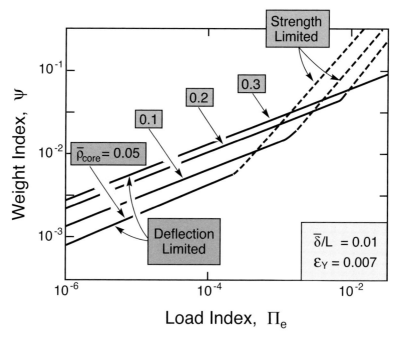

Fig. 2 Variations in the weight index with load capacity for a stochastic-core panel in bending, subject to uniformly distributed pressure on one side. Note that for all core densities, the panel is stiffness-limited at low loads but becomes strength-limited at large load capacities. Here $\bar{\delta}/L$ is the specified end deflection.

been used,[15] applicable for loadings that allow transverse shear to be neglected. This index is the ratio of the stiffness, S_B, of a beam in pure bending, having moment of area, I, to that for a solid circular beam, S_o, at the equivalent weight. It can be expressed as:[2,15]

$$\phi \equiv \frac{S_B}{S_o} = \frac{4\pi I}{A^2} \quad (3)$$

where A is the area of the cross-section. Note that values of ϕ are dictated solely by the cross section.[2,15]

Minimum weight designs, based on these indices, are found by *identifying the failure modes*, specifying the stiffness or load capacity and then varying the dimensions to determine the lowest weight for each mode.[1] The preferred topology depends on the configuration and the loading: whether bending or compression, and whether flat or curved.

When load capacity governs the design, indices (1) and (2) fully-specify the comparison between options. Plots for three cases are presented on Fig. 3[10,18,22,23] (note that $\sqrt{\pi_e}$ is commonly used as the abscissa).

(i) For flat panels subject to bending, honeycomb cores represent the performance benchmark (Fig. 3a).
(ii) For flat panels in axial compression, configurations that *align the core elements with the load* have the lowest weight. Hat-stiffened panels define the benchmark (Fig. 3b).
(iii) The preferences differ for curved panels and shells, because the loading on the core is bi-axial. Greater benefits accrue from cores with *isotropic topologies*. The reference system is a shell with distributed axial stiffeners and circumferential stiffeners at the ends (Fig. 3c).

While the goal is to identify topologies that outperform these benchmarks, designs with inferior performance can still have implementation opportunities, based on cost, durability and other performance criteria, such as strength retention after impact.[1,24]

Effects of Shape

For beams, an optimum (minimum weight) exists in *pure bending*[2,15] because of their explicit cross-sectional profile. The result is applicable at load indices sufficiently small that transverse shear is negligible. Conversely, for panels in pure bending, lower and lower weights emerge upon systematically increasing the core thickness and there is no realisable minimum weight. The discriminator that sets the optimum is the response in transverse shear.

(a)

(b)

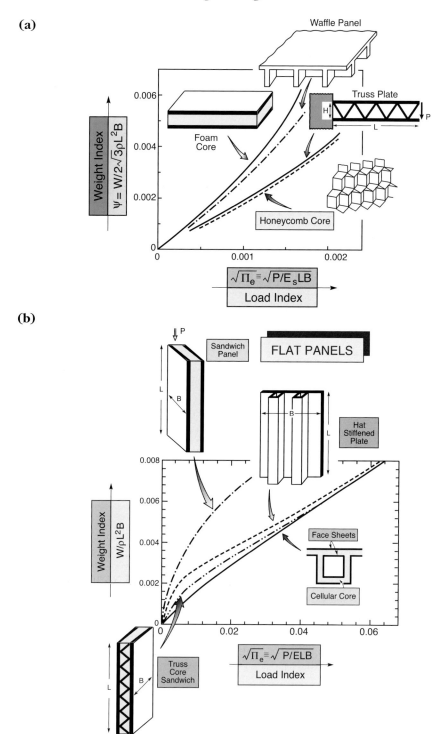

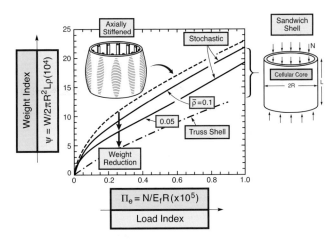

Fig. 3 Plots of minimum weight against load index for fully-optimised, strength-limited panels subject to bending and compression. (a) Bending of flat panels: the honeycomb panel is the benchmark design that must be equalled or surpassed from a performance perspective. (b) Axial compression of flat panels: the hat-stiffened design is the benchmark. The isotropic nature of the truss core is a disadvantage for this loading. Nevertheless, it is as light as the benchmark design. (c) Axial compression of curved panels. Now biaxial strain is introduced into the core because of the hoop expansion. Accordingly, the isotropic nature of the stochastic and truss cores offers benefits and all of these cores are lighter than the stiffened panel.

Beam failure mechanisms applicable in pure bending have been documented.[15] An example for a *box* illustrates the responses and provides a benchmark. First, note that, absent failure mode considerations, the stiffness and strength can be increased indefinitely by thinning the wall. A minimum weight only emerges when yielding and buckling responses are introduced. These failure domains are illustrated using a cross plot of $\phi(\Pi_p)$ for representative yield strain, ε_Y^s, and for an allowable non-dimensional curvature, κL (Fig. 4). Contours of constant weight index, Ψ, are superposed.

The highest realisable load factors, at specified mass, occur at the boundaries between mechanisms,[1,18,20] as evident from Fig. 4a. The characteristics near these boundaries dictate lightweight design. Since buckling is a catastrophic mode,[15] the transition from stiffness to buckling-failure is most important. This occurs at a shape factor:[15]

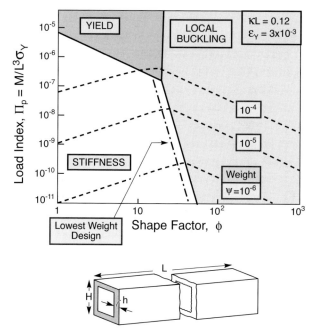

Fig. 4 Design factors for box beams in pure bending, neglecting the role of transverse shear. The map of load index against shape factor indicates the domains of dominance based on stiffness, local buckling and yielding. The objective is a curvature not to exceed, $\kappa L = 0.12$, for a material with yield strain, $\varepsilon_Y = 3 \times 10^{-3}$ (typical of steel). Contours of the weight index are superposed. Note that the highest load index for given weight occurs at the transition between stiffness and buckling-dominated responses. Accordingly, by using a safety factor, 2/3, to avert the catastrophic response upon buckling, the lowest weight designs are just to the left of the mechanism transition, as shown by the dotted line.

$$\phi_{opt} \equiv \frac{\pi c}{6b} = 0.6(\kappa L)^{-1/3}(\Pi_p)^{-1/9} \qquad (4a)$$

where a safety factor of 2/3 has been used to assure that buckling is averted. The weight at this optimum is:[15]

$$\Psi_{opt} = 2.52 \frac{(\Pi_e/\varepsilon_Y^s)^{5/9}}{(\kappa L)^{1/3}} \qquad (4b)$$

The shape applicable to each load is given by (4a). Note that, to attain the optimum, the shape must be changed as the load index changes. The result (4a) is nominally applicable in the load index range prior to failure by yielding:

$$\Pi_e^b \leq \frac{\varepsilon_Y^{7/2}}{3(\kappa L)^3} \qquad (4c)$$

In practice, transverse shear effects are likely to govern the design at higher load index.

Basic Topologies

The design of minimum weight panels with all-metallic constituents requires an open core attached to dense faces. The faces provide the bending resistance and the core the shear/crushing resistance. Performance is most strongly affected by the shear and compressive properties of the core, as affected by its attachment to the faces. Cores with stochastic topologies are *bending-dominated*, resulting in power law dependence of the properties on the relative density, $\bar{\rho}_{core}$.[1,27] Cores with periodic topologies can be designed such that the trusses experience tension/compression, no bending. These *stretch-dominated* cores exhibit a linear dependence of properties on $\bar{\rho}_{core}$.[18,20]

For isotropic, *stochastic cores*, the shear modulus, G^c is[1] (Fig. 5):

$$G^c/E_s = \alpha_{13}[1/2(1+\nu)]\bar{\rho}_{core}^2 \qquad (5a)$$

where Poisson ratio, $\nu \approx 3/8$. The coefficient α_{13} is a quality factor. It is about unity for the best commercial materials, but can be smaller when manufacturing defects are present, particularly at the lowest achievable densities ($\bar{\rho}_{core} \approx 0.05$). It increases for thin cores, because of stiffening enabled by a boundary layer interaction with the faces.[1]

The shear yield strength τ_Y^c and compressive strength, σ_Y^c are similar and given by:[1]

$$\tau_Y^c/\sigma_Y^s \approx \sigma_Y^c/\sigma_Y^s \approx 0.3\beta_{13}\bar{\rho}_{core}^{3/2} \qquad (5b)$$

The coefficient β_{13} is another quality factor, also about unity for the best materials, but again diminished by manufacturing imperfections and elevated by a thinness factor.

Among all possible truss toplogies, only a small subset has the characteristic that the trusses stretch and compress, without bending.[18,20,28] Some have been identified by topology optimisation, others by search procedures. Three are

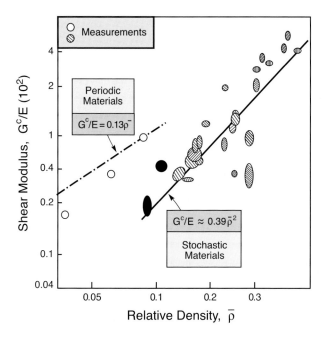

Fig. 5 A plot of shear modulus against relative density for stochastic and periodic cores. The line for the stochastic cores is a fit to experimental data for Al alloys: with exponent, $\bar{\rho}^2$ appropriate to bending-dominated structures.[1,27] The line for the periodic cores is that for the tetragonal or pyramidal trusses. The circles are experimental measurements. Note that the relative density for optimised truss core panels is about, $\bar{\rho}_{core} \approx 0.03$. In this range, the trusses have appreciably greater performance than the stochastic cores. This is reflected in the superior panel performance shown on Fig. 3.

discussed: tetragonal and pyramidal trusses (Figs 6a,b)[18,20] and a diamond plain-weave[24,29] (Fig. 6c). Others exist, such as the Kagomé truss,[28] but have yet to be fully-characterised. The theoretical results express the best achievable performance. The experimental manifestations include manufacturing-based degradation factors.

The theoretical properties for *stretch-dominated cores* are represented by:[18,20]

$$G/E_s = A_{ij}\bar{\rho}_{core} \qquad (6a)$$

$$\tau_Y/\sigma_Y^s = B_{ij}\bar{\rho}_{core} \qquad (6b)$$

$$\sigma_Y^c/\sigma_Y^s = C_{ij}\bar{\rho}_{core} \qquad (6c)$$

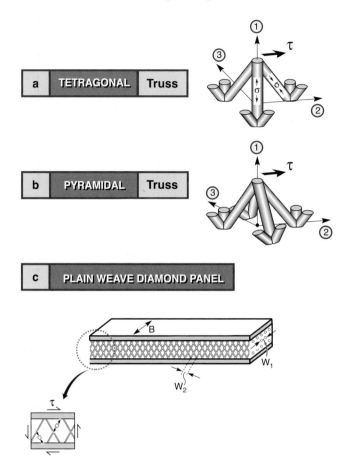

Fig. 6 Three of the periodic core designs wherein, upon in-plane shear loading, the trusses experience either tension or compression, with negligible bending.

The coefficients A_{ij}, B_{ij} and C_{ij} are functions of truss architecture, loading orientation and node design. They can be derived upon specifying a stress/strain, $\sigma(\varepsilon)$, representation for the truss/face material: a Ramberg-Osgood representation is convenient:[23]

$$\varepsilon = \sigma/E_s + \alpha(\sigma_Y^s/E_s)(\sigma/\sigma_Y^s)^N \tag{7}$$

where N is the strain hardening exponent and is a fitting coefficient.

For trusses, there is an associated tension/compression asymmetry. In tension, they strain harden after yielding in accordance with eqn (7). When compressed, beyond yield, they are susceptible to plastic buckling, at stress σ_{pb}, given by the implicit formula:[20]

$$\left(\frac{\pi k R_c}{2L_c}\right)^2 \varepsilon_Y^{-1} = \left(\frac{\sigma_{pb}}{\sigma_Y}\right) + \alpha N \left(\frac{\sigma_{pb}}{\sigma_Y}\right)^N \qquad (8)$$

where k is a measure of the rotational stiffness of the nodes (it ranges between 1 and 2),[18,20] L_c the truss length and R_c its radius. *The system softens once σ_{pb} is reached.*

The following results are presented for situations wherein the trusses fail by yielding/plastic buckling. These are the common failure modes for optimised metallic panels. Failure by elastic buckling is only expected for panels with sub-optimal relative densities ($\bar{\rho} < 1\%$) and, conceivably, in polymer systems having high yield strain.[30]

(i) Tetragonal

For a tetragonal core (Fig. 6a) having included angle, $\omega = \pi/4$, the relative density is related to the dimensions by:[18,20]

$$\bar{\rho}_{core} = 4\pi(\sqrt{2/3})(R_c/L_c)^2 \qquad (9)$$

The transverse shear modulus in the (1, 2) orientation, defined on Fig. 6, is:[20]

$$G^c/E_s = (\pi/\sqrt{6})(R_c/L_c)^2 \qquad (10a)$$

such that,

$$A_{12} = 1/8 \qquad (10b)$$

The coefficient is the same in the (1,3) orientation.

The in-plane shear yield strength is slightly anisotropic. The minimum value is:[20]

$$\tau_Y^c/\sigma_Y = (\pi\sqrt{2/3})(R_c/L_c)^2 \qquad (11a)$$

such that,

$$B_{12} = 1/4 \qquad (11b)$$

The maximum value is 15% larger.[20] The corresponding compressive strength coefficient is:[20]

$$C_{33} = 1/2 \qquad (11c)$$

The relatively large C_{33} indicates that this core would not normally fail by crushing: *it is shear limited.*

This topology has one truss that sustains a larger stress than the other two (see Fig. 6a). When the imposed shear causes these trusses to be in tension, they strain hardening, allowing the load capacity to increase with deformation. For converse loadings that cause these trusses to be in compression, they buckle at σ_{pb} eqn (8) and the load capacity diminishes upon further deformation.[20,23] This difference leads to the asymmetry in panel bending described below.

(ii) *Pyramidal*

At included angle, $\omega = \pi/4$, the pyramidal core (Fig. 6b) has relative density:[20]

$$\bar{\rho}_{core} = 4\pi(\sqrt{2})(R_c/L_c)^2 \tag{12a}$$

The shear modulus is:[20]

$$G^c/E_s = (\pi/\sqrt{2})(R_c/L_c)^2 \tag{12b}$$

Accordingly, the stiffness coefficient is the same as that for the tetragonal core: $A_{12} = A_{13} = 1/8$.

The shear strength, τ_Y^c, while more anisotropic than that for the tetragonal core, has the same minimum: $B_{12} = 1/4$.[20] The maximum is 41% larger.[20]

(iii) *Textile Cores*

Cores with plain weave architectures (Fig. 6c) have been evaluated.[23,24,29] Such weaves have weak shear response when oriented as squares, because of unconstrained truss bending. The same weaves exhibit very attractive characteristics when oriented into a *diamond cross-section* (Fig. 6c). In this case, the ligaments are subject to tension/compression with minimal bending. The following beam theory results characterise the shear response. The relative density is:[29]

$$\bar{\rho}_{core} = (2\pi)R_c^2/W_1 W_2 \tag{13a}$$

where W_1 and W_2 are the cell dimensions depicted on Fig. 6c. Subject to in-plane shear, τ, each truss is subject to axial stress, $\sigma = \pm\tau$. The consequence is a shear yield strength coefficient given by:[29]

$$B_{12} = 1/2 \tag{13b}$$

Note that this strength is twice that for plates with either the tetragonal or pyramidal truss cores. The shear strength in the other (1, 3) plane has yet to be ascertained, but is expected to be lower and performance limiting in bi-axially loaded panels. The shear modulus coefficient is also twice that for the truss cores: $A_{12} = 1/4$.[29]

This topology would have crushing limitations *except for the constraint provided by the faces.* The constrained compressive strength is expressed through the coefficient:[29]

$$C_{33} = \frac{1/2}{1 + (9/64)\nu\bar{\rho}_{core}^2 W_2/W_1} \tag{13c}$$

This is essentially the same as the in-plane shear strength and also equal to the compressive strength of the tetragonal and pyramidal cores.

Panel Properties

The relevance of the core properties is addressed in the context of panels subject to three-point bending (Fig. 7). The relative trends between core designs are the same for all other bending modes as well as for axial compression. Analytic results highlight the connections. The overall panel *bending stiffness*, S_B, is given by:[1,5,27]

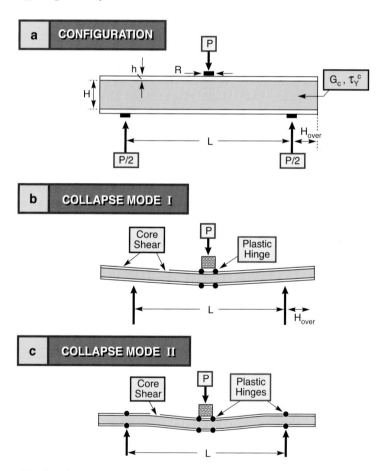

Fig. 7 (a) The three point loading configuration, defining the dimensions. The two core shear responses referred to in the text are shown on (b) and (c).

$$\frac{B}{S_B} = \frac{2L^3}{C_1 E_s hc^2} + \frac{HL}{C_2 c^2 G^c} \qquad (14)$$

where h is the face thickness and H the core thickness, with $c = H + h$, and, for three-point bending, $C_1 = 48$ and $C_2 = 4$.[1,2] The first term is the bending contribution governed by the face sheet properties. The second is the shear contribution, dominated by the core. Substituting G^c from above into eqn (14) gives S_B.

The *load capacity* depends on whether the panel fails by face yielding or core shear. When *core shear* predominates, the response at the outer supports has an important influence. If the core is sufficiently weak that plastic hinges develop (Fig. 7), the overhang does not contribute and the limit load index is given by:[1,20,23]

$$\Pi_{max}^{core} = 4(h/L)^2 + 2(c/L)B_{12}\bar{\rho}_{core} \qquad (15a)$$

The first term is the contribution from the plastic hinges formed in the faces and the second due to shear yielding of the truss core. Absent hinges at the supports, the corresponding result is:[1,20,23]

$$\Pi_{max}^{core*} = 2(h/L)^2 + 2(c/L)B_{12}\bar{\rho}_{core}[1 + 2H_{over}/L] \qquad (15b)$$

where H_{over} is the overhang (see Fig. 7). The transition between these two modes occurs when:

$$H_{over}/c \leq (1/2)(h/c)^2(C_{33}/B_{12}) \qquad (15c)$$

When the panel fails by *face yielding*, the limit load is:[1,20,23]

$$\Pi_{max}^{face} = \frac{4h(c+h)}{L^2} + (c/L)^2 C_{33}\bar{\rho}_{core} \qquad (15d)$$

The corresponding panel weight is:[1,20]

$$\Psi = 2(h/L) + (c/L)\bar{\rho}_{core} \qquad (16)$$

An optimisation based on these formulae leads to the cross-plots, $\Pi(\Psi)$, shown on Fig. 3. These plots reveal that, in flat configurations subject to either bending or axial compression, panels with tetragonal and pyramidal cores have about the same weight as the benchmark systems (honeycombs for bending and hat-stiffeners for compresssion). Preferences for truss core systems are based on their multifunctionality attributes, as well as their cost and durability. However, for curved panels, because of the associated strain biaxiality,[32] the isotropy of the truss cores results in substantial weight benefits relative to the (axially and peripherally-stiffened) benchmark.

The diamond core system remains to be optimised. It is expected that, for in-plane bending, it would be lighter than the tetragonal and pyramidal core panels, because of the larger β_{12}.

Practical Experience

Several of the periodic structures discussed above have been fabricated and tested. The tests reveal phenomena not addressed in the analytic models and establish knock-down factors associated with manufacturing imperfections. Panels with tetragonal cores have been made by the rapid prototyping of a polymer template, followed by investment casting with Al and Cu alloys.[23] One of the overarching findings is that relatively high ductility (about 20%) is needed to realise the expected performance, otherwise failure occurs prematurely at the nodes. A second major discovery is that the performance is strongly influenced by the design of the nodes. The theoretically expected response is achievable if a 'gap' design is used, wherein the axes of the trusses converge at the mid-plane of the face sheet.[23] For other designs, particularly when the axes converge beneath the face, failure occurs prematurely by shearing of the nodes.[20] Results for panels with a 'gap' design fabricated with a ductile Cu/Be alloy, summarised in Figs 8 and 9, are compared with

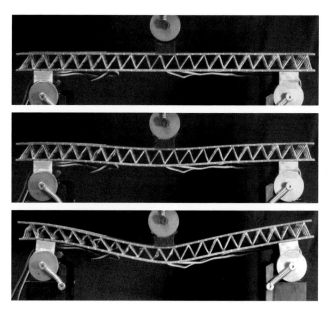

Fig. 8 A sequence of optical images of a tetragonal truss core panel subject to three point loading. The truss core has a relative density, $\bar{\rho} \approx 0.03$, close to that for the optimised panel. Note the asymmetry, especially apparent in (c). This happens as a consequence of plastic buckling on the left side.

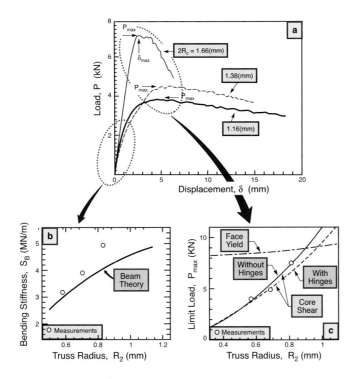

Fig. 9 The response characteristics of tetragonal truss core panels of the type depicted on Fig. 8 showing: (a) load/deflection curves, (b) a comparison between the measured and theoretically predicted stiffness and (c) the measured and predicted limit loads. Note that the measured stiffness is slightly larger than the theoretical values.

the formulation from Basic Topologies section.[23] The shear modulus and the in-plane shear strength are essentially the same as those expected analytically, with negligible knock-down effect (Fig. 9). This is true despite a significant proportion of defective nodes, embodying casting porosity. This robustness arises due to efficient load redistribution in this redundant structure. Note the asymmetric response on the two sides of the configuration (Fig. 8). On the right side, the trusses subject to the largest stress are in tension. As already noted, these trusses strain harden and sustain load. On the left side, the most highly stressed trusses are in compression. Once they yield, they buckle plastically and then soften. At this stage, the deformations occur preferentially on the left side, resulting in the observed asymmetry.[23] The full bending response of a panel designed to be close to the optimum is shown in Fig. 10. The numerical simulation is also shown. The simulation agrees well with the elastic stiffness and the limit load, but underestimates the strain hardening, for reasons not yet understood. It is concluded that the theoretical

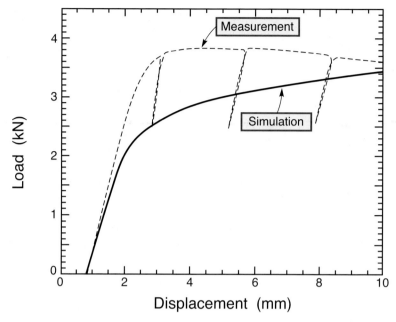

Fig. 10 A comparison of measured and simulated load/deflection curves for the panel shown on Fig. 8.

results summarised in Basic Topologies section may be used as a conservative basis for establishing minimum weight designs.

The limitations of the investment casting approach include: a relatively high fabrication cost and the limited range of lightweight casting alloys having high strength and adequate ductility. To obviate these limitations, two alternative fabrication approaches have been devised:[24,29,31] one based on textile technology, the other on stamping and hot forming. These alternatives have two clear benefits: lower fabrication cost plus the inherently superior properties of wrought alloys. On a relative basis, their performance is not as good as that for cast systems, because of manufacturing imperfections and anisotropy. On an absolute basis, the load capacity is excellent, because of the high yield strengths achievable in wrought alloys (large σ_Y in eqn 2) and the considerable ductility. The cores constructed from wrought wires with plain weaves illustrate the performance characteristics and also provide a visual depiction of the influence of topology.[24,29] The panels with square weaves have core members susceptible to bending. They fail by core shear with inferior load capacity (Fig. 11a). Panels with the same weave in the diamond orientation have a much lower susceptibility to truss bending. Accordingly, they have a high resistance to core shear and fail by face yielding with superior load capacity (Fig. 11b). Beyond distinguishing bending and stretch dominated

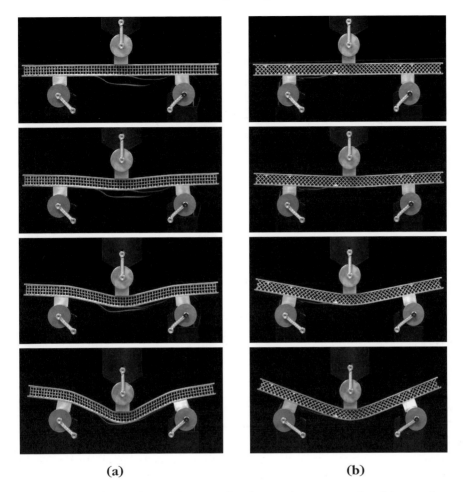

Fig. 11 Sequences of optical images for the textile core panels subject to three point loading. (a) The square core fails by core shear with plastic hinges at the inner and outer platens (refer to Fig. 7). (b) The diamond topology does not fail in core shear. Instead, the face sheets fail by stretching and bending beneath the inner platen.

core designs, the measurements indicate a relatively small (0.7) knock-down factor caused by weaving imperfections.[29] At lower densities and for other core failure modes, the degradation may be larger. Further studies are needed to establish the performance envelope.

The various truss designs and manufacturing concepts embody a cost/performance trade-off.[1–4] A value assessment is needed. Among the systems considered, the stochastic cores have demonstrably lower performance, but also lowest cost. Conversely, the truss systems with tetragonal and pyramidal cores made by investment casting have the best isotropic performance, but the highest manufacturing cost.[1] The trade-off between these concepts is application specific, and value sensitive.[2,4] The textile core materials and those based on plastic forming methods are intermediate, and while presently indeterminate from a value perspective, appear to have greatest promise. The exceptional performance of the plain weave core in the diamond configuration is particularly encouraging. This is the lightest core yet investigated, albeit within a limited design space explored experimentally.

Conclusion

This paper gives a brief overview of the opportunities provided by applying topology concepts, in conjunction with fabrication, to create new classes of robust, lightweight panels from structural alloys. Topologies have been identified that allow the truss members to resist bending, resulting in minimum possible weight. Models have been devised that relate the overall performance of the panel to the architecture of the core and to the properties of the constituent alloys. The models have been validated through measurements of the load capacity in bending, conducted on panels fabricated by investment casting and by textile-based approaches.

Metrics have been presented that allow competing concepts to be compared and, consequently, permit the specific weight benefits of the new systems to be defined.

References

1. M. F. Ashby, A. G. Evans, N. A. Fleck, L. J. Gibson, J. W. Hutchinson and H. N. G. Wadley, *Metal Foams: A Design Guide*, Butterworth-Heinemann, Boston, 2000.
2. M. F. Ashby, *Materials Selection and Mechanical Design*, 2nd edn, Butterworth-Heinemann, Oxford, 1999.
3. J. P. Clark, R. Roth and F. R. Field, Techno-Economic issues in material science, in *ASM Handbook*, **20**, Materials Science and Design, ASM International, 1997.
4. M. F. Ashby, *Acta Mater.*, 2000, **48**(1), 359–369.
5. H. G. Allen, *Analysis and Design of Structural Sandwich Panels*. Pergamon Press, Oxford, 1969.
6. B. L. Agarwal and L. H. Sobel, *AIAA J.*, 1977, **14**, 1000–1008.
7. W. T. Koiter *Koninkl. Nederl. Akademie van Wetenschappen. Ser. B*, 1963, **66**, 265–279.
8. J. W. Hutchinson, *Adv. Appl. Mech.*, 1974, **14**, 67–144.
9. B. Budiansky and J. W. Hutchinson (eds), *Buckling of Circular Cylindrical Shells Under Axial Compression, Contributions to the Theory of Aircraft Structures*, Delft University Press, Netherlands, 1972, 239–260.

10. B. Budiansky, *Int. J. Solids Struc.*, 1999, **36,** 3677–3708.
11. E. J. Catchpole, *J. Royal Aeronautical Society*, 1954, **58,** 765–768.
12. D. J. Farrar, *J. Royal Aeronautical Society*, 1949, **53,** 1041–1052.
13. G. Gerard, *Minimum Weight Analysis of Compression Structures*, New York University Press, New York, 1956.
14. V. Tvergaard, *Int. J. Solids Struct.*, 1973, **9,** 177–192.
15. P. M. Weaver and M. F. Ashby, *Progress in Materials Science*, 1997, **41,** 61–128.
16. J. G. Parkhouse, Structuring: A Process of Material Distribution, in *Proc. 3rd Int. Conf. On Space Structures*, H. Nooschin (ed.), Elsevier, London, 1984, 367–374.
17. P. M. Weaver and M. F. Ashby, *Journal of Engineering Design*, 1996, **7,** 129.
18. N. Wicks and J. W. Hutchinson, *Int. J. Solids .Struct.*, 2001, **38,** 5165–5183.
19. A. G. Evans, J. W. Hutchinson, N. A. Fleck, M. F. Ashby and H. N. G. Wadley, *Progress in Materials Science*, 2001, **46,** 309–327.
20. V. K. Deshpande and N. A. Fleck, *Int. J. Solids Struct.*, in press.
21. R. B. Fuller, U.S. Patent, 2,986,241, 1961.
22. A. G. Evans, J. W. Hutchinson and M. F. Ashby, *Progress in Materials Science*, 1999, **43,** 171–221.
23. S. Chiras, D. R. Mumm, A. G. Evans, N. Wicks, J. W. Hutchinson, K. Dharmasena, H. N. G. Wadley and S. Fichter, *Int. J. Solids Struct.*, in press.
24. D. J. Sypeck and H. N. G. Wadley, *J. Mater. Res.*, 2001, **16,** 890–897.
25. J. C. Wallach and L. G. Gibson, *Int. J. Solids Struct,.* submitted.
26. H. Bart-Smith, J. W. Hutchinson, N. A. Fleck, and A. G. Evans, *Int. J. Solids Struct.*, to be published.
27. L. J. Gibson and M. F. Ashby, *Cellular Solids: Structure and Properties*, 2nd edn, Cambridge University Press, Cambridge, 1997.
28. S. Hyun and S. Torquato, *J. Mater. Res.* 2001, **16,** 280.
29. D. R. Mumm, S. Chiras, A. G. Evans, J. W. Hutchinson, D. J. Sypeck, and H. N. G. Wadley, *Acta Mater.*, to be published.
30. N. A. Fleck, personal communication.
31. H. N. G. Wadley and D. J. Sypeck, personal communication.
32. J. W. Hutchinson and M. Y. He, *Int. J. Solids Struct.*, 2000, **37,** 6777–6794.

The Engineering of Cells and Tissues

WILLIAM BONFIELD
Professor of Medical Materials, Department of Materials Science and Metallurgy, University of Cambridge

ABSTRACT

The twin demands of an ageing population and the need to treat younger patients have driven research on a second generation of biomaterials for medical implants, prostheses and devices, with emphasis on an enhancement of implant lifetime *in vivo*. Particular attention has been given to the innovation of biomaterials which mimic the biological template, such as the laboratory synthesis of hydroxyapatite (HA) (a major constituent of bone), which has produced improved skeletal-defect filling and cementless fixation of total hip joint prostheses. Another advance has been from the development of hydroxyapatite reinforced polyethylene composite (HAPEX) as a bone analogue. Following a successful clinical application as a middle ear prosthesis. Derivative implants based on this technology are under development for a range of clinical applications, including maxilla-facial reconstruction, spinal implants and revision-hip prostheses.

HA and HAPEX are examples of bioactive materials in which the implant surface is designed to provide a favourable site for the recruitment of cells and biological factors *in vivo* from the adjacent bone (*in situ* tissue engineering). An alternative approach more suited to soft tissues, such as skin, cartilage, ligament and tendon, is to recruit and multiply the cells externally in culture, followed by incorporation into a suitable scaffold and then reimplantation.

Such tailor-made biomaterials and tissue-engineered substitutes provide the basis for a spectrum of new clinical procedures with the potential to improve the quality of life for many patients in the coming years.

A Genetic Basis for Biomedical Materials

LARRY L. HENCH, IOANNIS D. XYNOS, ALASDAIR J. EDGAR,
LEE D. K. BUTTERY and JULIA M. POLAK
Department of Materials and the Tissue Engineering Centre,
Imperial College of Science, Technology and Medicine, University of London, UK

Introduction

Most of the materials used in medical applications have been developed by trial and error. Structural stability under physiological conditions has usually been the primary criterion of success. This approach has been successful from the 1960s to the present and has resulted in the development of different materials for replacement of 40 different parts of the body. However, many people now outlive their replacement parts, called implants or prostheses, require revision surgery and two, three or even four devices during their lifetime. This is enormously expensive and painful. For our progressively ageing population an alternative to replacement of tissues with non-living parts must be found. This paper describes such an alternative based upon gene activation.

Recent results from the Imperial College Tissue Engineering Centre and the Department of Materials suggest that a new paradigm is possible for repairing or replacing diseased or damaged parts of the human body. It is our hypothesis that bioactive materials can be used to activate genes that stimulate regeneration of living tissues. Two alternative routes of repair are available in this new paradigm.

1. Tissue Engineering. In this approach progenitor cells are seeded on to bioactive resorbable scaffolds outside the body where the cells grow and become differentiated; e.g. specialised, to mimic naturally occurring tissues. The tissue engineered constructs are then implanted from the culture vessels, called bioreactors, into the patient to replace the diseased or damaged tissues. With time the scaffolds are resorbed and replaced by living tissues that include a viable blood supply and nerves. The living tissue engineering constructs adapt to the physiological environment and can provide a repair with long survivability.
2. *In Situ* Tissue Regeneration. This approach involves use of bioactive materials in the form of powders, fibres or solutions to stimulate localised tissue repair. The bioactive materials release inorganic chemicals in the form of ionic dissolution products at controlled rates

that activate the genes of the cells in contact with the bioactive stimuli. The cells produce growth factors that in turn stimulate multiple generations of growing cells to self-assemble into the required tissues *in situ*, under the biochemical and biomechanical gradients that are present.

The advantage offered by both new routes is genetic control of the tissue repair process. The result is ideally equivalent to natural tissues in that they are adaptable to the physiological environment, unlike all the biomedical materials presently used clinically.

Until 1969 it was accepted by the medical community that placing any man-made material in the body would result in a foreign body reaction and formation of non-adherent scar tissue at the interface with the material. Thus, the emphasis in biomaterials research and clinical application was on materials that were as inert as possible when exposed to a physiological environment. This understanding was irreversibly altered when a special composition of soda-lime-phosphate-silicate glass was synthesised by the lead author and implanted in the femurs of rats.[1,2,3] This glass composition contained only 45% SiO_2, in weight %. The network modifiers were 24.5% Na_2O and 24.5% CaO. In addition 6% P_2O_5 was added to simulate the Ca/P constituents of hydroxyapatite, the inorganic mineral phase of bone. The glasses did not cause interfacial scar tissue. Instead, they bonded to the living bone and could not be removed from their implant site.[1] This discovery led to a new class of materials, called *bioactive* materials, for use in implants or prostheses and repair or replacement of bones, joints and teeth.

Bioactive materials, including bioactive glasses[1-12] and glass-ceramics,[13-19] form a mechanically strong bond with bone. The bond to bone occurs at different rates depending upon composition of the material.[1-3,7,15-17,20-25] The glass compositions with the fastest rates of bone bonding also bond to soft tissues.[12,26]

Bioactive materials are used as bulk implants to replace bones or teeth,[27-35] coatings to anchor orthopaedic or dental devices[6] or in the form of powders to fill various types of bone defects.[28,36,37] When a particulate of bioactive glass, ceramic, or glass-ceramic is used to fill a bone defect, both the rate and quantity of bone regeneration depend on composition.[38]

More bone formed in one week in the presence of bioactive glass 45S5 than is formed when synthetic hydroxyapatite (HA) or other calcium-phosphate ceramic particulates are placed in the same type of defect. After several weeks of repair there is almost twice as much new bone in the defect containing bioactive glass. The amount of bone that is regenerated matches that originally present in the site. The architecture of the trabecular bone is also equivalent to the original bone[39] and the mechanical properties have been restored by the regenerated bone.[40] These large differences in rates of *in vivo* bone regeneration and extent of bone repair indicate that there are two classes of bioactive materials (Table 1).[38,41]

Class A bioactivity leads to both osteoconduction and osteoproduction[36,38] as a consequence of rapid reactions on the bioactive glass surface.[1,4,6,22–24,42–47] The surface reactions involve ionic dissolution of critical concentrations of soluble Si, Ca, P and Na ions that give rise to both intracellular and extracellular responses at the interface of the glass with its physiological environment.

Class B bioactivity occurs when only osteoconduction is present; i.e. bone migration along an interface, due to slower surface reactions, minimal ionic release and only extracellular responses occuring at the interface.[38]

Details are reviewed in references 6, 38 and 41. Compositions of several Class A and Class B bioactive materials are given in Table 1.

Glass Surface Reactions

The first five reaction stages shown in Fig. 1 occur at the surface of a Class A bioactive glass very rapidly and go to completion within 24 hours. The effect is rapid release of soluble ionic species and formation of a high surface area

Table 1 Composition and Properties of Bioactive Glasses and Glass-Ceramics Used Clinically.

Composition (wt%)	45S5 Bioglass	S53P4 (AbminDent 1)	A-W Glass-ceramic (Cerabone)
Na_2O	24.5	23	0
CaO	24.5	20	44.7
CaF_2	0	0	0.5
MgO	0	0	4.6
P_2O_5	6	4	16.2
SiO_2	45	53	34
Class of Bioactivity	A	B	B
Phases	Glass	Glass	Apatite Beta-wollastonite Glass
Density (g/cc)	2.7		3.07
Vickers Hardness (HV)	458 ± 9		680
Compressive Strength (MPa)			1080
Bending Strength (MPa)	42		215
Young's modulus (GPa)	35		218
Fracture toughness (MPam 1/2)			2
Slow crack growth (n) (unitless)			33

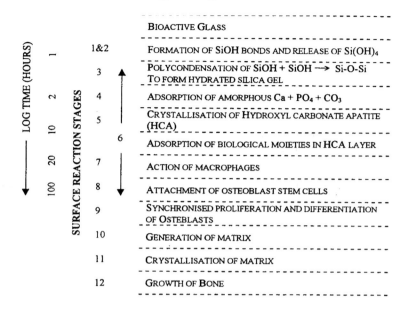

Fig. 1 Time dependence (log scale) of sequence of surface reaction stages (1–5) and biological reaction stages (6–11) of a Class A bioactive glass (45S5 Bioglass®) in bone.

hydrated silica and polycrystalline hydroxy carbonate apatite (HCA) bi-layer on the glass surface (Stages 1–5).[42–45] The reaction layers enhance adsorption and desorption of growth factors (Stage 6) and influence the length of time macrophages are required to prepare the implant site for tissue repair (Stage 7).[7] Attachment (Stage 8) and synchronised proliferation and differentiation of the cells that grow bone, osteoblasts, (Stage 9) rapidly occurs on the surface of Class A bioactive materials.[48–51] Several weeks are required for similar cellular events to occur on the surface of Class B bioactive materials. Osteoprogenitor cells, cells capable of forming new bone, colonise the surface of Class A bioactive materials within 48 hours and begin production of various growth factors which stimulate cell division, mitosis, and production of extracellular matrix proteins, (Stage 10). Mineralisation of the matrix follows soon thereafter and mature osteocytes, encased in a collagen-HCA matrix, are the final product by 6–12 days *in vitro* and *in vivo*.[48–49]

Control of Bone Cell Cycle

Formation of a surface HCA layer (Stages 1–5) is a useful but not the critical stage of reaction for bone regeneration. Xynos *et al.* compared the effect of Class A bioactive glass (45S5 Bioglass®) vs. a bioinert control (Thermanox®

plastic) on the cell cycle of human osteoblasts (hOBs).[48–49] The hOB cells were primary cultures obtained from excised femoral heads from patients aged 50–70 years, undergoing total hip replacements. Details are given in references 48 to 51. Various assays were used to quantify the percentages of cells in specific segments of the cell cycle.

In order for new bone to form it is essential for osteoprogenitor cells to divide, a process called cell division or mitosis. There are very few cells in the bones of older people that are capable of dividing and forming new bone. The osteoprogenitor cells that are present must receive the correct chemical stimuli from their local environment that instruct them to enter the active segments of the cell cycle. Figure 2 summarises the sequence of cellular events that comprise a cell cycle. Every new cell cycle begins after a cell has completed the preceding division into two daughter cells.

During step 1 in Fig. 2, called the G1 phase, the cell grows and metabolises normally.[49] Osteoblasts synthesise phenotype specific cellular products including tropocollagen macromolecules, which self assemble into type I collagen, the predominant collagen molecule present in the bone matrix. This process is listed in Fig. 1 as Stage 9.

If the local chemical environment is suitable, and following a critical period of growth in the G1 phase, the cell enters the S phase (step 2 in Fig. 2), when DNA synthesis begins. The S phase eventually leads to duplication of all the chromosomes in the nucleus. Next (step 3 in Fig. 2) the cell is ready to undergo division with a second phase of growth termed the G2 phase. During G2 the cell checks its replication accuracy using DNA repair enzymes. A critical increase in mass and synthesis and activation of various growth factors is necessary for the G2-M transition. Details are reviewed in ref. 49. If the local chemical environment does not lead to the full completion of either

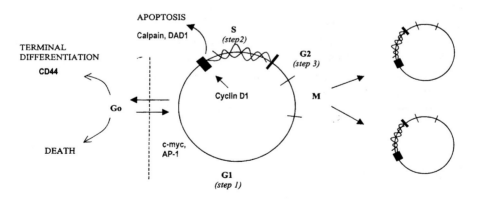

Fig. 2 Eukaryotic cell cycle leading to terminal differentiation, apoptosis (left) or mitosis (right) to create two daughter cells.

the G1 phase or the G2 phase then the cell proceeds to a programmed cell death, called apoptosis, also shown in Fig. 2.

Bioinert materials or Class B bioactive materials do not produce the local chemical environment to enable the few osteoprogenitor cells present in the tissues of older people to pass through these cell cycle checkpoints. Thus, only Class A bioactive materials produce rapid new bone formation *in vivo*.

The life dependent consequence of the checkpoints in the osteoblast cell cycle is the formation of two daughter cells, each nucleus receives a complete and equivalent complement of genetic material. However, the checkpoints in the cell cycle also mean that as we age we produce progressively fewer and fewer progenitor cells that can enter into the M phase unscathed. Our built-in protection from multiplication of damaged genes means that fewer cells are available to repair dying bone cells. The cumulative effect is a progressive decrease in bone density with age.

Effect of Bioactive Glass on Cell Cycle

Figure 3 is a summary of the differences in surface chemical, cell biology and tissue response of Class A bioactive materials vs. Class B bioactive materials. The rapid release of soluble inorganic species, especially hydrated silica and calcium ions, gives rise to rapid nucleation and crystallisation of an amorphous calcium-phosphate layer to form a polycrystalline HCA layer (Fig. 3a). The high surface area HCA layer provides many binding sites for osteoprogenitor cells. Xynos *et al*. showed that as early as day two osteoblasts grown on the bioactive glasses were maintained in a shape characteristic of cell activation.[48] In contrast, at day two osteoblasts grown on a bioinert substrate had flattened, divided and started to form confluent sheets of cells. Such behaviour is characteristic of cells in culture that tend not to differentiate.

The percentage of osteoblasts that are in the synthesis (S) phase of the cell cycle at day two shows a dramatic 100% increase for cells growing on the bioactive glass substrate (Fig. 3b). Likewise, the percentage of osteoblasts that are in the G2-M phase on the bioactive substrate is increases more than 100% over the control substrate.[48] The largest population shift occurs for cells in apoptosis. At two days, five times as many osteoblasts are apoptotic when grown on bioactive glasses. The cells capable of becoming mature bone are favoured to survive the first few days of growth on the bioactive glass substrates.

These shifts in the population of cells in various stages of the cell cycle are very important. They lead to enhanced growth of new bone in the presence of Class A bioactive glasses. Osteoblasts exposed to bioinert substrates quickly

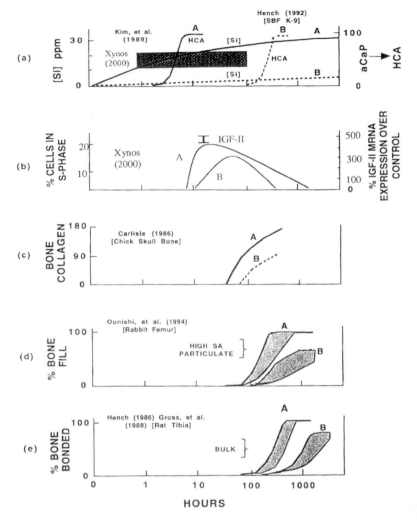

Fig. 3 Time dependence of materials reactions, and cellular and tissue responses to Class A and Class B bioactive materials.

attach and enter into a cell cycle that leads to formation of a confluent sheet of cells that are seldom capable of differentiating and forming bone matrix. They are much closer in phenotype to the cells of scar tissue typically found at the interface of bioinert materials and Class B bioactive materials at the earliest time of implantation.

The importance of the bioactive control of the osteoblast cell cycle becomes apparent by six days in culture, just as it does at six days *in vivo*. Scanning electron microscopy (SEM) analysis of the human osteoblast cultures shows that osteoblasts growing on the bioactive substrate for six days have already organised, in a process called self-assembly, into a three-dimensional structure composed of cells and mineralised extracellular matrix.[48] This is called a bone nodule and it has an organisational complexity similar to natural bone grown *in vivo*, although without a blood supply. The time for formation of collagen on bioactive substrates *in vitro* is similar to the kinetics of collagen formation *in vivo* summarised in Fig. 3c. The rate of forming mineralised bone nodules *in vitro* is similar to the kinetics of bone growth *in vivo* summarised in Fig. 3d,e.

The results confirm that human osteoblasts growing in culture, in the presence of a bioactive glass self-assemble into a three-dimensional architecture and create a mineralised matrix that is characteristic of mature osteocytes in living bone. In order for this architecture to be created by the osteoblasts there must be release of critical concentrations of the soluble ionic constituents of the bioactive glass. Approximately 17 ppm of soluble Si and 88 ppm of soluble Ca ions are required. The ions can be provided by controlled dissolution of a bioactive glass substrate. It is also possible to react bioactive glass powders in tissue culture medium and create the critical concentrations of soluble inorganic ions in the medium. When osteoblasts are grown in this ionically conditioned medium they differentiate and form a mineralised extracellular matrix and create bone nodules.

At six days osteoblasts grown on bioinert substrates show no evidence of formation of bone nodules.

At longer times the number of bone nodules growing on the bioactive substrates continues to increase and the organisation of the nodules becomes increasingly more complex with large numbers of osteocytes within them. At twelve days there are still no bone nodules present on the bioinert substrates although the osteoblast-like cells are still healthy. The cellular markers suggest that the cells growing on the inert substrates are not capable of forming new mineralised bone, but are more similar to the fibroblast-like cells found in scar tissues.

Genetic Control of the Osteoblast Cell Cycle

Recent findings by Xynos *et al.* have shown that the bioactive shift of osteoblast cell cycle described above is under genetic control.[50,51] Within a few hours exposure of human primary osteoblasts to the soluble chemical extracts of 45S5 Bioglass, several families of genes are activated including genes encoding nuclear transcription factors and potent growth factors, especially IGF-II, along with IGF binding proteins and proteases that cleave IGF-II from their binding proteins. Figure 3b shows that there is a 200 to 500%

increase in the expression of these genes over those of the control cultures. Activation of several immediate early response genes and synthesis of growth factors is likely to modulate the cell cycle response of osteoblasts to Bioglass®, as shown in Fig. 3. These findings indicate that Class A bioactive glasses enhance new bone formation (osteogenesis) through a direct control over genes that regulate cell cycle induction and progression.

The cell cycle represents a highly synthetic phase in the life of an osteoblast. During this phase the cell produces proteins and nucleic acids that eventually result in the formation of two daughter cells, as illustrated in Fig. 2. Mistakes in the synthesis of proteins and nucleic acids are quite likely, especially in the mitosis of progenitor cells of older people. In order to avoid such mistakes being passed on during cell division the cell possesses an arsenal of mechanisms that can determine whether damage is present, evaluate its extent and correct it.

The upregulation of DNA repair proteins by the ionic products of bioactive glass dissolution indicates that these mechanisms are activated in human osteoblasts. At least four important genes involved in DNA synthesis, repair and recombination are differentially expressed at levels of >200% over control osteoblasts. When the damage is beyond repair the cell voluntarily exits the mitotic cell cycle through death by apoptosis, programmed cell death. Apoptosis thereby prevents the creation of abnormal cells and represents a means to regulate the selection and proliferation of functional osteoblasts; i.e. osteoblasts capable of synthesising the complex array of extracellular proteins and mucopolysaccharides required to form a mineralised matrix that is characteristic of mature osteocytes. The treatment of the osteoblast cultures with the bioactive glass stimuli induced the expression of several important genes involved in apoptosis.

As discussed earlier, the cell cycle does not merely provide the framework for cell proliferation but also determines to some extent cell commitment and differentiation. Bone cells cover a broad spectrum of phenotypes that include predominately the osteoblast, a cell capable of proliferating and synthesising bone cell specific products such as Type I collagen. However, a vital cellular population in bone consists of osteocytes. Osteocytes are terminally differentiated osteoblasts and are usually not capable of cell division. They are capable of synthesising and maintaining the mineralised bone matrix wherein they reside. Thus, osteocytes represent the cell population responsible for extracellular matrix production and mineralisation, the final step in bone development (Fig. 1 Stages 10–12) and probably the most crucial one given the importance of collagen-hydroxyl carbonate apatite (HCA) bonding in determining the mechanical function of bone. Therefore, it is important to observe that the end result of the cell cycle activated by the ionic products of bioactive glass dissolution is the upregulation of numerous genes that express growth factors and cytokines and extracellular matrix components.

It is important to determine whether the bioactive glass activation leads to a significant increase of the secretion of the IGF-II protein by the cells and whether it is present in a biologically active state. Xynos et al.[50] confirmed the IGF-II mRNA upregulation using quantitative real time PCR and also showed that the unbound IGF-II protein concentration was increased. The results indicate that the ionic dissolution products of Bioglass 45S5 may increase IGF-II availability in osteoblasts by inducing the transcription of the growth factor and its carrier protein and also by regulating the dissociation of this factor from its binding protein. The unbound IGF-II is likely to be responsible for the increase in cell proliferation observed in the cultures. Similar bioactive induction of the transcription of extracellular matrix components and their secretion and self-organisation into a mineralised matrix is responsible for the rapid formation and growth of bone nodules and differentiation of the mature osteocyte phenotype.

Conclusions

Thirty years ago it was discovered that bioactive glasses bond to bone. We now know that the same glass compositions will lead to regeneration and repair of bone in both young and old people. This unique material has been used clinically for more than 15 years with many thousands of successful cases. Research has shown that the mechanisms of bone bonding and bone regeneration and repair (osteogenesis) involve rapid ion exchange reactions on the glass surface, nucleation and growth of biologically active surface reaction layers and release of critical concentrations of ionic dissolution products composed of soluble silicon and calcium ions.

The molecular biological mechanisms involved in the behaviour of bioactive glasses are finally beginning to be understood. The bioactive response appears to be under genetic control. Class A bioactive glasses that are osteoproductive enhance osteogenesis through a direct control over genes that regulate cell cycle induction and progression. Cells that are not capable of forming new bone are eliminated from the cell population, a response that is missing when osteoblasts are exposed to bioinert or Class B bioactive materials. The biological consequence of genetic control of the cell cycle of osteoblast progenitor cells is the rapid proliferation and differentiation of osteoblasts. The result is rapid regeneration of bone. The clinical consequence is rapid fill of bone defects with regenerated bone that is structurally and mechanically equivalent to normal, healthy bone.

Understanding the genetic basis for the reactions of bioactive glasses provides an important opportunity for glass research. It should now be feasible to design a new generation of gene-activating glasses tailored for specific patients and disease states. The new generation of gene activating glasses can also be fabricated into bioactive resorbable scaffolds for tissue engineering of bone constructs for patients with large bone defects. Perhaps of even more importance is the possibility that bioactive ionic dissolution products can be used to activate genes in a preventative treatment to maintain the health of our bones as we age. Only a few years ago this concept would have seemed impossible. We need to remember that only thirty years ago the concept of a material that would not be rejected by living tissues was considered to be impossible. If we can activate genes by use of glasses it is certainly possible that we may one day be able to use glasses to control genes.

Acknowledgements

The authors gratefully acknowledge the support of the UK Engineering and Physics Research Council, the UK Medical Research Council, US Biomaterials Corp., the Golden Charitable Trust, and the Julia Polak Transplant Fund, and editing by Dr. June Wilson Hench who has helped for twenty five years in developing clinical applications of bioactive glasses.

References

1. L. L. Hench, R. J. Splinter, W. C. Allen and T. K. Greenlee, Jr., Bonding Mechanisms at the Interface of Ceramic Prosthetic Materials, *J. Biomed. Mater. Res.*, 1971, **2**(1), 117–141.
2. C. A. Beckham, T. K. Greenlee, Jr. and A. R. Crebo, Bone Formation at a Ceramic Implant Interface, *Calc. Tiss. Res.*, 1971, **8**, 165–171.
3. T. K. Greenlee, Jr., C. A. Beckman, A. R. Crebo and J. C. Malmborg, Glass Ceramic Bone Implants, *J. Biomed. Mater. Res.*, 1972, **6**, 235–244.
4. L. L. Hench and E. C. Ethridge, *Biomaterials: An Interfacial Approach*, Academic Press, New York, 1982.
5. S. F. Hulbert, J. C. Bokros, L. L. Hench, J. Wilson and G. Heimke, Ceramics in Clinical Applications, Past, Present and Future, in *High Tech Ceramics*, P. Vincenzini (ed.), Elsevier Science Pub. B.V., Amsterdam, 1987, pp. 189–213.
6. L. L. Hench and June Wilson, *An Introduction to Bioceramics*, World Scientific, London, 1993.
7. U. Gross, R. Kinne, H. J. Schmitz and V. Strunz, The Response of Bone to Surface Active Glass/Glass-Ceramics, *CRC Critical Reviews in Biocompatibility*, D. Williams (ed.), 1988, **4**, 2.

8. L. L. Hench, Bioceramics: From Concept to Clinic, *J.Am. Ceram. Soc.*, 1991, **74**(7), 1487–1510.
9. L. L. Hench, Bioactive Ceramics, in *Bioceramics: Materials Characteristics Versus In Vivo Behavior*, Vol. 523. P. Ducheyne and J. Lemons (eds), Annals N.Y. Acad. Sci., 1988, 54.
10. L. L. Hench and J. W. Wilson, Surface-Active Biomaterials, *Science*, 1984, **226**, 630.
11. U. Gross and V. Strunz, The Interface of Various Glasses and Glass-Ceramics with a Bony Implantation Bed, *J. Biomed. Mater. Res.*, 1985, **19**, 251.
12. J. Wilson, G. H. Pigott, F. J. Schoen and L. L. Hench, Toxicology and Biocompatibility of Bioglass, *J. Biomed. Mater. Res.*, 1981, **15**, 805.
13. T. Nakamura, T. Yamamuro, S. Higashi, T. Kokubo and S. Itoo, A New Glass-Ceramic for Bone Replacement: Evaluation of its Bonding to Bone Tissue, *J. Biomed. Mater. Res.*, 1985, **19**, 685.
14. T. Kokubo, S. Ito, S. Sakka and T. Yamamuro, Formation of a High-Strength Bioactive Glass- Ceramic in the System $MgO–CaO–SiO_2$-P_2O_5, *J. Mater. Sci.*, 1986, **21**, 536.
15. T. Kitsugi, T. Yamamuro and T. Kokubo, Bonding Behavior of a Glass-Ceramic Containing Apatite and Wollastonite in Segmental Replacement of the Rabbit Tibia Under Load-Bearing Conditions, *J. Bone Jt. Surg.*, 1989, **71A**, 264.
16. S. Yoshii, Y. Kakutani, T. Yamamuro, T. Nakamura, T. Kitsugi, M. Oka, T. Kokubo and M. Takagi, Strength of Bonding Between A-W Glass Ceramic and the Surface of Bone Cortex, *J. Biomed. Mater. Res.*, 1988, **22A**, 327.
17. T. Yamamuro, J. Shikata, Y. Kakutani, S. Yoshii, T. Kitsugi and K. Ono, Novel Methods for Clinical Applications of Bioactive Ceramics, in *Bioceramics: Material Characteristics Versus In Vivo Behavior*, P. Ducheyne and J. E. Lemons (eds), New York Academy of Science, New York, 1988, 107.
18. T. Yamamuro, L. L. Hench and J. Wilson (eds), *Handbook on Bioactive Ceramics: Bioactive Glasses and Glass-Ceramics*, Vol. I. CRC Press, Boca Raton, FL, 1990.
19. L. L. Hench and J. K. West, Biological Applications of Bioactive Glasses, *Life Chemistry Reports*, 1996, **13**, 187–241.
20. L. L. Hench and H. A. Paschall, Direct Chemical Bonding Between Bio-Active Glass-Ceramic Materials and Bone, *J Biomed. Mater. Res. Symp.*, 1973, **4**, 25–42.
21. L. L. Hench and A. E. Clark, Adhesion to Bone; Chap. 6 in *Biocompatibility of Orthopedic Implants*, Vol. 2, D. F. Williams (ed.), CRC Press, Boca Raton, FL, 1982.
22. Ö. H. Andersson, G. Liu, K. H. Karlsson, L. Niemi, J. Miettinen and J. Juhanoja, In Vivo Behavior of Glasses in the SiO_2–Na_2O–CaO–P_2O_5–Al_2O_3–B_2O_3 System, *J. Materials Sci., Materials in Medicine*, 1990.
23. U. M. Gross and V. Strunz, The Anchoring of Glass Ceramics of Different Solubility in the Femur of the Rat, *J. Biomed. Mater. Res.*, 1980, **14**, 607.
24. U. Gross, J. Brandes, V. Strunz, J. Bab and J. Sela, The Ultrastructure of the Interface Between a Glass Ceramic and Bone, *J. Biomed. Mater. Res.*, 1981, **15**, 291.
25. K. Kangasniemi and A. Yli-Urpo, Biological Response to Glasses in the SiO_2–Na_2O–CaO–P_2O_5–B_2O_3 System, in *Handbook of Bioactive Ceramics*, Vol. I. T. Yamamuro, L. L. Hench and J. Wilson (eds), CRC Press, Boca Raton, FL, 1990, 97–108.

26. June Wilson and D. Nolletti, Bonding of Soft Tissues to Bioglass, in *Handbook of Bioactive Ceramics*, Vol. 1, T. Yamamuro, L. L. Hench, and J. Wilson (eds), CRC Press, Boca Raton, FL, 1990, 283–302.
27. J. Wilson, A. E. Clark, E. Douek, J. Krieger, W. K. Smith and J. S. Zamet, Clinical Applications of Bioglass® Implants, in *Bioceramics 7,* Andersson, Happonen, Yli-Urpo (eds), Butterworth-Heinemann, Oxford, 1994, 415.
28. C. A. Shapoff, D. C. Alexander and A. E. Clark, Clinical Use of a Bioactive Glass Particulate in the Treatment of Human Osseous Defects, *Compendium Contin. Educ. Dent*, 1997, **18**(4), 352–363.
29. R. Reck, S. Storkel and A. Meyer, Bioactive Glass-Ceramics in Middle Ear Surgery: An 8-Year Review, in *Bioceramics: Materials Characteristics Versus In Vivo Behavior*, Vol. 523, P. Ducheyne and J. Lemons (eds), *Annals NY Acad. Sci.*, 1988, 100.
30. G. E. Merwin, Review of Bioactive Materials for Otologic and Maxillofacial Applications, in *Handbook of Bioactive Ceramics*, Vol I, T. Yamamuro, L. L. Hench and J. Wilson (eds), CRC Press, Boca Raton, FL, 1990, 323–328.
31. E. Douek, Otologic Applications of Bioglass® Implants, in *Proceedings of IVth International Symposium on Bioceramics in Medicine*, W. Bonfield (ed.), London, Sept. 10–11, 1991.
32. K. Lobel, Ossicular Replacement Prostheses, in *Clinical Performance of Skeletal Prostheses*, Hench and Wislon (eds), Chapman and Hall, Ltd, London, 1996, 214–236.
33. H. R. Stanley, M. B. Hall, A. E. Clark, J. C. King, L. L. Hench and J. J. Berte, Using 45S5 Bioglass® Cones as Endosseous Ridge Maintenance Implants to Prevent Alveolar Ridge Resorption – A 5 Year Evaluation, *Int. J. Oral Maxillofac. Implants*, 1997, **12**, 95–105.
34. H. R. Stanley, A. E. Clark and L. L. Hench, Alveolar Ridge Maintenance Implants, in *Clinical Performance of Skeletal Prostheses*, L. .L. Hench and J. Wislon (eds), Chapman and Hall, Ltd., London, 1996, 255–270.
35. T. Yamamuro, A/W Glass-Ceramic: Clinical Applications, Chap. 6 in ref. 6, p. 89.
36. J. Wilson and S. B. Low, Bioactive Ceramics for Periodontal Treatment: Comparative Studies in the Patus Monkey, *J. Appl. Biomaterials*, 1992, **3**, 123–169.
37. A. E. Fetner, M. S. Hartigan and S. B.Low, Periodontal Repair Using Perioglas® in Non-Human Primates: Clinical and Histologic Observations, *Compendium Contin. Educ. Dent.*, 1994, **15**(7), 932–939.
38. L. L. Hench, The Challenge of Orthopaedic Materials, *Current Orthopaedics*, 2000, **14**, 7–16.
39. H. Oonishi, L. L. Hench, J. Wilson, F. Sugihara, E. Tsuji, S. Kushitani and H. Iwaki, Comparative bone growth behaviour in granules of bioceramic materials of different particle sizes. *J. Biomed. Mater. Res.*, 1999, **44**(1), 31–43.
40. L. L. Hench, D. L. Wheeler and D. C. Greenspan, Molecular Control of Bioactivity in Sol-Gel Glasses, *J. Sol-Gel Sci. and Tech.*, 1998, **13**, 245–250.
41. L. L. Hench, Bioceramics., *J. Am. Ceram. Soc.*, 1998, **81**(7), 1705–1728.

42. M. Ogino, F. Ohuchi and L. L. Hench, Compositional Dependence of the Formation of Calcium Phosphate Films on Bioglass, *J. Biomed. Mater. Res.*, 1980, **14**, 55–64.
43. A. E. Clark, C. G. Pantano and L. L. Hench, Auger Spectroscopic Analysis of Bioglass Corrosion Films, *J. Am. Ceram. Soc.*, 1976, **59**(1–2), 37–39.
44. L. L. Hench, H. A. Paschall, W. C. Allen and G. Piotrowski, Interfacial Behavior of Ceramic Implants, *National Bureau of Standards Special Publication*, 1975, **415**, 19–35.
45. L. L. Hench and H. A. Paschall, Histo-Chemical Responses at a Biomaterials Interface, *J. Biomed. Mater. Res.*, 1974, **5**(1), 49–64.
46. T. Kokubo, Surface Chemistry of Bioactive Glass-Ceramics, *J. Non-Cryst. Solids*, 1990, **120**, 138–151.
47. T. Kokubo, Bonding Mechanism of Bioactive Glass-Ceramic A-W to Living Bone, in *Handbook of Bioactive Ceramics*, Vol I, T. Yamamuro, L. L. Hench and J. Wilson (eds), CRC Press, Boca Raton, FL, 1990, 41–50.
48. I. D.Xynos, M. V. J. Hukkanen, J. J. Batten, I. D. Buttery, L. L. Hench and J. M. Polak, Bioglass® 45S5 Stimulates Osteoblast Turnover and Enhances Bone Formation *In Vitro*: Implications and Applications for Bone Tissue Engineering, *Calcif. Tiss. Int.*, 2000, **67**, 321–329.
49. L. L. Hench, I. D. Xynos, L. D. Buttery and J. M. Polak, Bioactive Materials to Control Cell Cycle, *J. Mater. Res. Innovations*, 2000, **3**, 313–323.
50. I. D. Xynos, A. J. Edgar, L. D. Buttery, L. L. Hench and J. M. Polak, Ionic Products of Bioactive Glass Dissolution Increase Proliferation of Human Osteoblasts and Induce Insulin-like Growth Factor II mRNA expression and Protein Synthesis, *Biochem. and Biophys. Res. Comm.*, 2000, **276**, 461–465.
51. I. D. Xynos, A. J. Edgar, L. D. K. Buttery, L. L. Hench and J. M. Polak, Gene Expression Profiling of Human Osteoblasts Following Treatment with the Ionic Products of Bioglass® 45S5 Dissolution, *J. Biomed. Mater. Res*, 2001, **55**(2), 151–157.

Controversial Concepts in Alloy Theory Revisited*

D. G. PETTIFOR

Department of Materials, University of Oxford, Parks Road, Oxford OX1 3PH, UK

ABSTRACT

Hume-Rothery strongly believed that electron theory would help industrial metallurgists develop new and better alloys by providing concepts that were underpinned by quantum mechanics rather than empiricism. This talk will focus on three concepts that aroused confusion and controversy amongst academics in Hume-Rothery's day: firstly, the relevance of Jones' theory of Brillouin zone touching to the experimental Hume-Rothery electron per atom rule; secondly, the relevance of the Engel-Brewer theory to the structural stability of metals and alloys; and thirdly, the relevance of Pauling's theory of resonant bonds to transition metal cohesion. I will end by asking whether the advent of quantitative electron theory will indeed help industrialists design new and better alloys in the twenty-first century.

Introduction

'... if metallurgists of this country do not learn about the electron theories they may discover, too late, that the industry in other countries has benefited from knowledge of modern theories and is developing new alloys.'[1]

Hume-Rothery was a great believer that electron theory would place the metallurgical industry on firm scientific foundations just as the fundamental principles of chemistry had underpinned the chemical industry.[2] This feeling is perhaps most eloquently expressed in the quotation above from H-R's young protegé, Geoffrey Raynor, who had been assigned the task by the UK's Institute of Metals of presenting H-R's *Atomic Theory for Students of Metallurgy*[3] in a language understandable to the 'older metallurgist'. Alas, as is well known, this post war optimism in the utility of electron theory in the metallurgical industry was to be short lived.

---

*This talk was first given at the TMS symposium in October 2000 celebrating William Hume-Rothery's work . It appeared in *The Science of Alloys for the 21st Century: a Hume-Rothery Symposium Celebration*, P.Turchi, R.Shull and A.Gonis eds, TMS 2000, 121–150, and is reprinted here with permission of the publishers. Hume-Rothery made his famous prediction that the beta-phase of Cu_3Al would take the bcc structure because it had an average electron per atom ratio of 3/2 in 1925, whilst a graduate student at the Royal School of Mines. A companion chapter about his life and science appears in the above TMS book.

The modelling and simulation of materials is, in fact, a daunting task because linking the world of the electron theorist to the world of the industrial metallurgist requires spanning at least 12 orders of magnitude in both length and time. This is illustrated schematically in Fig. 1 which has divided this wide range of modelling into four definite hierarchies.[4,5] At the electronic level we are solving the Schrödinger equation of quantum mechanics in order to predict the behaviour of the electrons. As an example the upper left-hand panel of Fig. 2 shows the sp^3 hybrids that are responsible for the occurrence of the diamond structure amongst group IV elements. However, even with the largest supercomputer, only about one thousand non-equivalent atoms can be simulated from first principles, which corresponds to a three-dimensional cell size of about 1 nm. Therefore, in order to simulate larger systems, the electronic degrees of freedom must be removed by imagining the atoms are held together by some sort of glue or interatomic potential. This allows large scale molecular dynamics to be performed on fracture at crack tips, for example.[6] However, even with the largest supercomputer, only about one thousand million atoms can be treated, which corresponds to a cell size of about 0.1 µm. Therefore, in order to compute the evolution of microstructure, the atoms must be coarse grained into cells which interact via deterministic or

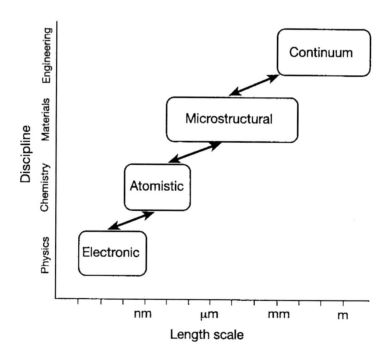

Fig. 1 Hierarchies of models.

stochastic rules, thereby allowing, for example, the simulation of the grain structure during the growth of single crystal turbine blades.[7] Finally, the continuum world of the industrial metallurgist or engineer is reached by averaging over the microstructure and describing the material through a set of constitutive relationships.[8]

Hume-Rothery and his contemporaries, however, were faced with the immediate problem that the solutions of the quantum mechanical equations at the electronic level were not accurate enough even to begin thinking about trying to link up the modelling hierarchies in Fig. 1. The 1936 free-electron Brillouin zone model of Jones[9] that seemed to explain the Hume-Rothery electron per atom rule for alloys of the noble metals had been discredited in 1955 by the anomalous skin effect measurements of Pippard, which showed that the Fermi surface of copper had already made contact with the zone boundary.[10] The success of the 1933 paper of Wigner and Seitz[11] on the cohesion of sodium had not been repeated for non-monovalent metals because the self-consistent Hartree approximation led to severe underbinding and the Hartree-Fock approximation led to the anomalous vanishing of the density of states at the Fermi level. It was not surprising that Hume-Rothery

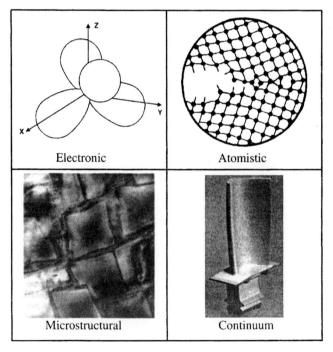

Fig. 2 Upper left-hand panel: sp^3 hybrids. Upper right-hand panel: a crack tip. Lower left-hand panel: superalloy microstructure. Lower right-hand panel: a turbine blade.[4]

wrote in the preface to the 1962 fourth reprint of his *Atomic Theory for Students of Metallurgy*: 'The work of the last ten years has emphasised the extreme difficulty of producing any really quantitative theory of the electronic structure of metals, except for those of the alkali group.'

Although unrecognised at the time, the breakthrough came two years later with the publication of the two papers by Hohenberg, Kohn and Sham on density functional theory.[12,13] They proved that the ground state energy of a many-electron system is a unique functional of the electron density,[12] which allowed them to derive an effective one-electron Schrödinger equation that was similar to Hartree's except that it also contained an additional attractive contribution to the potential arising from the so-called exchange-correlation hole.[13] It required, however, the advent of fast computers and improved algorithms during the late 1970s and early 1980s before the local approximation to the exact shape of the exchange-correlation hole was demonstrated numerically to be surprisingly accurate in predicting intrinsic ground state properties. Theorists then realised that the concept of the exchange-correlation hole was particularly robust; even though the shape of the hole might be poorly represented by the Local Density Approximation (LDA), the attractive potential that the electron sees at its centre is usually well described because the approximate and exact holes both exclude precisely one electron each.

In 1998 Walter Kohn, a theoretical physicist, shared the Nobel Prize in Chemistry with John Pople for helping to establish quantitative electron theory. In this paper we will show how the effective one-electron description of density functional theory allows us to clarify three concepts that aroused considerable confusion and controversy in Hume-Rothery's day: firstly, the relevance of the Jones theory of Brillouin zone touching to the experimental Hume-Rothery electron per atom rule;[9] secondly, the relevance of the Engel-Brewer theory to the structural stability of metals and alloys;[14] and thirdly, the relevance of Pauling's theory of resonant bonds to transition metal cohesion.[15]

Jones' Brillouin Zone Theory

> 'If the zone faces are already touched, then the Jones theory which requires contact only after the addition of more electrons per atom by alloying becomes, on the face of it, untenable.'[16]

Jones' 1936 theory, which was published in Mott and Jones,[9] showed that the spherical Fermi surface of a free electron gas first made contact with the Brillouin zone (or Jones zone) boundary at the electron per atom counts of 1.36, 1.48 and 1.54 for the α-, β- and γ-phases respectively. This correlated very well with Hume-Rothery's empirical electron per atom ratios for the solid solubility limit of the α-phase and the narrow stability regions of the β- and γ-phases of alloys of the noble metals with group B sp valent elements.[17] The reason for this correlation between Brillouin zone touching and structural stability has most lucidly been explained by Raynor: 'Let us now consider the effect of dissolving another metal of higher valency in such a monovalent metal. . . . The number of electrons present increases, so that the energy of the electron of maximum energy increases. This increase in the number of electrons is therefore represented . . . by an increase in the radius of the sphere in k-space which represents the occupied energy levels. . . . Eventually, however, a stage is reached at which the surface of the expanding sphere comes near to some of the planes bounding the zone. This produces . . . a rise in the number of energy states per very small energy range. After the surface of constant energy (somewhat distorted at this stage) has touched one set of bounding planes, a steep fall takes place in the curve showing the distribution of energy states.'[1]

This behaviour in the density of states (DOS) is illustrated in the left hand panel of Fig. 3. This represents the DOS associated with only those states within the first Brillouin zone. 'We may suppose the zone to be "filled" up to the energy represented by p. It will be seen from this figure that the further addition of a number of electrons represented by the area $pqrs$ leads to an increase in energy of the most energetic electron from p to s. A second further addition of the same number of electrons, represented by the area $srut$ which is equal to the area $pqrs$, in the region where the curve is falling leads to the much greater rise in energy from s to t. Since the total energy of the electrons depends upon the value E_{max}, the energy of the most energetic electron, it will be appreciated that the energy of the whole structure begins to rise sharply as soon as the zone has been filled up to a point represented by the peak of the

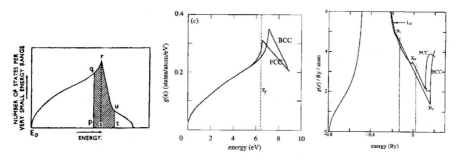

Fig. 3 Left-hand panel: Jones' 1936 DOS.[1] Middle-panel: Jones' 1937 DOS.[18] Right-hand panel: LDA copper DOS where the vertical lines show the energies at which the Fermi levels in the fcc and bcc lattices are identical.[22]

curve in the figure. It is when we reach the stage at which the addition of a small number of electrons causes a large increase in energy that we may expect, on theoretical grounds, that the structure will become unstable with regard to possible alternative structures. The zone theory, therefore, leads to the expectation that the fcc structure of the solid solution of a polyvalent metal in a monovalent metal will become unstable when the energy of the most energetic electron has the value corresponding roughly with the peak in the (density of states). This condition . . . occurs approximately when the surface of the sphere in k-space . . . touches a set of bounding planes . . . so that the (fcc) structure becomes unstable at 1.36 electrons per atom.'[1]

Raynor was, therefore, right to believe that Jones' simple theory became 'untenable' when Pippard's 1957 paper demonstrated conclusively that the Fermi surface of copper, with only 1.00 valence electrons per atom, had already made contact with the fcc Brillouin zone. And here the confusion multiplied, because 20 years earlier in 1937 Jones had published an additional paper,[18] in which he looked at the relative stability of the α- and β-phases of Cu–Zn alloys in the presence of the very large {111} zone boundary gap of 4.1 eV that had recently been determined experimentally. It is clear from the middle panel of Fig. 3 that in this new model the Fermi surface of fcc copper has nearly touched the zone boundary, requiring only the addition of a further 0.03 electrons per atom to do so. As the number of electrons increases above 1.03, the fcc DOS begins to fall whereas the bcc DOS continues to rise up to meet the van Hove singularity at the {110} zone boundary. Jones used a closed form expression for the band structure that was free-electron-like throughout most of the Brillouin zone, deviating from it, however, as the wave vector approached the Bragg reflecting planes. Using the cone approximation for performing the integration over the Brillouin zone, he calculated a transition from fcc α-phase stability to bcc β-phase stability at an electron per atom ratio of 1.41 that was in good agreement with experiment.

The subsequent confusion in the literature arose from the fact that in Jones' 1937 model there is absolutely no connection between stability and the Fermi energy falling near the maximum in the peak of the DOS, as there had been in his 1936 model. However, after Pippard's experimental bombshell, Jones readdressed the conditions for structural stability within his 1937 model.[19] He assumed a rigid-band approximation in which the bands of fcc and bcc copper are taken to remain unchanged (or rigid) on alloying, so that the structural energy difference between the α- and β-phases arising from the electronic structure contribution could be written

$$\Delta U = \Delta \left[\int^{E_F} En(E)dE \right] \tag{1}$$

where

$$N = \int^{E_F} n(E)\,dE \qquad (2)$$

with E_F, $n(E)$ and N the Fermi energy, DOS and number of valence electrons per atom respectively and with $\Delta U = U_\beta - U_\alpha$. It immediately follows from equations (1) and (2) that

$$\frac{d}{dN}(\Delta U) = \Delta E_F \qquad (3)$$

and

$$\frac{d^2}{dN^2}(\Delta U) = \Delta\left[\frac{1}{n(E_F)}\right] \qquad (4)$$

Jones used these last two equations for the slope and curvature to show that the minimum in the structural energy difference curve ΔU occurred for an electron per atom ratio N that lay well to the right of the peak in the fcc DOS in the middle panel of Fig. 3.

The properties of copper are, of course, very far from those of a monovalent sp metal such as sodium or potassium. It has been known for a long time[20] that the narrow $3d$ valence band in copper hybridises strongly with the broad free-electron-like $4s$ and $4p$ band to provide the additional cohesion, shortness of bond length, and increase in bulk modulus that the metal displays. Remarkably, however, although masking the first van Hove singularities at $L_{2'}$ in fcc and N_1 in bcc (see the right-hand panel of Fig. 3), the presence of the d band does not alter fundamentally Jones' 1937 and 1962 Brillouin zone arguments regarding the relative stability of the fcc and bcc lattices, at least for electron per atom ratios away from $e/a = 1.0$.[21] This is demonstrated by the upper panel in Fig. 4 which shows the relative structural stability of the fcc, bcc and hcp lattices for Cu–Zn alloys within the rigid band approximation.[22] These curves were computed using local density functional theory with the structural energy differences evaluated directly from the bandstructure term alone as justified by the frozen-potential or force theorem of Pettifor.[24,25]

The shape of the structural energy difference curves in Fig. 4 can be understood in terms of the relative behaviour of the DOS in the middle panel using a Jones-type analysis. In particular, from equation (3), the stationary points in the upper curve correspond to band occupancies for which ΔE_F vanishes in the lower panel (see the two vertical lines in the right-hand panel of Fig. 3). Moreover, whether the stationary point is a local maximum or minimum depends on the relative DOS at the Fermi level through equation (4). Thus, the bcc–fcc energy difference curve has a minimum around $N = 1.6$ where

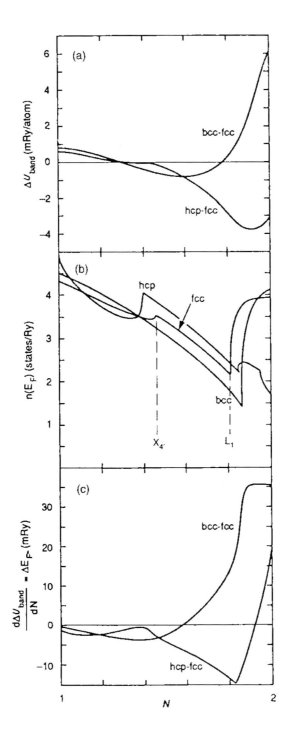

Fig. 4 Analysis of fcc, bcc and hcp relative structural stabilities within the rigid-band approximation for Cu–Zn alloys. Upper panel: The difference in band energy as a function of band filling N with respect to elemental rigid copper bands. Middle panel: DOS at the Fermi level as a function of band filling, $X_{4'}$ and L_1 marking the bottom and top of the band gaps across the {100} and {111} Brillouin zone faces respectively. Lower panel: The difference in the Fermi energies as a function of band filling.[22,23]

the bcc DOS is lowest, whereas the hcp–fcc curve has a minimum around $N = 1.9$ where the hcp DOS is lowest. The fcc structure is most stable around $N = 1$ where $\Delta E_F \approx 0$ and the fcc DOS is lowest. Thus, the top panel of Fig. 4 displays the succession of phases from α(fcc) to β(bcc) to ε(hcp) as found in the Cu–Zn phase diagram. The γ-phase has not been included in the above discussion because its structure is too distorted for the first order frozen-potential or force theorem[24,25] to be valid. However self-consistent LDA calculations by Paxton et al.[22] confirm Massalski and Mizutani's results[26] of a dramatic drop in the DOS for an e/a greater than 1.5 due to scattering across the $\{330\}$ and $\{441\}$ zone boundaries as first discussed by Jones.[27]

Copper and silver are the two best conductors of electricity at room temperature due to their carriers at the Fermi level being free-electron-like.[9] This is illustrated in the left-hand panel of Fig. 5 which shows the total DOS and the partial sp and d contributions to the conductivity of copper within an Anderson-like disorder scattering model.[28] We see from the middle panel that at the Fermi energy almost all the current is being carried by the sp electrons

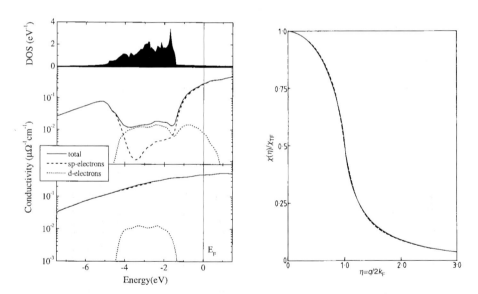

Fig. 5 Left-hand panel: The total DOS (upper panel) and the partial sp and d contributions to the total conductivity of copper in the presence and absence of hybridisation between the valence sp and d bands (middle and lower panels respectively).[28] Note that the vertical conductivity scale is logarithmic. Right-hand panel: The wave-vector dependence of the normalised Lindhard response function. The dashed curve shows the rational polynomial approximation of[31] that does not include the weak logarithmic singularity in the slope at $q/2k_F = 1$.

just as predicted by the unhybridised sp-d band model of Mott and Jones in the lower panel. Importantly, we also see from the middle panel that the mixture of the 3d band into the 4sp band rapidly falls to zero as the Fermi energy or the electron per atom ratio increases. Thus, away from $e/a = 1.0$, the crystal structure trend is being driven by essentially free electrons in the system.

The success of Jones' original theory on the importance of the spherical free-electron Fermi surface just touching the Brillouin zone boundary may, therefore, be understood by considering a free electron gas that is perturbed by the presence of an underlying lattice. Within second-order perturbation theory the resultant change in the band structure energy can be written:[29,23]

$$U^{(2)} = \frac{1}{2} V \sum_{q \neq 0} |S(\mathbf{q})|^2 [\chi(\mathbf{q})/\varepsilon(\mathbf{q})] \left| v_{ps}^{ion}(\mathbf{q}) \right|^2 \tag{5}$$

where V, $S(\mathbf{q})$ and $v_{ps}^{ion}(\mathbf{q})$ are the atomic volume, structure factor and Fourier transform of the ionic pseudopotential respectively. $\chi(\mathbf{q})$ and $\varepsilon(\mathbf{q})$ are the density response function and dielectric constant respectively which are dependent on the Lindhard response function for a non-interacting free electron gas, namely

$$\chi_o(\eta = q/2k_F) = \left[\frac{1}{2} + \frac{1-\eta^2}{4\eta} \ln\left|\frac{1+\eta}{1-\eta}\right|\right] \chi_{TF} \tag{6}$$

χ_{TF} is the Thomas-Fermi response function, $-\kappa_{TF}^2/8\pi$, where $\kappa_{TF} = (4k_F/\pi)^{\frac{1}{2}}$. The right-hand panel of Fig. 5 shows the normalised Lindhard response function, χ_o/χ_{TF}, plotted as a function of $q/2k_F$. It is seen from equation (6) that the slope diverges logarithmically at $q = 2k_F$, the spanning wave vector for the spherical Fermi surface.

In 1971 Stroud and Ashcroft[30] used this second-order perturbation theory to study the relative stability of the fcc, bcc and hcp lattices as a function of e/a. They made the virtual-crystal approximation, in which all ions are treated the same, and characterised the ionic pseudopotential by the local Ashcroft empty core form.[32] They found structural energy difference curves that are qualitatively similar to the LDA results in the top panel of Fig. 4 (apart from predicting copper with $e/a = 1.0$ to be hcp). This structural behaviour can be shown to be driven directly by the weak logarithmic singularity at $q = 2k_F$ in the Lindhard response junction. This singularity is well-known to be responsible for the Friedel oscillations in the electronic screening cloud which fall off asymptotically with distance from the centre of the ions as $\cos 2k_F r/r^3$.[33] Pettifor and Ward[31] replaced the Lindhard function by the dashed curve in the right-hand panel of Fig. 5 that is correct to third order in q^2 and $1/q^2$ but which has no singularity in slope at $q = 2k_F$. As can be seen from the upper panel in Fig. 6, the removal of the logarithmic singularity leads to the absence of the rapid oscillation in the bcc–fcc structural energy difference curve about $e/a = 1.48$ and hence the destruction of the narrow domain of bcc stability in this vicinity.[34]

Thus, the passage of a reciprocal lattice vector through $2k_F$ is marked by special structural significance as Jones hypothesised in his original 1936 theory. The critical condition, $|\mathbf{G}| = 2k_F$, depends only on the electron per atom ratio, not on any other properties of the atom such as its core size. For this reason these Hume-Rothery alloys are correctly termed electron phases. We should note, in conclusion, that the virtual crystal approximation (like the rigid band approximation) is not able to predict correctly the heat of formation which depends, as expected, on the chemical difference between the constituents.[35] This requires the addition of the second order

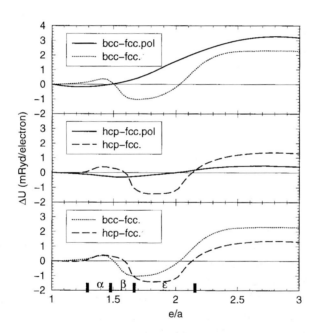

Fig. 6 Upper panel: the bcc–fcc structural energy difference curves using the exact (dotted curve) and the approximate polynomial (solid curves) Lindhard response function. Middle panel: the hcp–fcc structural energy difference curves using the exact (dashed curve) and the approximate polynomial (solid curve) Lindhard response function. Lower panel: A comparison of these exact results, showing the transition from α(fcc) to β(bcc) to ε(hcp) as the electron per atom ratio increases.[34] Note the similarity of these second-order perturbation theory curves in the lower panel with those of the LDA results in the upper panel of Fig. 4 (apart from the copper-rich end).

contribution $(\nu_{ps}^A - \nu_{ps}^B)^2$ as first discussed by Inglesfield.[36] However, the difference in energy between two structures is still mainly determined by the second-order contribution $[(\nu_{ps}^A + \nu_{ps}^B)/2]^2$ which we have presented in equation (5). The actual first principles prediction of phase diagrams requires a proper treatment of the total free energy of the different phases.[37] Ten years ago this was carried through for the α- and β-phases of the brass system by Turchi et al.[38]

Engel–Brewer Structural Stability Theory

'Hume-Rothery said that it would be of great interest if the present meeting would discuss the Engel–Brewer correlations, and say whether they are sound or not.'[39]

The Engel–Brewer theory correlated the crystal structure of metals with the number of valence s and p electrons in the system.[14] In particular, the bcc, hcp and fcc structures are characterised by 1, 2 and 3 valence sp electrons respectively, the archetypical example being bcc Na,[40] hcp Mg and fcc Al. This led to the postulate that the crystal structure of the transition elements is due solely to the valence s and p electrons (according to the above rule) even though their cohesion results from the bonding between all the valence s, p and d electrons.

We saw in the previous section that the structures of the Hume-Rothery electron phases really do depend just on their electron per atom ratio, the chemistry of the constituents not playing a significant role. This arises from the weak logarithmic singularity in the slope of the Lindhard screening function at the spanning vector $|\mathbf{G}| = 2k_F$ across the free electron Fermi sphere. However, for e/a occupancies away from this condition, the long-range Friedel oscillations will interfere destructively and the structural stability will depend also on other properties such as the core size. We, therefore, expect the rational polynomial approximation to the Lindhard function to provide the correct structural trends in this case. Rather than using the reciprocal-space representation of the previous section which is ideal for treating singularities in k-space, we switch now instead to the real-space representation of second-order perturbation theory.[23]

The real-space representation is very beautiful because it separates out the small structure-dependent contribution to the total binding energy as a single sum over a density-dependent pair potential $\phi(R)$. This can be written

$$\phi(R) = \frac{4Z^2}{\pi R} \int_0^\infty \left\{ [\hat{v}_{ps}^{ion}(q)]^2 / \varepsilon(q) \right\} [(\sin qR)/q] \, dq \qquad (7)$$

where the normalised ion-core pseudopotential matrix element

$$\hat{v}_{ps}^{ion}(q) = [(Vq^2)/8\pi Z]v_{ps}^{ion}(q) = \cos qR_c \tag{8}$$

The last identity follows for a simple Ashcroft empty core pseudopotential with core radius R_c. For the particular level of approximation that is illustrated in the right-hand panel of Fig. 5, we find

$$\phi(R) = (2Z^2/R) \sum_{n=1}^{3} A_n \cos(2k_n R + \alpha_n) e^{-\kappa_n R} \tag{9}$$

The wave vector k_n and the screening length κ_n^{-1} depend only on the density of the free electron gas through the poles of the approximated inverse dielectric response function, whereas the amplitude A_n and the phase shift α_n depend also on the nature of the ion-core pseudopotential. The behaviour of these three contributions is illustrated in the left-hand panel of Fig. 7 for the

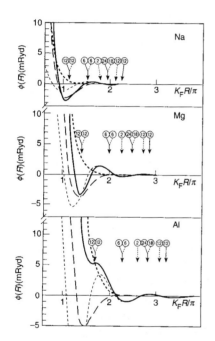
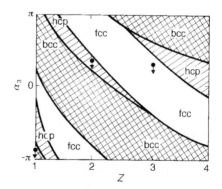

Fig. 7 Left-hand panel: The analytic pair potentials (full curves) for Na, Mg and Al together with the short-ranged (bold short-dashed curves), medium-ranged (long-dashed curves) and long-ranged (light short-dashed curves) contributions. The full (dotted) arrows mark the positions of the first four (five) nearest-neighbour shells in fcc (ideal hcp).[31] Right-hand panel: The structure map (Z, α_3) predicted using the long-range oscillatory pair potential $\phi_3(R)$. The three dots indicate the value of the phase shifts for Na, Mg and Al corresponding to $Z = 1$, 2 and 3 respectively, the arrows indicating the direction the phase shift changes under pressure.[42,23]

archetypical case of Na, Mg and Al using Ashcroft empty core pseudopotentials with core radius 1.75, 1.31 and 1.11 au respectively. We see that all three metals are characterised by a short-ranged repulsive contribution, a medium-ranged attractive contribution, and a long-ranged oscillatory contribution. The total analytic pair potentials in Fig. 7 follow closely those for Na, Mg and Al that have been predicted numerically by McMahan and Moriarty using their generalised pseudopotential theory within the density functional formalism.[41]

The relative structural stability can be shown to be driven by the long-range oscillatory contribution to the pair potential (see Fig. 6.11 of Ref. 23). At metallic densities this takes the form

$$\phi_3(R) = (2Z^2/R) A_3 \cos(1.92 k_F R + \alpha_3) e^{-0.29 k_F R} \qquad (10)$$

By introducing the Wigner–Seitz radius S, $k_F R$ can be written as

$$k_F R = (3\pi^2 Z/V)^{\frac{1}{3}} R = (9\pi/4)^{\frac{1}{3}} Z^{\frac{1}{3}} (R/S) \qquad (11)$$

Thus, for a given crystal structure, $k_F R$ takes fixed values proportional to $Z^{\frac{1}{3}}$ as R/S is independent of atomic volume. The volume dependence and core-size dependence enters through the phase shift α_3 which determines how the peaks and troughs in the oscillatory pair potential fall with respect to the underlying lattice. The phase shift is responsible for changes in structure both under pressure and on going down a group in the periodic table.

This is illustrated in the right-hand panel of Fig. 7 by the structure map (Z, α_3) that results from summing over the long-range oscillatory potential $\phi_3(R)$.[42,23] We see that Na, Mg and Al at zero pressure fall in bcc, hcp and fcc domains respectively (with Na lying close to an hcp domain as expected from its experimental Sm-type ground state structure). Under pressure the phase shift α_3 decreases, driving hcp Mg and fcc Al into the bcc phase, as first predicted by Moriarty and McMahan[43] in 1982 and later confirmed experimentally for Mg by Olijnyk and Holzapfel[44] in 1986. Thus, the structural stability of the Engel-Brewer archetypical metals Na, Mg and Al is determined not only by the electron per atom ratio through $k_F R$ in equation (11), but also by the phase shift α_3 that depends on both core size and density.[45]

The Engel-Brewer theory[14] postulates that the number of sp electrons will vary across the transition metal series, since hcp Y and Zr, for example, would have two sp electrons, bcc Nb and Mo one sp electron, hcp Tc and Ru two sp electrons, and fcc Rh and Pd three sp electrons. Self-consistent density functional calculations have, of course, shown that this hypothesis is incorrect in the metallic state. For example, the left-hand panel of Fig. 8 plots the predicted behaviour of the number of d electrons as a function of the Wigner–Seitz radius across the non-magnetic 4d transition metal series.[46] We see from the open circles in Fig. 8 that at their equilibrium volumes they all have approximately one valence electron each in the sp band. Thus, the structural trend from hcp → bcc → hcp → fcc must be being driven by the change in the number of valence d electrons across the series.

This is confirmed numerically in the right-hand panel of Fig. 8, which compares the fcc, bcc and hcp d band energies as a function of d band filling using equations (1) and (2) with the total DOS $n(E)$ replaced by the unhybridised tight-binding d band DOS $n_d(E)$.[47] We find that apart from the noble metal end of the series, the curves predict correctly the hcp → bcc → hcp → fcc structural trend. Interestingly, a change in the number of d electrons also drives the observed structural trend across the rare earths and certain structural phase changes in the early transition metals under pressure. Both are related to the sensitivity of the number of d electrons to the fraction of the atomic volume occupied by the ion core, since this controls the relative position of the nearly-free-electron (NFE) sp band with respect to the tight-binding (TB) d band.[23] Calculations predict that the number of d electrons decreases across the trivalent rare earths as the ion core contracts, thereby accounting for the structural trend from $d$hcp(La) → Sm-type → hcp(Lu).[48] Under pressure the number of d electrons

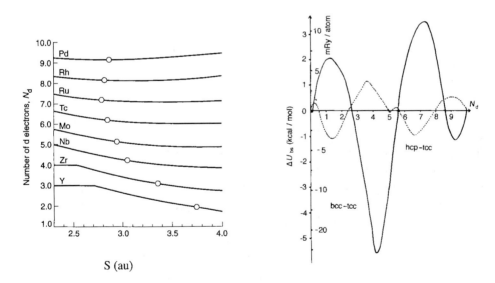

Fig. 8 Left-hand panel: The predicted d band occupancy versus the Wigner-Seitz radius for the non-magnetic $4d$ series. The open circles mark the equilibrium values.[46] Right-hand panel: The d band energy differences between bcc and fcc (solid line) and hcp and fcc (dotted line) as a function of the d band filling.[47]

312 Materials Science and Engineering: Its Nucleation and Growth

in the early transition metals increases as the fraction of the atomic volume occupied by the ion core grows (as seen in the left-hand panel of Fig. 8) thereby leading to the structural phase change in Mo, for example, from bcc → hcp.[49]

Following a Jones-type analysis, the behaviour of the band energy difference curves in the right-hand panel of Fig. 8 may be related directly to the structure in the transition metal DOS. Figure 9 shows an early histogram DOS for bcc, fcc and hcp structures.[47] They were calculated by diagonalising

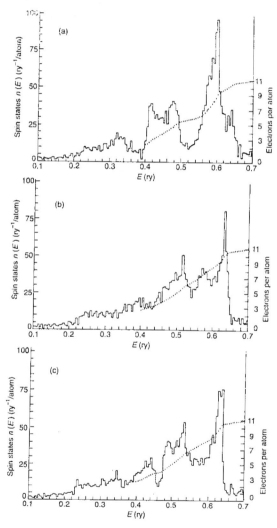

Fig. 9 The DOS for the three structures (a) bcc, (b) fcc, and (c) hcp for a model transition metal. The dotted curves represent the integrated DOS.[47]

a hybrid NFE-TB secular equation that can be derived directly from scattering theory under the physically-motivated assumption that the *d* electrons undergo resonant scattering, whereas the *sp* electrons remain unscattered.[50] This allows all the matrix elements in the hybrid NFE-TB secular equation to be expressible in terms of just the two fundamental parameters, E_d and Γ, the position and width of the *d* resonance respectively. Figure 9, therefore, represents a model transition metal with $E_d = 0.5\,\text{Ry}$ and $\Gamma = 0.06\,\text{Ry}$. (This is to be compared with the parameters for a real transition metal, Fe, with $E_d = 0.540\,\text{Ry}$ and $\Gamma = 0.088\,\text{Ry}$). We see at once that the strong stability in Fig. 9 of the bcc lattice for a half-full *d* band arises from the deep minimum that separates the bonding from antibonding peaks in the DOS. Moreover, even though the conventional band structures of fcc and hcp appear very different as the latter has a two atom rather than one atom basis, their DOS are very similar. The main differences such as the peak (trough) in the fcc (hcp) DOS around $N = 7.5$ may be linked to the oscillatory behaviour of the hcp–fcc curve in the right-hand panel of Fig. 8, whose amplitude of oscillation is seen to be about a factor of five smaller than that for the bcc–fcc curve.

The origin of the structure in the *d*-band TB DOS of the transition metals is most directly understood within a real-space representation in which the moments of the eigenspectrum are related to self-returning hopping or bonding paths within the lattice, as first shown by Cyrot-Lackmann in 1967.[51] This is in direct contrast to the discussion in the previous section where the structure in the *sp*-band NFE DOS of the noble metals above their Fermi energies was interpreted within a reciprocal-space representation in which energy gaps and van Hove singularities appeared at the Brillouin zone boundaries. The link between moments and path counting is easy to demonstrate starting from the usual definition of the *p*th moment of a spectrum of eigenvalues E_n, namely

$$\mu_p = \frac{1}{\mathcal{N}} \sum_n E_n^p \tag{12}$$

where $\mathcal{N}$ is the total number of states. Since the eigenfunctions $|n\rangle$ diagonalise the Hamiltonian, equation (12) can be written

$$\mu_p = \frac{1}{\mathcal{N}} \sum_n (H^p)_{nn} = \frac{1}{\mathcal{N}} TrH^p \tag{13}$$

But, since the trace is invariant with respect to the choice of basis functions that are related by a unitary transformation, we can work instead with the basis of TB atomic orbitals $|i\rangle$, so that

$$\mu_p = \frac{1}{\mathcal{N}} \sum_{i_1} (H^p)_{i_1 i_1} = \frac{1}{\mathcal{N}} \sum_{i_1, i_2, \ldots, i_p} H_{i_1 i_2} H_{i_2 i_3}, \ldots, H_{i_p i_1} \qquad (14)$$

That is, the pth moment of an eigenspectrum is given by the sum over all self-returning bonding paths of length p. We see at once, therefore, that the centre of gravity of the TB d band, μ_1, is simply the atomic energy level E_d, corresponding to the on-site diagonal element H_{ii}.[23]

The d band energy can be written in terms of the moments as an exact many-atom expansion,[52,53,54] namely

$$U = -\left\{\hat{\chi}_2(N_d) + \hat{\chi}_3(N_d)(\mu_3/\mu_2^{\frac{3}{2}}) + \hat{\chi}_4(N_d)\left(\mu_4/\mu_2^2 - \mu_3^2/\mu_2^3 - 2\right) + \ldots\right\}\mu_2^{\frac{1}{2}} \qquad (15)$$

where the centre of gravity of the d band, E_d, has been taken as the zero of energy, so that $\mu_1 = 0$. $\hat{\chi}_n(N_d)$ are reduced response functions that in the simplest approximation[52] depend only on the band filling N_d as

$$\hat{\chi}_n(N_d) = \frac{2}{\pi}\left[\frac{\sin(n-1)\phi_F}{n-1} - \frac{\sin(n+1)\phi_F}{n+1}\right] \qquad (16)$$

where ϕ_F is related to the band filling N_d through

$$N_d = (10\phi_F/\pi)[1 - (\sin 2\phi_F)/2\phi_F] \qquad (17)$$

These response functions are plotted in the left-hand panel of Fig. 10 where we see that the number of nodes (excluding the end points) is $(n-2)$.

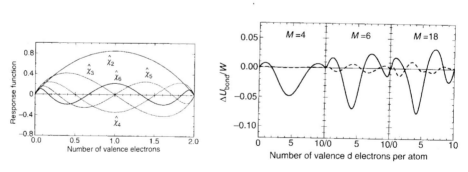

Fig. 10 Left-hand panel: The normalised response functions $\hat{\chi}_n$ as a function of the number of valence electrons per orbital for the case of the shape parameter $S = 1$.[52] Right-hand panel: Convergence of the bcc–fcc d band energy (full curves) and hcp–fcc d band energy (dashed curves) with respect to the number of moments retained in the many-atom expansion (15).[54]

The first term in the expansion is proportional to the root mean square width of the eigenspectrum, which sets the energy scale. It lies behind the square root embedding function which was used by Finnis and Sinclair[55] in their choice of interatomic potential. The second moment itself is pairwise in character because the second moment about atom i can be written from equation (14) as

$$\mu_2^i = \sum_j H_{ij} H_{ji} = \sum_{j \neq i} \beta^2 (R_{ij}) \qquad (18)$$

where $\beta(R_{ij})$ is the TB hopping or bond integral. The second term in expansion (15) is dependent on the third moment, μ_3, and represents the skewing or asymmetry of the band. This arises from the presence of three-membered rings in the lattice so that the electron can hop from $i \to j \to k \to i$. The third term in expansion (15) is proportional to $(S - 1)$, where

$$S = \mu_4/\mu_2^2 - \mu_3^2/\mu_2^3 - 1 \qquad (19)$$

If $S < 1$, the spectrum is said to show bimodal behaviour, whereas if $S > 1$, it shows unimodal behaviour. Two level systems such as the dimer H_2 or the tetrahedral tetramer H_4 display perfect bimodal behaviour with $S = 0$.[23] The importance of the fourth moment contribution to the properties of the group V and VI transition metals has been demonstrated by Carlsson,[55] Foiles[56] and Moriarty.[57]

The convergence of this many-atom expansion for the TB d band energy is demonstrated in the right-hand panel of Fig. 10. We see that the structural trend from hcp $\to$ bcc $\to$ hcp $\to$ fcc is reproduced correctly provided contributions up to μ_6 are included. We find that the fourth moment contribution is indeed responsible for the bcc stability of groups V and VI transition metals, whereas the sixth moment is required for differentiating between cubic and hexagonal close-packed structures. The numbers of inner nodes displayed by the bcc–fcc fourth moment curve (namely 2) and the hcp–fcc sixth moment curve (namely 4) are consistent with a theorem by Ducastelle and Cyrot-Lackmann.[58] They proved that if two eigenspectra have moments that are identical up to some level p_o (that is $\Delta\mu_p = 0$ for $p \leq p_o$), then the two band energy curves must cross at least $(p_o - 1)$ times as a function of electron count. This theorem is reflected in the behaviour of the response function curves in the left-hand panel of Fig. 10.

The origin of the bcc DOS being more bimodal than either fcc or hcp (as seen in Fig. 9) can be tracked down to differences in the four-member ring contribution to the fourth moment. From equation (13) this can be written

$$\mu_4^{ring} = \text{Tr}\, H_{12} H_{23} H_{34} H_{41} \qquad (20)$$

where the H_{ij} are the 5 × 5 TB matrices that link the five d orbitals on atom i to those on the neighbouring atom j. For the particular case of equilateral planar rings[59] with bond angle θ

$$\mu_4^{ring} = 5\left[\left(1757 - 60460x^2 + 327870x^4 - 563500x^6 + 300125x^8\right)/5792\right]\beta^4 \quad (21)$$

where $x = \cos\theta$ and $\beta^4 = (dd\sigma^4 + 2dd\pi^4 + 2dd\delta^4)/5$. This analytic expression was derived by Moriarty[60] assuming canonical TB parameters $dd\sigma : dd\pi : dd\delta = -6 : 4 : -1$. The left-hand panel of Fig. 11 shows the resultant rapid oscillations as a function of bond angle. They reflect the interference between the angular lobes of the d orbitals as the electron hops around the ring from $1 \to 2 \to 3 \to 4 \to 1$. The angular prefactor in the square brackets above would, of course, be totally absent for s orbitals since then the four-member ring contribution would be simply $ss\sigma^4$. The bcc and fcc nearest-neighbour planar rings are shown in the right-hand panel of Fig. 11. We see that the bcc and fcc bond angles of 70.5° (109.5°) and 90° correspond to negative and positive contributions respectively to the fourth moment. Thus, it follows from expansion (15) that since $\hat{\chi}_4(N_d)$ is negative for nearly half-full bands this ring contribution will stabilise the bcc lattice but destabilise the fcc lattice, as reflected in the right-hand panel of Fig. 10.

Density functional calculations have confirmed these oscillatory energy difference curves. Figure 12 shows Skriver's results[61] across the non-magnetic $4d$ series where we see that both the hcp–fcc and the bcc–fcc curves mirror the TB d-band results in the right-hand panel of Fig. 8.[47] However, these theoretical values differ markedly from the original values used by the

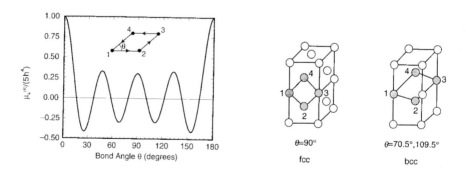

Fig. 11 Left-hand panel: the four-membered ring contribution as a function of bond angle.[60] Right-hand panel: a four-membered ring contribution in fcc and bcc lattices respectively. Note that from the upper panel the fcc and bcc rings shown contribute positive and negative contributions respectively to the fourth moment.

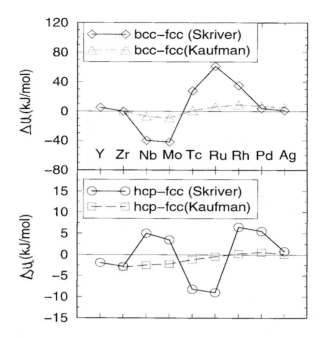

Fig. 12 The bcc–fcc (upper panel) and hcp–fcc (lower panel) energy difference curves across the $4d$ series predicted[61] by density functional theory (solid curves) and recommended[62] by Calphad (dashed curves).

Calphad community for predicting multicomponent phase diagram behaviour.[62] Firstly, as is seen in the lower panel of Fig. 12, the hcp–fcc Calphad curve does not oscillate between the two experimental domains of hcp stability. The predicted oscillation is not a numerical error in the density functional calculations since we expected such an oscillation from the sixth moment TB arguments discussed above. Alas, wave-mechanical, oscillatory functions are not easy to handle within an extrapolation scheme such as that used by Calphad! Secondly, the magnitudes of the energy differences predicted by theory are often nearly five times larger than the Calphad values.[63] Density functional calculations of the non-magnetic transition metals are known to give excellent agreement with the experimental structural energy differences when these are known, such as for example hcp–bcc zirconium.[63] However, where the metastable structures are not accessible directly to experiment, large discrepancies can be found between theory and Calphad. This can be traced back to the assignment of positive melting temperatures to metastable phases, which theoretically are predicted to be mechanically unstable at absolute zero.[64] These differences between theoretical predictions and empirical extrapolation is leading to a reassessment of the Calphad values.[65] Interestingly, Brewer already felt in 1966 that 'there is no justification for the assumption that all structures will have closely similar energies'.[66] The Engel–Brewer theory[14] predicts, for example, bcc–hcp energy differences for Nb and Mo that are in the same ball-park as the theoretical values plotted in Fig. 12.

Nevertheless, even though there are large discrepancies between the Calphad values and theory far away from experimental accessible metastable domains, density functional theory is invaluable for combining with the Calphad method for metastable phases that are accessible experimentally, for example, by rapid solidification. This is illustrated in Fig. 13 which shows the metastable phase equilibria between the β and ω phases of Ti–Al–V alloys at 673 K which were calculated[67] by the Calphad method using as input the structural energy differences between the metastable ω phase and the hcp, fcc or bcc ground states predicted by density functional theory.[68] We see good agreement between the resultant Calphad metastable phase diagram and experiment.

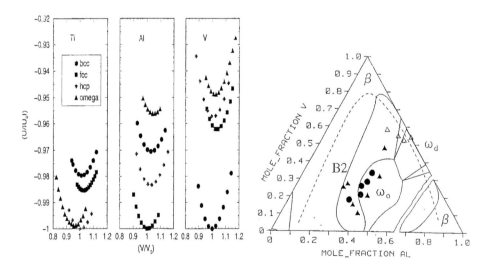

Fig. 13 The density functional binding energy curves for Ti, V and Al with respect of the ω, fcc, hcp and bcc structures (left-hand panel). The resultant Calphad metastable phase diagram for Ti–Al–V at 673 K (right-hand panel), where the open triangles represent mixed disordered bcc and ω phase experimental points, whereas the solid symbols represent mixed ordered regions.

In this section, we have shown that the structural stability of elemental metals is controlled by at least four factors: (i) the electron per atom ratio, (ii) the ionic core size, (iii) the angular character of the valence orbitals (whether they are s or d, for example), and (iv) the electron density (which affects the screening, thereby causing the atomic orbitals to contract or shrink in the metallic state). In addition, the structural stability of binary systems is influenced by the electronegativity difference or atomic energy level mismatch.[69] The lesson to be learned, therefore, is that it is very dangerous to conclude that because one has found a particular physical coordinate (such as the number of valence sp electrons per atom) that appears to correlate with structure that this is, in fact, the underlying reason for the structural trends observed. It was this caveat that led the author to propose his phenomenological 'string' for providing the best possible structural separation of binary $A_m B_n$ compounds within a single two-dimensional structure map.[70,71]

Pauling's Resonant Bond Theory

'Hume-Rothery's general conclusion was that the collective electron treatment was to be preferred for the general discussion of transition and other metals and their alloys, though it was admitted that the Pauling treatment provided a self-consistent scheme for the correlation of interatomic distances and structures in a wide variety of alloys and intermediate phases.' [72]

This success of density functional theory in predicting the cohesion and structure of transition metals has led to a reconciliation between the two apparently diametrically opposed viewpoints which are represented by the collective band model of Mott and Jones[9] and the covalent bond model of Pauling.[15] In this section we will show that the effective one-electron description that underpins density functional theory allows a transparent, direct link to be forged between the electron bands (expressing the global behaviour of the eigenstates) and the resultant bonds (expressing a local description of the energetics).

An immediate link between the behaviour of the energy bands and the resultant binding energy curves of transition metals is provided by the frozen-potential or force theorem of Pettifor.[24] This proved that within the classic approximation of replacing the Wigner–Seitz cell by a sphere, the pressure (or volume derivative of the total density functional energy) could be broken down uniquely into individual s, p and d valence band contributions. For the particular case of transition metals, in which the valence sp electrons are best describable within a NFE framework and the valence d electrons within a TB framework, the partial pressures take the physically intuitive form:[24,73,74]

$$3P_{sp}V = 3N_{sp}(B_s - \varepsilon_{xc})/m_s + 2U_{sp}^{ke} \qquad (22)$$

$$3P_dV = 2N_d(C_d - \varepsilon_{xc})/m_d + 5U_d^{bond} \qquad (23)$$

where

$$U_{sp}^{ke} = \int^{E_F}(E - B_s)n_{sp}(E)dE \qquad (24)$$

$$U_d^{bond} = \int^{E_F}(E - C_d)n_d(E)dE \qquad (25)$$

B_s and C_d are the bottom of the sp band and the centre of the d band respectively, which are defined in terms of the logarithmic derivatives at the Wigner–Seitz radius within Andersen's atomic sphere approximation.[5] m_s and m_d are the corresponding effective masses. ε_{xc} is the value of the exchange-correlation energy density at the Wigner–Seitz radius S and takes the value $\frac{3}{4}\phi(S)$ for the simple Slater $\rho^{\frac{1}{3}}$ exchange-correlation approximation, where $\phi(S)$ is the value of the potential at the Wigner–Seitz sphere boundary.[76]

The left-hand panel of Fig. 14 shows the behaviour of the energy bands of technetium as the volume changes about equilibrium. Technetium was chosen as the representative example because it lies in the middle of the $4d$ transition metal series. We see that the bottom of the NFE sp band, B_s, satisfies very closely a Bardeen-Fröhlich-type expression

$$B_s = -\frac{3Z_c}{m_sS}\left[1 - \left(\frac{R_c}{S}\right)^2\right] \qquad (26)$$

where Z_c and R_c can be interpreted as the core charge and ionic radius of an Ashcroft empty core pseudopotential.[46] The values of R_c, which are fitted to reproduce the density functional behaviour of the bottom of the sp band across the $4d$ series, are found to vary with the outer node in the free-atom $5s$ wave function, R_{node}, as

$$R_c = 1.89R_{node} \qquad (27)$$

to within 1%. This demonstrates unambiguously the importance of the repulsive core orthogonality constraints on the behaviour of the NFE sp band. We also see that the centre of gravity of the TB d band, C_d, moves up under compression as the electronic charge is confined into yet smaller Wigner–Seitz sphere volumes. However, at high compressions we see that the

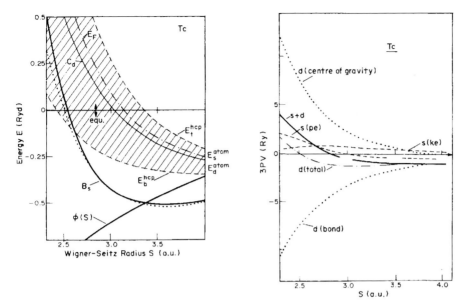

Fig. 14 Left-hand panel: the energy bands of Tc as a function of Wigner–Seitz radius S; the vertical arrow marks the observed equilibrium radius. Right-hand panel: the sp and d contributions to the virial 3PV as a function of the Wigner–Seitz radius.[77]

bottom of the sp band moves up much faster than the centre of gravity of the d band, causing the transfer of electrons from sp to d as observed in the left-hand panel of Fig. 8.

The right-hand panel of Fig. 14 shows the contributions to the sp and d partial pressures that result from equations (22) and (23) respectively. We see that about equilibrium the total sp pressure is *positive*, since we not only have the expected positive kinetic energy term, but we also find a positive contribution from the 'potential' energy term since $B_s > \frac{3}{4}\phi(S)$ (see the left-hand panel). At equilibrium this is countered by the *negative* total d pressure, which is driven by the large attractive d bond term. Integration over these pressure volume curves provides individual binding energy contributions that are consistent with the renormalised atom description of the cohesive energies, which are shown in Fig. 15.[78]

Thus, the main contribution to the cohesive energy of transition metals is the d bond term, which is illustrated by the solid black lines in Fig. 15. This is well represented by the rectangular d band model of Friedel,[79] in which the d band DOS, $n_d(E)$, is assumed to take the constant value of $10/W$ throughout its bandwidth W. Substituting this into equation (25) results in the d bond contribution

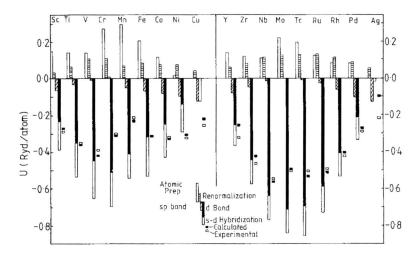

Fig. 15 The contributions to the cohesive energy of the $3d$ and $4d$ transition metals. (After Gelatt et al.[78])

$$U_d^{bond} = -\frac{1}{20} WN_d(10 - N_d) \tag{28}$$

whose variation with d occupancy is reflected in the parabolic shape of the black histogram in Fig. 15. Since the bandwidth scales as the root mean square width of the DOS, $\sqrt{\mu_2}$, it follows from equations (28) and (18) that

$$U_d^{bond} \propto \sqrt{z} \tag{29}$$

where z is the coordination of the lattice. The band/bond model of transition metals, therefore, predicts that the cohesive energy per atom varies less strongly than the linear coordination dependence expected within a saturated covalent bond description. As Heine[80] has stressed this is indicative of the correctness of Pauling's resonant bond model of metallic cohesion.[81]

The rectangular d band model may also be generalised to predicting the heats of formation of equiatomic transition metal alloys.[82,83] The lower part of the left-panel of Fig. 16 shows the formation of the metallic bond between two elemental transition metals A and B, in which a common band of width W_{AB} is formed in the alloy. Using the relation between the second moment of the total alloy DOS and hopping paths of length 2, we find that

$$\frac{1}{12} W_{AB}^2 = z\beta^2 + \left(\frac{1}{2}\Delta E\right)^2 \tag{30}$$

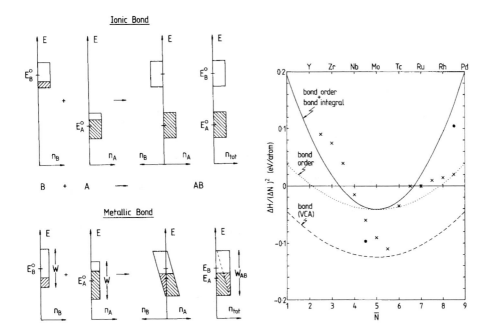

Fig. 16 Left-hand panel: Schematic representation of ionic and metallic bond formation in binary AB systems. Right-hand panel: The different contributions to the normalised heats of formation $\Delta H/(\Delta N_d)^2$ for the case of $4d$ transition metal alloys. The solid dots and crosses represent the experimental and Miedema values[85] respectively for disordered alloys with $\Delta N_d = 1$ or 2 where the second order approximation is valid.

where z is the coordination of the lattice, β is the covalent nearest neighbour bond integral, and $\Delta E = E_B - E_A$ is the energy level mismatch. Thus, the alloy band width may be related to the elemental bandwidth $W = \sqrt{12z}\,\beta$ through

$$W_{AB} = (1 + 3\delta^2)^{\frac{1}{2}} W \approx \left(1 + \frac{3}{2}\delta^2\right) W \tag{31}$$

where $\delta = \Delta E/W$.

This metallic model of the binary alloy is very different from the ionic model which is illustrated in the upper part of the left-hand panel in Fig. 16. In the ionic picture the binary alloy DOS is assumed to be a rigid superposition of the elemental ones, so that charge flows from the B site to the A site setting up an ionic bond through the electrostatic Madelung interaction. In the metallic picture, on the other hand, the binary alloy DOS comprises skewed partial DOS located on the A and B sites as shown in the figure. The amount of skewing is such as to guarantee local charge neutrality (LCN) as is required for a metal with its perfect screening. It follows from the model that

$$\delta_{LCN} \approx -\Delta N_d / \left[\frac{3}{5}\bar{N}_d(10 - \bar{N}_d)\right] \qquad (32)$$

Hence, in the vicinity of a half-full common band where the average number of d electrons $\bar{N}_d \simeq 5$, the atomic energy level mismatch for local charge neutrality will be given by

$$\Delta E_{LCN} \approx -\frac{1}{5}W\Delta N_d \simeq -\frac{2}{3}\Delta N_d \text{ eV} \qquad (33)$$

choosing a value of 10 eV for the bandwidth. This is to be compared with the difference in the d energy levels of free transition metal atoms of about -1 eV per valence difference across the $4d$ series.

The heat of formation arising from the bond contribution may now be obtained by comparing the bond energy of the disordered AB alloy with that of the elemental A and B transition metals, namely

$$\Delta H^{bond}_{order} = U^{bond}_{AB} - \frac{1}{2}\left(U^{bond}_A + U^{bond}_B\right) \qquad (34)$$

where the use of the label 'bond order' will become apparent later (cf. equation (41)).

It follows from the definition of the bond energy as the energy relative to the on-site atomic energy levels that the first term can be written

$$U^{bond}_{AB} = -\frac{1}{20}W_{AB}\bar{N}_d(10 - \bar{N}_d) - \frac{1}{4}(\Delta N_d)(\Delta E_{LCN}) \qquad (35)$$

Therefore, substituting equations (28), (31), (32) and (35) into equation (34) we find to second order that

$$\Delta H^{bond}_{order} = \left\{-\frac{1}{80} + \frac{5}{24}[\bar{N}_d(10 - \bar{N}_d)]^{-1}\right\} W(\bar{N}_d)(\Delta N_d)^2 \qquad (36)$$

where the explicit parabolic dependence of the d bandwidth across the series is inferred by the argument in $W(\bar{N}_d)$ (see, for example, Fig. 7.12 of Ref. 23). This is plotted as the dotted curve in the right-hand panel of Fig. 16 and comprises two contributions: the first contribution in equation (36) is attractive and corresponds to the virtual crystal approximation (VCA) in which the atomic energy level mismatch is neglected; the second contribution is repulsive and arises from the loss in covalent bond energy due to the mismatch in the on-sites energies E_A and E_B. We see that the bond order contribution to ΔH is negative for band fillings such that $2.11 < \bar{N}_d < 7.89$, but positive otherwise. The total heat of formation comprises an additional repulsive term arising from the difference in size of the A and B atoms. This is reflected in the different bond integrals $\beta_{AA}(R)$ and $\beta_{BB}(R)$ and leads to the term[83,84]

$$\Delta H^{bond}_{integral} = \frac{1}{100}(5 - \bar{N}_d)^2(\Delta N_d)^2 \tag{37}$$

The total heat of formation is now negative in the much smaller region about the centre of the band, namely for average band fillings such that $3\frac{1}{3} < \bar{N}_d < 6\frac{2}{3}$.

We see that the rectangular d band model reproduces the behaviour found by experiment and predicted by Miedema's semi-empirical scheme.[85] However, we must stress that the TB model does not give credence to any Pauling-like ionic model[86] since the atoms are perfectly screened in the metallic state and hence locally charge neutral. Instead the TB model supports the earlier suggestion by Brewer[87] that the most stable transition metal alloys would comprise elements from groups at the opposite ends of the transition metal series, such as Y and Pd. These groups have very few bonding electrons, since they have nearly empty or full d shells. Mixing these elements together results in a dramatic increase in their bond order as the electrons would be shared in the bonding states of the alloy corresponding to a half-full band, thereby leading to a sizeable lowering of their covalent bond energy. Subsequent, detailed density functional calculations have established the validity of the above simple rectangular d band model for the heats of formation of transition metal alloys.[88]

Pauling[86] had found empirically that there was a simple logarithmic relationship between the bond length R and the bond order Θ, namely

$$R_\Theta = R_1 - A \log_{10} \Theta \tag{38}$$

where A was a constant that depended on the particular element. For example, as is well known, the single, double and triple carbon–carbon bond lengths decrease from 1.51 to 1.34 to 1.20 Å as the bond order increases from 1 to 2 to 3. In order to account for fractional bonds that arise due to resonance in carbon systems or metals, Pauling defined the resonant bond order by $\Theta = N/z$, where N is the number of valence electrons per atom available for forming resonating electron-pair bonds between the z neighbours. Pauling showed that the above logarithmic relationship was valid not only for organic systems but also for transition metals where A took the value of 0.6 Å (provided N was suitably chosen across the transition metal series!).

This valence-bond concept of the bond order was quantified within the molecular-orbital framework by Coulson,[89] who defined the bond order between atoms i and j by

$$\Theta_{ij} = \sum_{n\ occ.} \left(c_i^{(n)*} c_j^{(n)} + c_j^{(n)*} c_i^{(n)} \right)/2 \qquad (39)$$

where $c_i^{(n)}$ are the eigenvectors associated with the atomic orbital $|i>$ in the eigenstate $|n>$. It is trivial to show (see, for example, sec. 4.6 of ref. 23) that the bond order is just one-half the difference between the number of electrons in the bonding state $(|i> + |j>)/\sqrt{2}$ compared to the antibonding state $(|i> - |j>)\sqrt{2}$, namely

$$\Theta_{ij} = \frac{1}{2}\left(N_{ij}^+ - N_{ij}^- \right) \qquad (40)$$

Thus, molecular hydrogen with both electrons in the bonding state has a bond order of unity corresponding to a saturated bond, whereas metallic hydrogen would display unsaturated bonds with a bond order much less than unity.

The beauty of Coulson's approach is that it allows the global description of energy bands with an associated total band energy to be broken down into a local description of covalent bonds with associated individual bond energies. In particular, for the case of a single orbital per site, the bond energy defined by equation (25) can be written

$$U^{bond} = \frac{1}{2}\sum_{i \neq j}\sum_{j} \left[2\beta(R_{ij})\Theta_{ij} \right] \qquad (41)$$

Thus, we recover the intuitive result that the bond energy between a given pair of atoms i and j is simply the product of the bond integral and the bond order (the prefactor 2 accounting for the spin degeneracy). If we now assume that the bond integral falls off exponentially with distance as $\beta = -B\exp(-\nu R)$ and the attractive bonding term is countered by a repulsive pair potential[90] $\phi = A\exp(-\mu R)$, then at equilibrium the bond length will satisfy the equation

$$R_\Theta = R_1 - [1/(\mu - \nu)]\log_e \Theta \qquad (42)$$

That is, Coulson's definition of the bond order allows one to recover Pauling's empirical logarithmic relationship, equation (38).

However, although the energy has been broken down in terms of individual bond energies, these still require the diagonalisation of the global Hamiltonian matrix in order to recover the eigenvectors $c_i^{(n)}$ that enter the definition of the bond order in equation (39). During the past decade a theory has been developed that allows the bond order to be written as an exact expansion over local hopping paths about the bond.[52,53,54,91] This bond-order potential (BOP) theory relies on a novel theorem[53] that expresses the off-diagonal Green's function matrix element G_{ij} (whose imaginary part is required for the bond order) as the derivative of a diagonal element. The proof is straightforward. Consider an orbital $|U_o^\lambda>$ that is an admixture of the two orbitals $|i>$ and $|j>$, namely $|U_o^\lambda> = \frac{1}{\sqrt{2}}(|i> + e^{i\gamma}|j>$ where $\lambda = \cos\gamma$. Then

$$G_{oo}^\lambda(E) = <U_o^\lambda|(E-\hat{H})^{-1}|U_o^\lambda> = \frac{1}{2}[G_{ii}(E) + G_{jj}(E)] + \lambda G_{ij}(E) \quad (43)$$

Hence, it follows at once that

$$G_{ij}(E) = \frac{\partial}{\partial\lambda} G_{oo}^\lambda(E) \quad (44)$$

The theory of how to expand diagonal elements of Green's functions in terms of hopping paths or moments is well developed.[92] The above theorem allows an exact many-atom expansion for the off-diagonal element G_{ij} to be derived, which for symmetric eigenspectra takes the form

$$G_{ij}(E) = \sum_{n=0}^{\infty} [G_{on}^{\lambda=o}(E)]^2 \delta a_n \quad (45)$$

where G_{on} are Lanczos Green's functions. The δa_n are related to the interference terms $\zeta_{p+1} = (H^p)_{ij}$ that link atoms i and j together through paths of length p.[92]

Very recently the convergence of this expansion has been speeded-up dramatically by constraining the poles of the intersite Green's function G_{ij} to be the same as those of the average on-site Green's function $\frac{1}{2}(G_{ii} + G_{jj})$.[91] This has allowed the derivation of explicit analytic expressions for the σ and π bond orders that accurately reproduce the TB results from direct matrix diagonalisation. For example, the σ bond order can be written

$$\Theta_{ij,\sigma}^{BOP} = \frac{1}{\sqrt{1 + \frac{2\Phi_{2\sigma} + \hat{\delta}^2}{\left[1 + \sqrt{(\Phi_{4\sigma} - 2\Phi_{2\sigma}^2 + \Phi_{2\sigma}^i \Phi_{2\sigma}^j)/\Phi_{2\sigma}}\right]^2}}} \quad (46)$$

where $\delta = (E_p - E_s)$. $\Phi_{2\sigma}$ and $\Phi_{4\sigma}$ are the 2-hop and 4-hop contributions about the bond that are illustrated in Fig. 17.

The accuracy of this 'classical' interatomic potential is illustrated in Table 1. We see that not only are the single, double and triple carbon-carbon bond orders well reproduced, but also the conjugate bonds in benzene and graphite. In addition, it is clear from the last column in Table 1 that this is the first 'classical' interatomic potential that handles correctly the formation of radicals within its remit, C_2H_5 remaining essentially a singly bonded system on the abstraction of a H atom from C_2H_6. Thus, the bond order really can be

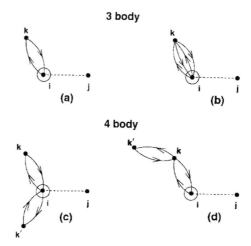

Fig. 17 Self-returning hopping paths of length 2 in (a) and length 4 in (b), (c) and (d), which contribute to the potential functions $\Phi_{2\sigma}$ and $\Phi_{4\sigma}$. There are an equivalent set of hopping paths starting from and ending on atom j.[91]

defined in terms of the local environment as Pauling had hoped with his simple expression for the resonant bond order, $\Theta = N/z$.[93]

Table 1 Comparison of the analytic bond-order potential (BOP) predictions for carbon-carbon bond orders with tight-binding (TB) matrix diagonalisation results.[91]

System	Local Coord.	β_σ^{CC} (eV) β_π^{CC} (eV)	Θ_σ BOP (TB)	Θ_{π_-} BOP (TB)	Θ_{π_+} BOP (TB)	Θ_{total} BOP (TB)
C_2	1	17.84	0.936	1.000	1.000	2.936
		2.76	(0.936)	(1.000)	(1.000)	(2.936)
C_2H_2	2	19.24	0.974	1.000	1.000	2.974
		2.98	(0.986)	(1.000)	(1.000)	(2.986)
C_2H_4	3	14.89	0.955	1.000	0.194	2.149
		2.30	(0.971)	(1.000)	(0.137)	(2.108)
C_6H_6	3	13.50	0.953	0.577	0.141	1.671
		2.09	(0.963)	(0.667)	(0.107)	(1.737)
C_{gr}	3	12.71	0.951	0.477	0.121	1.520
		1.97	(0.957)	(0.528)	(0.094)	(1.579)
C_2H_5	3.5	10.87	0.929	0.214	0.145	1.288
		1.68	(0.949)	(0.217)	(0.102)	(1.268)
C_2H_6	4	10.53	0.917	0.149	0.149	1.214
		1.63	(0.936)	(0.105)	(0.105)	(1.146)
C_6H_{12}	4	10.02	0.913	0.141	0.134	1.188
		1.55	(0.926)	(0.101)	(0.101)	(1.128)
C	4	10.02	0.915	0.126	0.126	1.167
		1.55	(0.912)	(0.103)	(0.103)	(1.118)

Conclusions

'This difference between Chemistry and Metallurgy is, of course, largely the result of the fact that the modern Chemical Industry has been developed chiefly from the scientific work of the laboratory, whereas Metallurgy has to a much greater extent grown from older traditional methods.'[94]

These words of Hume-Rothery, written 70 years ago, illustrate his hope that one-day the metallurgical industry would be underpinned by science rather than empiricism. He would be delighted to find today that the advent of quantitative electron theory has led to the 'Electronic Level' taking its rightful place at the base of the modelling hierarchies in Fig. 1. A glossy brochure from the Nippon Steel Corporation on *Computational Science for Material and Process Innovation* includes a section on the estimation of phase stabilities using the latest bandstructure techniques to solve the density functional equations.[95] Even though large gaps still exist between the modelling hierarchies, he would be encouraged by the rate at which these are being bridged by the many international, interdisciplinary, multiscale modelling programmes. In fact, the Accelerated Strategic Computational Initiative (ASCI) program in the United States to replace the experimental testing of weapons might lead him to feel that his worst nightmare had come true. As his daughter, Jennifer Moss, wrote to me recently: 'I am delighted to hear that there is to be a Hume-Rothery Symposium Celebration in St. Louis in October, and that the influence of his work continues to be felt. I should rather like to be a fly on the wall and hear what is said about H-R! At times I got the impression he feared that his very practical, experimental, approach to his subject might be upstaged by a more "modern" theoretical approach.' H-R need not have worried: experiment still remains as necessary as ever as materials modelling moves into the twenty-first century.

Acknowledgements

I should like to thank Dr Nguyen-Manh for plotting several of the figures, Andrew McKnight for photographic reproduction, and Kay Sims for high quality typing of the manuscript.

References

1. G. W. Raynor, *An Introduction to the Electron Theory of Metals*, Institute of Metals, London, UK, 1947.

2. W. Hume-Rothery, 'Researches on the structure of alloys', British Non-Ferrous Metals Research Association, Research Report No. 562, 1941.
3. W. Hume-Rothery, *Atomic Theory for Students of Metallurgy*, Institute of Metals, London, UK, 1946.
4. D. G. Pettifor, 'Computer-aided Materials Design: Bridging the Gaps Between Physics, Chemistry and Engineering', *Phys. Educ.*, May 1997, 164.
5. A. H. Cottrell and D. G. Pettifor, 'Models of Structure', *Structure*, W. Pullen and H. Bhadeshia eds, Cambridge University Press,, Cambridge, UK, 2000, chap. 2.
6. 'Atomistic theory and simulation of fracture', *MRS Bulletin*, 2000, **25**, 11.
7. Ch.-A. Gandin and M. Rappaz, 'A 3d Cellular Automaton Algorithm for the Prediction of Dendritic Grain Growth', *Acta Mater.* 1997, **45**, 2187.
8. Dierk Raabe, *Computational Materials Science: the Simulation of Materials, Microstructures and Properties*, Wiley VCH, Weinheim, FRG, 1998.
9. N. F. Mott and H. Jones, *The Theory of the Properties of Metals and Alloys*, 1936, Oxford University Press, Chapter V, Oxford, UK. (Mott writes in his Royal Society biography of Jones that 'my recollection is that his main contributions to our book were chapter I (§§ 6 and 7), chapter V, chapter VI (§6) and chapter VII (§§ 11 and 15)').
10. A. B. Pippard, 'An Experimental Determination of the Fermi Surface in Copper', *Phil. Trans. Roy. Soc. London*, 1957, **A250**, 325.
11. E. P. Wigner and F. Seitz, 'On the Constitution of Metallic Sodium', *Phys. Rev.*, 1933, **43**, 804.
12. P. Hohenberg and W. Kohn, 'Inhomogeneous Electron Gas', *Phys. Rev.*, 1964, **136**, B864.
13. W. Kohn and L. J. Sham, 'Self-Consistent Equations Including Exchange and Correlation Effects', *Phys. Rev.*, 1965, **140**, A1133.
14. N. Engel, Kemisk Maandesblad, 30 (1949) (5) 53; (6) 75; (8) 97; (9) 105; (10) 114 and L. Brewer, 'Thermodynamic Stability and Bond Character in Relation to Electronic Structure and Crystal Structure', *Electronic Structure and Alloy Chemistry of the Transition Elements*, P. A. Beck ed., Interscience Publishers, New York, NY, 1963, 221.
15. L. Pauling, 'Atom Radii and Interatomic Distance in Metals', *J. Amer. Chem. Soc.*, 1947, **69**, 542.
16. G. V. Raynor, 'The Next Stages in the Electron Theory of Metals', *Birmingham Metall. Soc. J.*, 1963, **43**, 16.
17. W. Hume-Rothery, *The Metallic State: Electrical Properties and Theories*, Clarendon Press, Oxford, UK, 1931.
18. H. Jones, 'The Phase Boundaries in Binary Alloys. II', *Proc. Phys. Soc.*, 1937, **A49**, 250.
19. H. Jones, 'Concentrated Solid Solutions of Normal Metals', *J. Phys. Radium*, 1962, **23**, 637.
20. J. Friedel, *Proc. Phys. Soc.*, 1952, **65**, 769.
21. A pure *sp* band model would not predict *fcc* to be the stable ground state structure of copper. Jones' prediction[18] that *fcc* is stable for $e/a = 1.0$ is probably due to errors in his approximations to the electronic structure and Brillouin zone integration.[22]

22. A. T. Paxton, M. Methfessel and D. G. Pettifor, 'A Bandstructure View of the Hume-Rothery Electron Phases', *Proc. Roy. Soc. Lond.*, 1997, **A453**, 1493.
23. D. G. Pettifor, *Bonding and Structure of Molecules and Solids*, Oxford University Press, Oxford, UK, 1995.
24. D. G. Pettifor, 'Pressure Cell-boundary Relation and Application to Transition-metal Equation of State', *Commun. Phys.*, 1976, **1**, 141.
25. D. G. Pettifor, 'Individual Orbital Contributions to the SCF Virial in Homonuclear Diatomic Molecules', *J. Chem. Phys.*, 1978, **69**, 2930.
26. T. B. Massalski and U. 'Mizutani, Electronic Structure of Hume-Rothery Phases', *Progr. Mat. Sci.*, 1978, **22**, 151.
27. H. Jones, *The Theory of Brillouin Zones and Electronic States in Crystals*, North-Holland, Amsterdam, The Netherlands, 1960.
28. E. Yu. Tsymbal and D.G. Pettifor, 'Effects of Bandstructure and Spin-independent Disorder on Conductivity and Giant Magnetoresistance in Co/Cu and Fe/Cr Multilayers', *Phys. Rev. B*, 1996, **54**, 15314.
29. V. Heine and D. L. Weaire, 'Pseudopotential Theory of Cohesion and Structure', *Solid State Physics*, 1970, **24**, 249.
30. D. Stroud and N. W. Ashcroft, Phase Stability in Binary Alloys, *J. Phys. F: Metal Phys.*, 1971, **1**, 113.
31. D. G. Pettifor and M. A. Ward, An Analytic Pairpotential for Simple Metals, *Sol. St. Com.*, 1984, **49**, 291.
32. Quantitative second-order perturbative calculations within the framework of density functional theory can be made for Cu and Zn by using non-local pseudopotentials which include the explicit presence of the $3d$-shell. (J. A. Moriarty, *Phys. Rev.*, 1988, **B38**, 3199 and references therein).
33. J. Friedel, 'The Distribution of Electrons Round Impurities in Monovalent Metals', *Phil. Mag.*, 1952, **43**, 153.
34. M. A. Ward, 'Analytic Pair Potentials for Simple Metals', Ph.D. thesis, 1985, Imperial College of Science, Technology and Medicine, London.
35. The heat of formation also depends on the volume-dependent free electron gas contribution to the energy (see Fig 6.8 of ref. 23).
36. J. E. Inglesfield, 'Pertubation Theory and Alloying Behaviour II. The Mercury-Magnesium System', *J. Phys.*, 1969, **C2**, 1293.
37. F. Ducastelle, *Order and Phase Stability in Alloys*, North-Holland, Amsterdam, The Netherlands, 1991.
38. P. E. A. Turchi, M. Sluiter, F. J. Pinski, D. D. Johnson, D. Nicholson, G. M. Stocks and J. B. Staunton, First-principles Study of Phase Stability in Cu–Zn Substitutional Alloys, *Phys. Rev. Lett.*, 1991, **67**, 1779.
39. J. Friedel and J. Stringer, 'Agenda Discussion: Pure Metals and Solid Solutions', *Phase Stability in Metals and Alloys*, P. S. Rudman, J. Stringer and R. I. Jaffee eds, 1967, 243, McGraw-Hill, Inc., New York, NY, 1967, 243.
40. In practice, Na transforms to the mixed chh close-packed stacking sequence of Sm below 5K.
41. A. K. McMahan and J. A. Moriarty, 'Structural Phase Stability in Third-period Simple Metals', *Phys. Rev.*, 1993, **B27**, 3235.

42. T. K. Wyatt, 'The Structural Stability of *sp*-bonded Metals', Third year undergraduate project, Imperial College of Science, Technology and Medicine, London, 1991.
43. J. A. Moriarty and A. K. McMahan, 'High-pressure Structural Phase Transition in Na, Mg and Al', *Phys. Rev. Lett.*, 1982, **48**, 809.
44. H. Olijnyk and W. B. Holzapfel, 'High-pressure Structural Phase Transition in Mg', *Phys. Rev.*, 1986, **B31**, 4682.
45. It is important not to confuse this phase shift α_3 that describes the *long-range* behaviour with the phase shifts from scattering theory that describe the *asymptotic* behaviour. The latter phase shifts are very small for the s and p electrons in nearly-free-electron metals.
46. D. G. Pettifor, 'Theory of Energy Bands and Related Properties of $4d$ Transition Metals: I. Band Parameters and their Volume Dependence', *J. Phys. F: Metal Phys.*, 1977, **7**, 613.
47. D. G. Pettifor, 'Theory of the Crystal Structures of Transition Metals at Absolute Zero', *Metallurgical Chemistry*, O. Kubaschewski ed., HMSO, London, UK, 1991, 191.
48. J. C. Duthie and D. G. Pettifor, 'Correlation Between d Band Occupancy and Crystal Structure in the Rare Earths', *Phys. Rev. Lett.*, 1977, **38**, 564.
49. J. A. Moriarty, 'Ultrahigh-pressure Structural Phase Transitions in Cr, Mo and W', *Phys. Rev.*, 1992, **B45**, 2004.
50. D. G. Pettifor, 'Accurate Resonance Parameter Approach to Transition Metal Band Structure', *Phys. Rev.*, 1970, **B2**, 3031.
51. F. Cyrot-Lackmann, 'On the Electronic Structure of Liquid Transition Metals', *Adv. Phys.*, 1967, **16**, 393.
52. D. G. Pettifor, 'New Many-body Potentials for the Bond Order', *Phys. Rev. Lett.*, 1989, **63**, 2480.
53. M. Aoki and D. G. Pettifor, 'Angularly-dependent Many-atom Bond-order Potentials Within Tight-Binding Hückel Theory', *Physics of Transition Metals*, P. M. Oppeneer et al. eds, World Scientific, Singapore, 1993, 299.
54. M. Aoki, 'Rapidly Convergent Bond-order Expansion for Atomistic Simulations', *Phys. Rev. Lett.*, 1993, **71**, 3842.
55. A. E. Carlsson, 'Angular Forces in Group VI Transition Metals: Application to W(100)', *Phys. Rev. B.*, 1991, **44**, 6590.
56. S. M. Foiles, 'Interatomic Interactions for Mo and W Based on the Low-order Moments of the Density of States', *Phys. Rev. B*, 1993, **48**, 4287.
57. J. A. Moriarty, 'Angular Forces and Melting in *bcc* Transition Metals: A Case Study of Molybdenum', *Phys. Rev. B*, 1994, **49**, 12431.
58. F. Ducastelle and F. Cyrot-Lackmann, 'Moments Developments. II. Application to the Crystalline Structure and Stacking Fault Energies of Transition Metals', *J. Phys. Chem. Sol.*, 1971, **32**, 285.
59. The general four-member ring expression in terms of bond angles and dihedral angles, is published in D. G. Pettifor, M. Aoki, P. Gumbsch, A. P. Horsfield, D. Nguyen-Manh and V. Vitek 'Defect Modelling: the Need for Angularly Dependent Potentials', *Mater. Sci. Eng. A*, 1995, **192/193**, 24.

60. J. A. Moriarty, 'Density-functional Formulation of the Generalized Pseudopotential Theory. III. Transition Metal Interatomic Potentials', *Phys. Rev. B*, 1988, **38**, 3199.
61. H. L. Skriver, 'Crystal Structure from One-electron Theory', *Phys. Rev. B*, 1985, **31**, 1909.
62. L. Kaufman and H. Bernstein, *Computer Calculations of Phase Diagrams*, Academic Press, New York, NY, 1970.
63. J. Hafner, 'Quantum Theory of Structure', *The Stuctures of Binary Compounds*, F. R. de Boer and D. G. Pettifor eds, North-Holland, Amsterdam, The Netherlands, 1989, 179.
64. P. J. Craievich, M. Weinert, J. M. Sanchez and R. E. Watson, 'Structural Instabilities of Excited Phases', *Phys. Rev. B*, 1997, **55**, 787.
65. A. P. Miodownik, 'Compatibility of Lattice Stabilities Derived by Thermochemical and First Principles', in *Structural and Phase Stability of Alloys*, J. L. Morn-López, F. Mejia-Lira and J. M. Sanchez eds, Plenum Press, New York, NY, 1992, 65.
66. L. Brewer, *Phase Stability in Metals and Alloys*, P. S. Rudman, J. Stringer and R. I. Jaffee eds, McGraw-Hill, 1967, 246, New York, NY.
67. G. Shao, D. Nguyen-Manh and D. G. Pettifor, 'On the Interplay of Omega Formation and Chemical Ordering in Titanium Alloys', in *Titanium '95*, B. A. Blenkinsop et al. eds, Institute of Materials, 1996, **3**, 2289.
68. D. Nguyen-Manh, D. G. Pettifor, G. Shao, A. P. Miodownik and A. Pasturel, 'Metastability of the ω-phase in Transition-metal Aluminides: First-principles Structural Predictions', *Phil. Mag. A*, 1996, **74**, 1385.
69. D. G. Pettifor and R. Podloucky, 'Microscopic Theory of the Structural Stability of *pd*-bonded *AB* Compounds', *Phys. Rev. Lett.*, 1984, **53**, 1080.
70. D. G. Pettifor, 'Structure Maps for Pseudo-binary and Ternary Phases', *Mat. Sci. Tech.*, 1988, **4**, 657.
71. D. G. Pettifor, 'Phenomenology and Theory in Structural Prediction', *J. Phase Equilibria*, 1996, **17**, 384.
72. G. V. Raynor, 'William Hume-Rothery', *Biographical Memoirs of Fellows of the Royal Society*, 1969, **15**, 126.
73. D. G. Pettifor, 'Theory of Energy Bands and Related Properties of $4d$ Transition Metals: III. s and d Contributions to the Equation of State', *J. Phys. F: Metal Phys.*, 1978, **8**, 219. (For simplicity we have omitted prefactors of order unity from the U_{sp}^{ke} and U_d^{bond} contributions in equations (22) and (23).)
74. The *first step* of reference 24 in deriving these equations was made by replacing the surface integral over the Wigner–Seitz cell by a surface integral over the Wigner–Seitz sphere in Liberman's virial expression (D. A. Liberman, *Phys. Rev.*, 1971, **B3**, 2081). Reference 24 pointed out that an identical expression would have been obtained by following Nieminen and Hodges' derivation of the first order change in total energy under change in volume, provided we keep a term linear in the logarithmic derivative which they argued vanished over the cell boundary (R. M. Nieminen and C. H. Hodges, Report TKK-F-A251, 1975).
75. O. K. Andersen, 'Simple Approach to the Bandstructure Problem', *Sol. State Commun.*, 1973, **13**, 133.

76. For a free electron gas, $B_s = \phi(S) = \frac{4}{3}\varepsilon_{xc}$, so that the first and second terms in equation (22) are just the potential energy, Φ, and twice the kinetic energy, T, respectively, thereby satisfying the viral theorem $3PV = 2T + \Phi$.
77. D. G. Pettifor, 'A Physicist's View of the Energetics of Transition Metals', *Calphad*, 1977, **1**, 305.
78. C. D. Gelatt, H. Ehrenreich and R. E. 'Watson, Renormalized Atoms: Cohesion in Transition Metals', *Phys. Rev. B.*, 1977, **15**, 1613.
79. J. Friedel, 'Transition Metals. Electronic Structure of the *d*-band. Its Role in the Crystalline and Magnetic Structures', *The Physics of Metals*, J. M. Ziman ed., Cambridge University Press, New York, NY, 1969, chapter 8.
80. V. Heine and J. Hafner, 'Volume and Pair Forces in Solids and Liquids', *Springer Proc. Phys.*, 1990, **48**, 14.
81. L. Pauling, 'The Nature of the Interatomic Force in Metals', *Phys. Rev.*, 1938, **54**, 899.
82. D. G. Pettifor, 'Theory of the Heats of Formation of Transition Metal Alloys', *Phys. Rev. Lett.*, 1979, **42**, 846.
83. D. G. Pettifor, 'A Quantum-mechanical Critique of the Miedema Rules for Alloy Formation', *Sol. State Phys.*, 1987, **40**, 43.
84. D. G. Pettifor, 'The Tight-binding Approximation: Concepts and Predictions', *Electron Theory in Alloy Design*, D. G. Pettifor and A. H. Cotrell eds, Institute of Materials, London, UK, 1992, chapter 4.
85. F. R. de Boer, R. Boom, W. C. M. Mattens and A. R. Miedema, *Cohesion in Metals: Transition Metal Alloys*, North-Holland, Amsterdam, The Netherlands, 1988.
86. L. Pauling, *The Nature of the Chemical Bond*, Cornell University, Ithaca, NY, 1960.
87. L. Brewer, 'Bonding and Structure of Transition Metals', *Science*, 1968, **161**, 115.
88. A. R. Williams, C. D. Gelatt and V. L. Moruzzi, 'Microscopic Basis of Miedema's Empirical Theory of Transition-metal Compound Formation', *Phys. Rev. Lett.*, 1980, **44**, 429.
89. C. A. Coulson, 'The Electronic Structure of Some Polyenes and Aromatic Molecules. VII, Bonds of Fractional Order by the Molecular Orbital Method', *Proc. Roy. Soc. Lond.*, 1939, **A169**, 413.
90. A. P. Sutton, M. W. Finnis, D. G. Pettifor and Y. Ohta, 'The Tight-binding Bond Model', *J. Phys. C: Solid State Phys.*, 1988, **21**, 35.
91. D. G. Pettifor and I. I. Oleinik, 'Bounded Analytic Bond-order Potentials for σ and π Bonds', *Phys. Rev. Lett.*, 2000, **84**, 4124.
92. D. G. Pettifor and M. Aoki, 'Bonding and Structure of Intermetallics: A New Bond Order Potential', *Trans. R. Soc. Lond. A*, 1991, **334**, 439.
93. The simplest form of BOP theory predicts that $\Theta = 2/\sqrt{z}$ for three-dimensional *sp* valent systems with $N = 4$, corresponding to bond orders of 1.00, 0.82 and 0.58 for diamond ($z = 4$), simple cubic ($z = 6$) and close-packed ($z = 12$) respectively. However, this simplest level of approximation does not predict any structural differentiation, the cohesive energies being independent of z under the realistic assumption that the repulsive pair potential falls off with distance as the square of the bond integral (cf. eqs 7.49 and 8.11 of Ref. 23).

94. W. Hume-Rothery, *The Metallic State. Electrical Properties and Theories*, Clarendon Press, Oxford, UK, 1931.
95. *Computational Science for Material and Process Innovation*, Nippon Steel Corporation, Japan, 1993.

Virtual Materials

ALAN WINDLE
Professor, University of Cambridge

ABSTRACT

Computational modelling is growing rapidly in capability and significance and its application to the simulation of the structure, properties and processing of materials is an important new area of our discipline. The inexorable increase in cheap computer power is one driver for the current growth, the other being the ability to model levels of structure as a hierarchy. Nevertheless, computer resources are finite, and the so called multi-scale modelling approach is a key which unlocks the door, particularly for systems which are rich in microstructure. Polymer systems will be used to illustrate these precepts, which are of course more widely applicable as will be shown by their extension to packed aggregates.

However, while it is enlightening and informative to simulate basic internal materials processes, such as atomic and molecular movement, liquid flow, microstructural development, and response to external fields, such as stress, the full future potential of modelling lies with the interpolation and extrapolation of experimentally determined structure–property space to give us materials which do not yet exist. Available data will be used in concert with modelling predictions, with the role of the former being to provide datums, or 'spot heights' while the 'contours' of the topography will be put in by the science driving the models.

Atomistic Simulation of Materials

ROBIN W. GRIMES, K. J. W. ATKINSON, MATTHEW O. ZACATE
and MOHSIN PIRZADA
*Department of Materials, Imperial College, Prince Consort Road,
London, SW7 2BP, UK*

Abstract
The aims and objectives of atomistic simulation are reviewed. These ideas are then illustrated by examining two different simulation studies. The first concerns transport of oxygen through a ceramic lattice. It employs energy minimisation and forces between ions are described using the ionic model. The second illustrates the evolution of a gas atom layer on a metal surface. This simulation is based on a cellular automata and forces are described using a short-range model. The results of both simulation studies emphasise the role that defects play in controlling the properties and structures of materials.

Overview

The result of carrying out an atomistic simulation is that the property of a material is predicted. Examples of properties include thermodynamic quantities such as total energy and specific heat or structural properties such as atomic configurations and kinetic energy distributions. Thus, the model on which the simulation is based is dictated by what property it is necessary to predict although nearly all involve the evaluation of an energy. In computer simulations the models are the sets of equations or algorithms. Of course, for the model to be atomistic, the equations must describe or imitate atomic scale interactions.

At the present time there are an enormous number of types of atomistic computer simulations covering all levels of approximations.[1,2] However, the majority of simulations have been carried out relatively recently. Prior to 25 years ago it had only been possible to simulate a few simple systems and even these were subject to approximations that may now seem crude.[3] Nevertheless, the models on which today's simulations are based often began their development much longer ago. Quantum mechanical simulations, for example, owe much to specific progress made between 1930 and 1960.[3,4] Likewise, Born and Mayer described the classical ionic model in 1932[5,6] and its use in predicting defect energies was suggested by Mott and Littleton in 1938.[7]

Of course developments in basic methodology continue apace but they are now driven by the demand to simulate increasingly complex systems. In fact, we are still faced by the same dilemma as were those who carried out the first simulations. That is, the desire to use the highest quality model while carrying out the simulation within a reasonable time scale given the computational facilities available. Practical experience suggests that this translates into each separate simulation taking between half an hour and two days. Simulations have never become quicker and, of course, we still cannot afford to wait any longer for the results!

The question is, why have simulations not become faster? The answer is, of course, that the increased processing power has been employed in increasing quality not increasing speed. Furthermore, processor power is still progressing via Moore's Law.[8] However, a cursory inspection of what is entailed in simulating a real material draws the inevitable conclusion that a universally reliable atomic scale model is unlikely to evolve within the foreseeable future.

The reasons why a universal model is not yet possible are well known to the perpetrators of simulations. First, the number of atoms that must be modelled in a real material is dauntingly large. Of course, for perfect crystals it is possible to circumvent the problem by using periodic boundary conditions.[9] Unfortunately, most real materials are full of atomic scale defects and it is the defects that control the majority of properties (this will be one focus of the present article). Consequently, techniques that can generate distributions of defects within a lattice are themselves becoming a focus for model development.[10–12]

The second issue that presently bodes against a universal model are the great differences between the description of forces in metals, ceramics and polymers. Each material type has its own advantages and disadvantages with respect to its simulation. For example, ceramics can be described using effective potentials but their Coulomb interactions are very long-range.[13] Conversely interactions in metals are shorter range but many-body.[14] Computational efficiencies therefore result when a model is developed for a specific class of material. Unfortunately this brings particular difficulties when simulating systems which involve more than one class of material such as metal ceramic interfaces.[15]

Finally, it is becoming clear that even the largest molecular dynamic simulations presently possible[16] (i.e. 10^6–10^7 species for ~ 100 ns) are still limited by specific approximations. Nevertheless, recent studies on grain growth[17,18] have shown that very useful and surprising results can be gleaned from present levels of simulation. Consequently it is through the simulation of dynamic processes that many important advances in the understanding of materials will occur over the next 20 years. In particular, we are beginning to observe complex non-linear mechanisms for bulk transport[19] and surface growth processes,[20] which hitherto have not been possible to simulate or observe experimentally.

Nevertheless, at the present time, the first crucial step in an atomistic simulation project is choosing the appropriate model. This will depend foremost on the experimental details available. In particular, is the system and property well defined? This means that the processes involved are clear. Consequently, an appropriate model can be identified and boundary conditions such as number of particles can be selected. The aim of such a problem may be, for example, to simulate different compositions and thereby predict an optimum composition. If the system is not well defined this implies that the rate determining process is not known; there may be the potential for multi-factorial issues. In the two cases, the relationship between the simulation and experiment becomes quite different. For a well defined system the prediction or recommendation can be tested. When the system is not well defined, it is hard to assess if the simulation is reproducing experiment. Consequently a more iterative process may be necessary.

In what follows we will consider two very different problems. In the first case, a well defined problem is addressed. This concerns the prediction of the activation energy for oxygen migration in a series of pyrochlore oxides. In the second case a poorly defined problem is addressed. This concerns the formation and growth of a monolayer of gas atoms on a metal surface.

Oxygen Migration in Pyrochlore Oxides

Introduction

$A_2B_2O_7$ pyrochlore materials are being considered for use in applications ranging from solid-oxide fuel cells[21,22] or catalysts[23] to potential phases for the immobilisation of actinides in nuclear waste.[24,25] The way in which ions are transported through the lattice is central to each of these applications. Consequently, oxygen ion conduction mechanisms have been investigated using atomic scale computer simulation techniques based on classical pair-potential models. The aim has been to identify A and B component cations that yield the lowest migration activation energy. The results are presented in the form of a contour map of activation energy so that combinations of elements that may yield a low energy can be easily selected.

Crystal Structure

This study deals with so-called (3–4) pyrochlore compounds, i.e. A is nominally a 3+ valence cation and B is a 4+ valence cation. The pyrochlore structure (see Fig. 1) is closely related to fluorite and can be considered as an ordered defective fluorite. It exhibits space group Fd$\bar{3}$m with eight formula units within the cubic unit cell. If we write the general formula as $A_2B_2O_6O'$, and choose to fix the origin on the B site,[26] the ions occupy four crystallographically nonequivalent positions: A at 16d, B at 16c, O at 48f and O' at 8b. There is an interstitial site that, within this space group, is at 8a. This will become important when we consider migration of oxygen ions between 48f sites.

Model

The atomic scale computer simulation technique used here is based on energy minimisation with a Born-like description of the lattice.[5,6] The interactions between ions are composed of two terms: long-range coulombic forces, which are summed using Ewald's method,[27] and short-range forces, which are modelled using parameterised pair potentials. The perfect lattice is described by defining a unit cell, which is repeated throughout space using periodic boundary conditions as defined by the usual crystallographic lattice vectors. The lattice energy is:

$$U_L = \frac{1}{4\pi\varepsilon_0} \sum_{i \neq j} \frac{q_i q_j}{r_{ij}} + Ae^{-r_{ij}/\rho} - \frac{C}{r_{ij}^6}$$

where A, ρ and C are the adjustable parameters, r_{ij} is the interionic separation and q is the charge on the ion. These adjustable parameters were chosen to accurately reproduce the lattice parameters of 54 pyrochlore structures.[28]

Since oxygen migration is mediated by defects, an approach is needed that is able to simulate the effect of a defect on the surrounding lattice ions. In other words, the calculations of defect energies must include the resulting

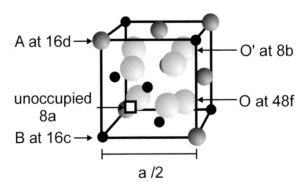

Fig. 1 Partial unit cell of the pyrochlore structure.

structural relaxation or displacement polarisation. This is achieved by partitioning the energy-minimised perfect lattice into three concentric spherical regions.[29] In region I ions are treated explicitly and relaxed to zero strain. The defect is positioned near the centre of this region. Region IIa is an interfacial region where forces between ions are determined via the Mott-Littleton approximation.[7] The interaction energies between ions in region IIa and region I are calculated explicitly. The outer region IIb extends to infinity and provides the Madelung field of the remaining crystal. The relaxation energy of ions in region IIb is determined using the Mott–Littleton approximation.[7]

Interestingly this type of simulation was first attempted by Tharmalingham in 1971, in this department.[30]

Migration Simulations

Oxygen ion conduction in pyrochlores, like in fluorites, proceeds via an oxygen vacancy mechanism.[31] The migration mechanism consists of sequential jumps of oxygen ions into vacant sites. The activation energy for migration is then defined as the difference between the energy of the system when the migrating oxygen ion is at the saddle point and the energy of the oxygen vacancy at equilibrium. The saddle point energy is calculated by introducing a fixed oxygen ion at the saddle point location and then relaxing the surrounding lattice. Evaluating the potential energy surface both parallel and perpendicular to the diffusion path identifies the configuration of the saddle point.

In the pyrochlore lattice, the lowest energy contiguous pathways for oxygen ion migration are provided by jumps from a $48f$ to a vacant $48f$ site along <100> directions.[32–34] The first step in this study is therefore to model the structure of the $48f$ oxygen vacancy.

Clearly, this is a well defined problem. The atomic configuration (in this case the crystal structure) is known in detail, the model parameters can be selected and the simulation will yield an energy that can be directly compared with experimental data, if it is known. However, this apparent simplicity conceals complex processes.

Results and Discussion

The 48f Vacancy

The structure of the $48f$ vacancy was determined for 54 $A_2B_2O_7$ compounds: A = Lu to La, B = Ti to Pb. In many cases, the lowest energy configuration is a single vacancy with a symmetric relaxation of neighbouring ions: towards the vacancy for anions, away from the vacancy for cations. However, some compounds exhibit a different lowest energy configuration. This is formed when a 48f oxygen ion adjacent to the vacancy relaxes considerably towards the unoccupied 8a site. This forms the so-called split vacancy[32] consisting of two 48f vacancies, orientated along <110>, and an interstitial oxygen ion. However, the interstitial ion never actually occupies the 8a site being symmetrically displaced away from 8a towards the two 48f sites (see Fig. 2).

Implication to the 48f–48f Migration Mechanism

For compounds that do not exhibit split vacancy formation, the 48f to 48f <100> mechanism will be a simple activated hopping process. All oxygen ion jumps will be along <100> directions. For those compounds that exhibit a split vacancy, the overall motion of the oxygen ions is not along <100>. In fact, what has occurred is the reorientation of a split vacancy associated with one 8a site to a new 8a site.[28] If we assume the oxygen interstitial occupies the 8a site, the overall mass transport is, on average, in a <111> direction.

Calculated Activation Energies

The activation energies for oxygen migration are presented in the form of an energy contour map (see Fig. 3). The ordinate shows A cation radius, the abscissa shows B cation radius. Equal activation energies are connected by contours. Thus compounds on the same contours are predicted to exhibit the same activation energies. The white points in Fig. 3 are compounds for which a calculation was made and from which the contours were constructed.

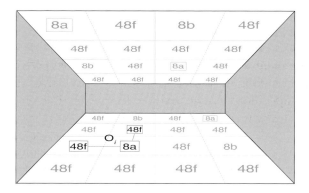

Fig. 2 The structure of the split vacancy in $A_2B_2O_7$ compounds. □ indicates an anion lattice site that is not occupied by an oxygen ion, O_i is the oxygen interstitial ion in the split vacancy. Two adjacent anion (001) sublattice planes are shown; $z = \frac{3}{8}$ and $z = \frac{5}{8}$.

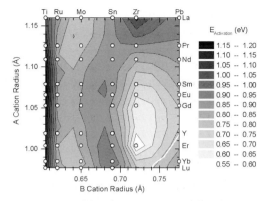

Fig. 3 Contour map of oxygen migration activation energy (eV).

To assess the reliability of these predictions we must compare the calculated activation energies for specific compounds with the available experimental data. Although data is limited, there is generally excellent agreement.[28] However, for $Gd_2Zr_2O_7$, the experimental values range from 0.73–0.90 eV[33,35,36] but the predicted activation energy is 0.58 eV. At first sight it seems that this prediction is somewhat low. However, it is well established that $Gd_2Zr_2O_7$ exhibits significant disorder[33,35] and this will increase the experimental activation energy.[34] Nevertheless, although the value presented in the present work assumes a fully ordered $Gd_2Zr_2O_7$ lattice, it does provide an estimate of the activation energy for local oxygen rearrangements even though it will not mirror the long-range migration value of a significantly disordered material.

Summary

Given the agreement with experiment described above, we may now use the activation energy contour plot (Fig. 3) in a predictive capacity to select pyrochlore compositions which exhibit low activation energies. However, the onset and effect of disorder makes prediction less straightforward. Compounds that exhibit higher degrees of disorder also display higher activation energies. Therefore, although $Y_2Zr_2O_7$ for example might seem to have an especially low activation energy, the extent of cation disorder in this compound and the difficulty in forming such a stoichiometry[37] causes a significant increase in activation energy. Conversely, $Sm_2Zr_2O_7$ exhibits a much lower disorder (<3%)[38] which is sufficiently low not to unduly increase the activation energy. As such $Sm_2Zr_2O_7$ will exhibit the lowest activation energy although compounds just above (i.e. $Nd_2Zr_2O_7$) and to the right (e.g. $Gd_2Sn_2O_7$) should also be considered. A guide to the extent of disorder expected in all these materials is provided in Ref. 38.

Cellular Automata Simulation of Surface Evolution

Introduction

The evolution of surface structure is very difficult to investigate experimentally. Consequently, there is considerable interest in developing an atomic scale predictive capability that allows the user to include, explicitly, the role of all types of chemical and crystallographic defects.

A number of atomistic-scale simulation techniques exist (e.g. molecular dynamics) that could potentially be used as the basis for predicting surface structure evolution. However, most become computationally challenging when the number of atoms reaches the point necessary to simulate the evolution of a large surface area. As such, we have developed a Monte Carlo-related methodology based on cellular automata (CA).

Formally, CA are discrete dynamical systems, the behaviour of which are completely specified in terms of local interactions. Typically, the evolution of a CA is controlled by updating the state of each discrete cell on the basis of a rule involving only the states of adjacent cells. The entire cell assembly is updated simultaneously in an update step that may be taken to represent a single step in time. Clearly such an approach is only useful if the interactions are short-range. This is the case in the system investigated here, which is comprised of a layer of argon-gas atoms on a calcium (111) surface. The simulation results in the formation and growth of domains, with the domain size increasing as a function of simulation temperature.

Crystallography

The (111) surface of a calcium exhibits a close packing of calcium atoms. These may be regarded as occupying c sites. When a gas atom attaches itself to the surface, it will reside in a site formed by three metal atoms since this maximises its interaction with the metal surface. Two sets of interstitial sites exist, which can be regarded as a sites and b sites (see Fig. 4).

If we examine the (111) surface more closely, it is clear that the distance between an interstitial b site and a nearest neighbour a site is rather short. Indeed, a b site gas atom in Fig. 4 overlaps considerably onto the adjacent a sites. As such, if a gas atom resides in a b site, it is not possible for a second gas atom to occupy a nearest-neighbour a site due to steric hindrance (as indicated in Fig. 4). It is important to realise that this is true only because of the

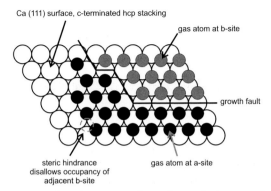

Fig. 4 Surface structure showing *c*-terminated metal plane and gas atoms at *a* and *b* sites.

specific matching of the size of the calcium (111) surface to the size of the argon atoms.

When the first few gas atoms impinge on the surface in the evolution of a gas-atom layer, they may reside in either *a* or *b* sites. As more atoms collect, patches or islands of *a*- and *b*-type atoms form. The distance between two neighbour *a*-type gas atoms (i.e. in next-nearest-neighbour sites) is the same as between two *b*-type gas atoms (see Fig. 4). However, the distance between allowed, non-hindered *a*- and *b*-type atoms is greater (i.e. between third or fourth neighbours). This results in a surface area with a lower density of gas atoms, termed a growth fault. In three dimensions, this would give rise to a stacking fault. Faults are important to the stability of a gas atom on the surface since the interaction energy of gas atoms is lower when they interact over longer distances. Therefore, atoms adjacent to growth faults have lower energy. This is the driving force that results in the formation and growth of islands of gas atoms that occupy only one type of site.

Methodology

Atomic Scale Interactions

In this study, two types of interactions are considered: the interaction between inert gas atoms, and the interaction between inert gas atoms and the metal substrate. Both are described by the Lennard-Jones potentials; if R is the distance between the atoms i and j, the interaction energy between two atoms, E_{ij}, is given by

$$E_{ij} = a_{ij}R^{-12} - b_{ij}R^{-6}$$

where *a* and *b* are constants specific to the atoms i and j. Detailed derivations have been provided previously.[39]

Using these potentials, it is possible to calculate the interaction energies that define the stability of an argon atom at all possible sites on the surface.[39] These will be a function of the number and positions of neighbouring gas atoms. They are then used to determine the probabilities of events occurring on a stochastic basis, which forms the basis for the time evolution of the surface. This link between the local atomic structure and the probability for adsorption, P_{ads}, or desorption, P_{des} is the key feature of the model.[39]

The CA Simulation

This is implemented in three stages per time step. The first stage randomly assigns thermal energies to atoms present on the surface and removes them according to their desorption probability, P_{des}. The probability of desorption decreases with more neighbours. The second stage randomly assigns lattice locations where incoming gas atoms interact with the surface and determines whether or not each incident gas atom adsorbs depending on their adsorption probability, P_{ads}. The probability of adsorption increases with more neighbours. The third stage updates the periodic boundary conditions of the edges of the surface. Further results can be found in previous publications.[39,40]

Surface Roughness

Here results are presented graphically for four groups (see Fig. 5). The first group of results concerns evolution on a perfectly flat (111) surface. Subsequent results consider various types of surface roughness; however, in each case, the degree of roughness, characterised by the number of additional calcium atoms on the (111) surface, is constant at 10% coverage. Thus, when it is assumed that the additional calcium atoms are randomly distributed, one in every 20 surface sites contains a calcium atom (the maximum number of surface sites that could be occupied is 50%, i.e. all *a* or all *b* filled, which would correspond to 100% coverage). Since the distribution is random, no possible aggregation of surface calcium atoms to form islands has been allowed. The opposite scenario was also considered, in which all additional calcium atoms aggregated to form a hexagonal (i.e. roughly spherical) island occupying only B sites. Nevertheless, such an island still has a very small radius, on the scale of a typical microstructure. Therefore, in the final group of results, the additional atoms form a canyon-like structure that has an effectively infinite radius of curvature. All simulations were carried out at 120 K. Results derived at other temperatures have been discussed separately.[39,40]

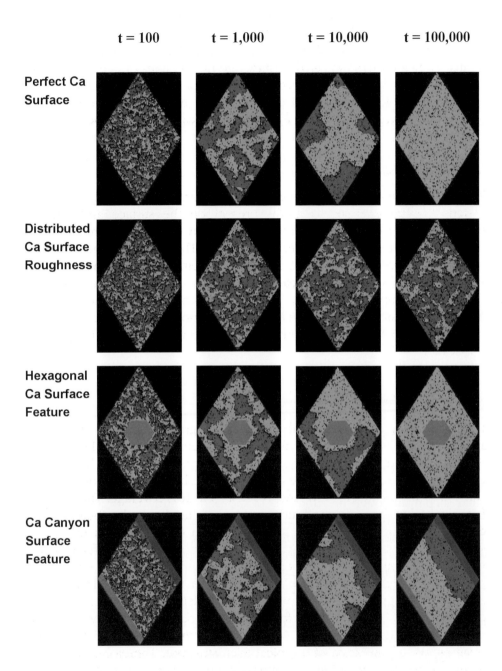

Fig. 5 Surface structure evolution subject to four different types of surface roughness.

Results and Discussion

Starting with an empty, perfectly flat surface, the coverage initially increases as a function of time as empty surface sites become occupied. Once the majority of sites are occupied, the surface begins to evolve. This is characterised by domain growth, with the smaller domains being consumed. After 100 000 time steps the surface is covered by gas atoms occupying only one type of site (see Fig. 5).

Figure 5 also presents data for the hexagonal feature. In this case, the time evolution for coverage is practically identical to that found for the perfect surface. However, it was observed that, to a certain extent, the growth proceeded from the hexagon. Thus the hexagon may affect the initial nucleation and growth, but this is a local process, which, after a certain point, becomes unimportant to the slower growth process.

Observations on the canyon feature (see Fig. 5) show that it has a number of characteristics in common with the hexagon. For example, the canyon causes nucleation and growth of same-site gas atom structures over the first 1000 time steps adjacent to the feature; this stabilises the growth of the single-site domains. However, since the canyon feature is composed of two different sites opposite each other, the result is growth of two different site domains. After 10 000 time steps the microstructure is that of an intergrowth. The remaining 90% of the simulation time (i.e. to 100 000 time steps) is required for a clear bimodal microstructure to develop.

Finally, Fig. 5 shows the surface on which the 10% roughness is randomly distributed. This example shows quite a different final behaviour to that of the perfect, hexagonal, or canyon surfaces. Nevertheless, after 100 time steps, the distributed rough surface morphology is very similar to that of the perfect surface. Even after 1000 time steps the rough and perfect surfaces are somewhat similar, although the domains of the perfect surface are a little larger. However after 10 000 time steps the surfaces are quite different, with the domain size of the distributed roughness surface showing little change. It is tempting to state that the domain size has been pinned, but by 100 000 time steps there is some indication that more growth has occurred, although the effect is small considering that the number of time steps has increased by an order of magnitude. It would clearly be interesting to increase the number of time steps by at least another order of magnitude to see if the surface is truly pinned or just strongly retarded.

An interesting observation is that for the distributed roughness surface, the domain size at 1000 time steps and beyond is greater than the distance between the additional surface calcium atoms. Consequently, each domain contains a number of additional calcium atoms. Furthermore, we observe that the surface calcium atoms do occupy both *a* and *b* sites within each domain;

clearly, the additional calcium atoms are not simply acting as classical nucleation points. The reason the microstructure becomes contained in the way that it does must be due to a complex interplay between the specific surface roughness distribution and the initial random deposition process over the first few hundred time steps.

Summary

Despite the apparent simplicity of this surface structure model these simulations reveal a complex interplay of physical processes that results in surface structural evolution. It would be interesting to further analyse the relationship between roughness distribution and the resulting microstructure, including varying the amount of surface roughness and the gas-atom deposition rate. Other distributions of surface roughness, intermediate between the totally random and clustered features presented in this work, would also be possible. Finally, this type of simulation can be expanded to three dimensions and other physical processes such as surface migration could be included.

Concluding Comments

The two atomic scale simulation studies presented here employ different simulation techniques to investigate quite different properties. The first study is concerned with migration activation energies in oxides, the second with surface evolution on a metal. However, both highlight the same broad issue. That is, our poor predictive understanding of the role that defects play in governing the properties of materials. Of course, there has been a broad appreciation of the importance of defects for many decades.[41] Unfortunately, in most real materials the defect populations are relatively small, pseudo-randomly distributed (pseudo- because they form strong associations or clusters) and more than one defect species is usually present. Consequently defects are often hard to evaluate experimentally. Also, it is necessary to simulate a large number of species in order to develop a meaningful atomic distribution. Therefore atomic scale simulations have also been challenging, although progress is being made.[10–12] One consequence of having to simulate a large number of particles is that for the foreseeable future many defect related problems will continue to calculate energies approximately. Models which employ state-of-the-art *ab initio* quantum mechanical techniques to determine the interaction energies will not be capable of simulating sufficient numbers of atoms.[1]

The ways in which interaction energies are used will also become more sophisticated. This will become particularly true when it is necessary to simulate systems where multiple physical processes are occurring (e.g. physorption, chemisorption and migration). With so many processes included it will become harder to identify the rate determining or controlling steps. Therefore, the analysis tools that interpret large volumes of simulation data will be as important as the simulation methods themselves. This is partly because, even as simulation methods become more proficient, it will still be important to extract physical understanding. That will continue to be aided by developments in computer graphics visualisation.

Acknowledgements

Computing facilities were also provided by the EPSRC, grant number GR/M94427.

References

1. E. Wimmer, *Mat. Sci. Eng.*, 1996, B, **37**, 72.
2. J. M. Newsam and E. Wimmer, *Curr. Opin. Solid St Mat. Sci.*, 1999, **4**, 491.
3. J. R. Chelikowsky, *J. Phys. D: Appl. Phys.*, 2000, **33**, R33.
4. P. O. Löwdin, *Lecture Notes in Chemistry*, Vol. 44, Springer-Verlag, 1986, 1.
5. M. Born and J. E. Mayer, Z. *Physik*, 1932, **75**, 1.
6. M. Born and J. E. Mayer, *J. Chem. Phys.*, 1933, **1**, 270.
7. N. F. Mott and M. J. Littleton, *Trans. Faraday Soc.*, 1938, **34**, 485.
8. G. E. Moore, *Electronics*, 1965, **38**, 114. See also, *http://www.intel.com/research/silicon/moorespaper.htm*.
9. N. W. Ashcroft and N. D. Mermin, *Solid State Physics*, Saunders College, 1976.
10. J. A. Purton, G. D. Barrera, N. L. Allan and J. D. Blundy, *J. Phys. Chem. B*, 1998, **102**, 5202.
11. N. L. Allan, G. D. Barrera, M. Yu. Lavrentiev, I. T. Todorov and J. A. Purton, *J. Mat. Chem.*, 2001, **11**, 63.
12. M. O. Zacate and R. W. Grimes, *Phil. Mag. A*, 2000, **80**, 797.
13. J. H. Harding, *Rep. Prog. Phys.*, 1990, **53**, 1403.
14. M. I. Baskes, R. G. Hoagland and A. Needleman, *Mat. Sci. Eng. A*, 1992, **159**, 1.
15. M. W. Finnis, *J. Phys.: Conds. Mat.*, 1996, **8**, 5811.
16. T. C. Germann and P. S. Lomdahl, *Comput. Sci. Eng.*, 1999, **1**, issue 2, 10.
17. P. Keblinski, D. Wolf, S. R. Phillpot and H. Gleiter, *Phil. Mag. A*, 1999, **79**, 2735.
18. P. Keblinski, D. Wolf, S. R. Phillpot and H. Gleiter, *Scripta Mat.*, 1999, **41**, 631.
19. V. L. Bulatov, R. W. Grimes and A. H. Harker, *J. Materials-e*, 1997, **49**, 2, *http://www.tms.org/pubs/journals/JOM/9704/Bulatov*.
20. F. H. Baumann, D. L. Chopp, T. Diaz de la Rubia, G. H. Gilmer, J. E. Greene, H. Huang, S. Kodambaka, P. O'Sullivan and I. Petrov, *MRS Bull.*, 2001, **26**, 182.
21. S. Kramer, M. Spears and H. L. Tuller, *Sol. State Ionics*, 1994, **72**, 59.

22. B. J. Wuensch, K. W. Eberman, C. Heremans, E. M. Ku, P. Onnerud, E. M. E. Yeo, S. M. Haile, J. K. Stalick and J. D. Jorgensen, *Solid State Ionics*, 2000, **129**, 111.
23. S. J. Korf, H. J. A. Koopmans, B. C. Lippens, A. J. Burggraaf and P. J. Gellings, *J. Chem. Soc. Faraday Trans.*, 1987, **83**, 1485.
24. S. X. Wang, B. D. Begg, L. M. Wang, R. C. Ewing, W. J. Weber and K. V. Godivan Kutty, *J. Mater. Res.*, 1999, **14**, 4470.
25. K. E. Sickafus, L. Minervini, R. W. Grimes, J. A. Valdez, M. Ishimaru, F. Li, K. J. McClellan and T. Hartmann, *Science*, 2000, **289**, 748.
26. M. A. Subramanian, G. Aravamudan and G. V. Subba Rao, *Prog. Solid State Chem.*, 1983, **15**, 55.
27. P. P. Ewald, *Ann. Phys. (Leipzig)*, 1921, **64**, 253.
28. M. Pirzada, R. W. Grimes, L. Minervini, J. F. Maguire and K. E. Sickafus, *Solid State Ionics*, 2001, **140**, 201.
29. C. R. A. Catlow and W. C. Mackrodt eds, *Computer Simulation of Solids*, Springer-Verlag, 1982.
30. K. Tharmalingham, *Phil. Mag.*, 1971, **23**, 199.
31. H. L. Tuller, in *Defects and Disorder in Crystalline and Amorphous Solids*, C. R. A. Catlow ed., Kluwer, 1994, 189.
32. M. P. van Dijk, A. J. Burggraaf, A. N. Cormack and C. R. A. Catlow, *Solid State Ionics*, 1985, **17**, 159.
33. M. P. van Dijk, K. J. de Vries and A. J. Burggraaf, *Solid State Ionics*, 1983, **9** & **10**, 913.
34. R. E. Williford, W. J. Weber, R. Devanathan and J. D. Gale, *J. Electroceram.*, 1999, **3**, 409.
35. P. K. Moon and H. L. Tuller, *Mat. Res. Symp. Proc.*, 1989, **135**, 149.
36. A. J. Burggraaf, T. van Dijk and M. J. Veerkerk, *Solid State Ionics*, 1981, **5**, 519.
37. V. S. Stubican, G. S. Corman, J. R. Hellmann and G. Senft, *Adv. Ceram.*, 1983, **12**, 96.
38. L. Minervini, R. W. Grimes, Y. Tabira, R. L. Withers and K. E. Sickafus, *Phil. Mag. A*, in press.
39. M. O. Zacate, R. W. Grimes, P. D. Lee, S. R. LeClair and A. G. Jackson, *Modelling Simul. Mater. Sci. Eng.*, 1999, **7**, 355.
40. K. J. W. Atkinson, R. W. Grimes, M. O. Zacate, P. D. Lee, A. G. Jackson and S. R. LeClair, *J. Materials-e*, 1999, **51**.
41. N. F. Mott and R. W. Gurney, *Electronic Processes in Ionic Crystals*, Oxford University Press, 1940.

Author and Subject Index

Note: References to Imperial College and RSM are so ubiquitous that they have been omitted from this index.

1851 Exhibition, Royal Commission for *xxii*
46*f* oxygen vacancy 343, 344

ABB PIUS reactor 117
ABWR 117
Acerinox 91
ACL *see* anterior cruciate ligament
Acta Metallurgica 75, 76
activation energies 344345
advanced gas-cooled reactor *see* AGR
advanced materials 103
AEI 68
AEI EM7 100 kv microscope 56, 57
   decommissioned 67–68
   used to observe chemical reactions 61–63
used to observe chemical reduction tunnelling 61–63
AEI MeV microscope 58–59, 61
AERE *xvii, xix,* 60, 113, 114
aero engines 103
aero engines *xix*
aerodynamics in Formula 1 176–177
aeronautics *xxi*
aerospace materials *xi*
AGR 114, 115
Agricola 4
AK Steel 91
Alani, Reza 69, 70
Albert, Prince, the Prince Consort x, 5
Alcan 94
Alcan xviii, 83, 96, 97
alchemy 4
Alcoa 96, 97
Alcock, Ben 56

Alcock, C. B. *xvii*, 41–51
Algoma Steel 205
Allegheny 91
alloy theory 297336
aluminium 103
   demand for 82
   first architectural use of 28
aluminium–beryllium alloys 172–176, 181
   banned 176
American Ceramic Society xvii, xix
American Institute of Metallurgical Engineers *xvii*
AMR head technology 105, 110
Andrade, E. N. da Costa *xi*
aneurysm, treatment of 141
angioplasty 141
anisotropic magneto-resistive head technology 105, 110
annealing 2
anodes 216–217
anterior cruciate ligament (ACL) repair 139
aorta 142
AP600 reactor 117, 119
Apollo 11 mission to Moon 56, 57
areal density and magnetic bit size 108, 110
areal density in magnetic media 105–111
Argonne National Laboratories 45
Arrows Formula 1 team *xxi,* 163
arsenates 3
arsenopyrite 2, 3
artefacts of electron microscopy 63–64
arterial bypass surgery 141142
arthritis 139
artificial cornea, PHEMA used for 144

356 Author and Subject Index

artificial cornea, PTFE used for 144
artificial limbs, materials for 135
arts, metallurgy and 13–40
Asano, Nagayuki 35
Ascot 90
ASM International *xix*
astronautics *xxi*
AstroPower 90, 97
atherosclerotic plaques 141
Atkinson, K. J. W. 339–353
atomic bomb 133
    first UK 134
Atomic Energy Commission (US) 74
atomic scale interactions 347–348
atomistic simulation *xix*, 339–354
Attlee, Clement 133
Audi 96
autografts 141
autologous endothelial cells 142
autosport engineering 155–183
AVLIS laser-based isotope separation 118
azurite 1

BAA 97
back-end fuel cycle 118
Bacon, Francis 5
Ball, J. G. *x, xiv,* 56, 57, 58, 60, 114
Banaras Hindu University *xviii*
Bangor, University of Wales, *xvii*
BAR Grand Prix *xxi*
Barbedienne 19
Bartlett, Paul Wayland 25
basic oxygen furnace 91
basic oxygen furnaces *see* BOF
basic topologies 267–272
Battelle *xviii*
Beche, Henry Thomas de la *ix,* 5
Bell Laboratories *xxi*, 74
Benetton 93
BEP0 experimental pile 114, 133, 134
Berkeley (UK) nuclear power station 134
Berkeley, University of California at *xix*
Bessemer Laboratory 10
Bessemer Medal *xxii*

Bessemer process 91
Bielby, R. C. 135–151
bioactive glass 145–146, 288, 29–293
bioactive materials 284, 289
biocompatibility 137, 142
biodegradable polymers for support of
    damaged ligaments 139
Bioglass® *xix,* 146
biomedical materials 283–296
bioprosthetic heart valve 142, 143
    degradation of 143
Birmingham (UK), University of *xvii,
    xviii, xx,* 58, 75
Bishop, Bill 70
BISRA 41
Black, Joseph *ix*
blast furnace 187, 188, 189, 194–197
BNFL 115
BOC 94
Boeing 93
BOF 91, 186, 188, 198, 200
Boiling Water Reactor *see* BWR
Bollmann, W. 53
bond order potential 326, 328
bone cell cycle 286–288
bone graft substitutes, ceramic 145
Bonfield, W. *xvii,* 281
Boulton, Matthew *ix*
Boulton–Watt steam engine *ix*
Boyle, Robert 5
Boys, Charles Vernon 11
Bradwell nuclear power station 134
Bragg reflecting planes 302
Bragg, William Lawrence 75
brand awareness and metals industries
    93, 94, 95
brass produced by calamine process in
    Roman times 4
Briers, Graham 70
Brillouin zone 299, 300308, 313
Bristol, University of *xxii*
British Airports Authority *see* BAA
British Energy 115
British Iron and Steel Research
    Association *see* BISRA

British Nuclear Fuels *xii, xx*
British Oxygen Company *see* BOC
British Steel *see also* Corus PLC 58
bronze 3, 17
    lead additions to in ancient Egypt 4
bronze age 3
Brook, R. *xvii*, 153
brownmillerite 238, 251
Buenos Aires, University of *xxi*
bump stops in Formula 1 178–180, 181
burns 138
Burrows, Sue *see* Ion, Sue
Butler, Paul 70
Buttery, L. D. K. 135–151, 283296
BWR 115

Cahn, John 74
Cahn, Robert W. *xvii*, 73–79, 114
    and *The Coming of Materials Science* 74
calcium, solubility of in molten halides 45
calcium aluminate cement clinker 237–239
Calder Hall reactors 114, 134
California, University of, at Los Angeles 73
Calphad 317, 318
Cambridge Instrument Company 56
Cambridge, University of *xvii, xviii, xix, xxi, xxii*, 53, 54, 56, 58, 75
Cambridge–MIT Institute *xxii*
Canada, HWRs in 115
CANDU reactor 115
capillary pores 236
carbon fibre in ACL replacement 139
carburisation 15
cardiovascular implants 141–143
Carlyle, W. A. *x, xiii*
Carnegie–Mellon University *xviii*
Carpenter Technology 91
Carpenter, H. C. H. *x, xi, xiv*, 11
cars, lightweight 96
Case Western Reserve University *xix*

cassiterite 3
cathodes 217–218
Cavendish Laboratory *xxi*, 56, 58
Cavendish, William 8
CEGB 60
cell cycle 288–289
cell engineering 281
cellular automata simulation 346–351
cement 229–242
cement paste microstructure 232
cement processes 230–231
Centennial Exposition of 1876 19
Centre for Ion Conducting Membranes 243
ceramic fuel for nuclear reactor 114, 119
ceramic ion conduction membrane devices 243–256
ceramic lattice 339
ceramic membrane reactors 243
ceramics *xix*
    as bone graft substitutes 145
    electrical *x*
    for hard tissue replacement 136
    in medical implants 137
    scientific approach to 76
    structural *xi*
chalcolithic period 2
chalcophyllite 2
Chaparral 91
Charles, J. A. *xviii*, 112
Chatterjee, A. *xviii*,185–211
Cheetham, Howard 70
Chelsea and Westminster Hospital *xxi*
chemical engineering 77
chemical reactions
    induced by electron microscope 63–64
    observed with AEI EM7 61–63
Chipman, John 42
Christodoulou, J. A. *xvii*, 227
Christodoulou, L. *xvii*, 70, 227
Christofle 19
chromium alloys in medical implants 137
cladding in nuclear reactors 113, 123–125
cloisonné enamel 19

coating implants 137
cobalt, oxidation of 44
cobalt–chromium–molybdenum alloy in artificial heart valve 142
Cobham 90
Coca-Cola® in electropolishing medium 58
Cockroft, John 133
coining 27
coke strength after reaction at high temperature *see* CSR
cokemaking 188, 189, 192
cold rolling 208
collage, type II 144
colouring of alloys 16–17, 27, 29–35
Columbian Exhibition of 1893
competition on price 84, 90
composites *xxi*, 103, 157165, 180
  comparison of mechanical properties with other materials 159
computational modelling 337
concrete 229–242
contact lens
  hard 143–144
  soft 143–144
continuous casting 201–205
copper 2
  pyritic 2
  solubility of in liquid silicates 44
copper–arsenic alloys 2, 3
corneal abrasion caused by hard contact lenses 143
corneal graft 144
coronary bypass 141–142
corrosion 64–67
  resistance in nuclear reactor cladding 122, 126
Corus PLC *xii, xviii,* 93, 95
Cottrell, A. H. 75, 113
Coulomb interactions 340
Coulson, C. A. 325, 326
crano-facial surgery 145
crash frond 170
crystal structure 341–342
crystallography 346–347

CSR 193, 194, 197
cuprite 3
cutting tool 27
CVD used to prepare nuclear fuels 120
Cyprus, bronze age metal-working 3
cytokines 137
cytotoxicity 137

Dacron for stent 141
dampers in racing cars 179
Dannatt, C. W. *x, xiv*
Dar es Salaam, University of *xxi*
DARPA *xviii*
data storage, magnetic 105–111
deep bed sintering 189–192
defect detection in nuclear industry 126
Delaware, University of xviii
Dennis, Bill 44
deposition of metal from molten salts 46
Derby Silver Co. 25
dermal substitutes 138
desulperisation 47
Devonshire, 7th Duke of *see* Cavendish, William
diffusion control 46, 47
diffusion in metals, solid-state 13, 15–16, 29–35
diffusion-controlled phase transformations 13
disk drive 105–111
dislocations, electron micrographs of 54
Disraeli, Benjamin 7
Doole, Ronald 70
Dresser, Christopher 19, 24
Ductal 239
Dupouy, Gaston 55, 56
dynamic recrystallisation *xx*

EAF 91, 186, 198, 198, 211
Edgar, A. J. 283–296
Edington, J. *xvii,* 81–102
Education Act of 1870 8
effects of shape 263–267

Egypt, investment casting in ancient 4
eigenspectrum 314, 315
eigenvalues 313
Elam, Constance *see* Tipper, Constance
electric arc furnaces *see* EAF
electrochemistry 218–221
electrodes 216–218
electromicrotome thinning of specimens 53
electron microscopy 53–71
electropolishing 57
electrostatic Madelung interaction 323
electro-thinning of specimens 53
Elkington & Co. 18
Ellingham, Dr 41
Engel–Brewer theory 297, 300, 308–319
Engineering and Physical Science Research Council *see* EPSRC
environmental impact of fossil fuels 116
enyme mimics 138
EPSRC 153
    Advanced Research Fellowship *xx*
Erkar 4
Eskom Pebble Bed Modular Reactor 117
ettringite 234, 235, 236
European Commission and 'energy gap' 117
European Union Science and Technology Committee *xx*
eutectics, binary and ternary 26
Evans, A. G. *xix*, 259–279
extractive metallurgy 213–224
Eyre, Brian 113

facial surgery 145
failure modes 263
failure of implants 139
Falize, Alexis 19
Falize, Lucien 19
Faraday Society 41, 42
Faraday, Michael 5
fast-breeder reactor 117
    liquid sodium cooled 114
Fermi energies 313

Fermi level 299, 303, 305
Fermi surface 300, 302, 306
Ferrari 176
ferrite, hot rolling of 205–207
FIA 157
fibre-reinforced composite 76
Finniston, Monty 114
Fischer Tropsch process 245
Fisher, Robert 56, 70
Florida, University of *xix*
Flower, Harvey 63, 70
fluorinated methacrylate for contact lenses 144
fluorocarbons in contact lenses 143
Ford Motor Company 97
Formula 1 *xxi*
    engineering 155–183
fossil fuels and environment 116
Foster, Clement Le Neve 11
France, gas-cooled reactors in 115
Fray, D. *xix*, 213–223
French threat to British manufacturing industry 8
Frenkel, Jack 73
Friedell oscillations 306, 307
Frisch, Otto Robert 133
Frishmuth, William 28
front-end fuel cycle 118
fuel reprocessing, non-aqueous 118
fully-integrated steel mills 91
furnace control in prehistoric times 2
fusion reactor 114

Gabor, Denis 53
gas centrifuge 118
gas trains made of Pyrex glass 45
gas-cooled reactor 115
gas-permeable contact lenses 143–144
Gatan Inc. *xxii*, 69–70
Gatan Ltd 60, 69
Gellner, O. H. 43
General Atomics Advanced Thermal Reactor 117

General Electric Company 73, 74, 75
Geological Museum 9
Geological Survey 5
German contribution to science in 19th century 8
German metallurgical education for Japanese students 35
giant magneto-resistive head technology 105
Gilbert, Alfred 26, 27–29
Gilchrist, Percy 11
gin-shibuichi (alloy) 17, 18, 20, 21, 25, 27
Gladstone, William Ewart 8
Glasgow, University of *ix*
glass surface reactions 285–286
GLEEP reactor 134
global warming 116
GMR head technology 105
gold 1, 2
   diffusion in 15
   solubility of in liquid silicates 44
   transmutation of base metal into 4
gold–aluminium alloys 16
gold-based alloys 16
gold–copper alloys 26
Goldschmidt, Viktor 77
Goodeve, Charles 41
Gorham 19, 24
Gowland, Willian *x, xiii,* 11, 27, 35
Graham, Thomas 14
grain boundaries 257
Grainger chair *xx*
graphite resistance furnace 44
graphite-moderated reactor 115
graphite-rich matrix 120
Great Exhibition (*see also* 1851 Exhibition) *x,* 1, 5, 9
Greek metallurgy 4
Green's function 326, 327
Greenwood, Geoffrey 114
Greey, Edward 18
Griffiths Medal *xvii, xix*
Grimes, R. W. *xix,* 339–353
growth factors 137
Grunfield Prize

HA 144
Hadfield Medal *xviii*
haematite 60–63
   reduction tunnelling in 6162
Hall & Miller Co. 25
Hall-Heroult cell 216, 217
hallmarking changes 25
Hammersmith Hospital *xxi*
Hanford, USA, nuclear waste stored at
Harboard, F. W. 11
hard disk 105111
Harris, J. E. 13
Harris, K. *xix,* 225
Harvard University *xix*
Harwell *see* AERE
heart bypass surgery 141–142
heart valve
   bioprosthetic 142, 143
   synthetic replacement for 141
hearth management 195–196
heat-cured concrete 233–237
heavy water as moderator and coolant 115
heavy water reactor *see* HWR
helium coolant in reactors 119
Hench, L. L. *xix,* 283–296
Hewlett Packard 90
high T_c superconductors 257
Hinton Medal *xx*
Hinton, Christopher 133
hip replacement 139–140
Hirsch, P. B. 53, 113
history of materials technology *xx*
history of metallurgy *xviii,* 1–12
Hitachi 58
Hitachi HU 650 kv microscope 58
Holland, iron nail in bronze age 3–4
Hollomon, Herbert 73
Hooke, Robert 5
Horne, R. W. 53
Howe, John 73
Hukin and Heath 18
Hull, Derek 113
Hume-Rothery, William 11, 75, 297, 299, 301, 307, 308, 329

Huxley Building 9
Huxley, T. H. *x*
Huxley, Thomas, the younger 9
HVEMs 5569
   Cambridge Instrument Company resists building 56
   Philips resists building 68
   shortcomings of 59–60
   UK decision to order from AEI 58, 68–69
HWR 114, 115
hyaluronan 144
hyaluronic acid (HA) 144
hydro-electricity 116
hydroxyapatite, synthetic 145

IAEA *xx*
IBM 90
ICI *xxi, xxi*, 133
immune response to implant 137
impact performance of composites vs aluminium 165–166
impact response of composites 166–167
impact structures in racing cars 170–176
Imperial Mint at Osaka 35
implant
   cardiovascular 141143
   coating to reduce inflammatory effects 137
   design 136–138
   integrity 137
   loosening 139
implant–tissue interface 137
India, CANDU reactor in 115
induction heating 42
inductive playback heads 110
industrial perspectives 229–242
Industrial Revolution 5
inorganic transport membrane 244
Institute of Materials *xii, xviii, xix, xx, xxi*
Institute of Metals 297
Institution of Mechanical Engineers *xxi*
Institution of Nuclear Engineers *xx*

integrated steel plants 185–212
Intel 93
International Atomic Energy Agency *xx*
Internet and the supply chain 94
investment casting in Late Kingdom Egypt 4
ion beam synthesis *xx*
ion milling machine 69
Ion, Sue *xx*, 70, 113–134
Ipitata Sponge Iron Company *xviii*
iron 3
iron arsenates 3
ironmaking 185–212
Ispat International 91

Japan, nuclear power generation in 116
Japanese Civil War 24
Japanese influence on metallurgical arts 13, 16, 17–27, 29
JEOL Ltd *xii*, 58
Jermyn Street ix, 5, 9
John Percy Process Group 46, 47
Johnson-Matthey 43
Jones, H. 75, 300–308
Journal of Materials 78
Journal of Materials Research 78

Kawasaki top blowing 201
Keele University *xix*
Kelly, Anthony 76
Kilner, J. A. *x, xv, xx*, 243–255
Kingery, David 76
Kiritsu Kosho Kaisha 19, 25
knee replacement 139, 140
Kobe Steel Ltd *xii*
Korea, CANDU reactor in 115
Krivanek, Ondrej 70
Kroll Medal *xviii*
Kroll process 220
Kyoto targets for greenhouse gas emissions 116, 117

Lafarge *xxii*
Lally, Scott 56
Larbalestier, D. *xx*, 257

laser-based isotope separation 118
Lausanne, École Polytechnique Fédérale
 de *xxii*
Laves, Fritz 77
lead
 diffusion in 15
 oxide 43
 silicates 43
 solubility of 44
Leeds, University of *xvii, xx,* 58
Lehigh University *xx*
lightweight cars 96
lightweight materials 259–280
lightweight structures 259–280
Linz and Donawitz 198
Linz–Donawitz process 46
liquid sodium cooled fastbreeder reactor
 114, 117
Lloyd, Tony 60, 69, 70
local charge neutrality 324
local density approximation 300
logistics 94
London, University of 7
Londros & Co. 24
Los Alamos National Laboratory *xix*
lost wax process in Late Kingdom
 Egypt 4
low carbon steel 205–207
low phosphorus steel 198–200
Luigi Losana Gold Medal *xvii*
LWR 117
 advanced 119

magnesium
 alloy *xx*
 solubility of in molten halides 45
magnetic data storage 105–111
magnetite 63
Magnox reactor 114, 115
malachite 1
Manchester, University of 58
Mannesmann 95, 99
March of Dimes 147
market forces vs. political decision-
 making 68–69

Martin Marrieta *xviii*
Massachusetts Institute of Technology
 see MIT
materials processing 227
Materials Research Society 78
materials science as discipline 73–79
Matthew Prize *xix*
Matthias Scholar *xix*
Maude Committee 133
Max Planck Institut für Metallforschung
 xvii
MBA 101
McCance, Andrew 11
McLaren International *xxi,* 163, 164
McLean, M. *ix–xv*
Mecedes 176
medical implants 135–151
medical materials 135–151
medical materials *in vitro* testing
 of 137
Medical Research Council 147
medium weight structures 260–263
Meiji Period 35
Meridien Britannia Co. 25
Mesopotamia 2
metal sulphides 46
metal/solder interfaces *xx*
metallography 13
metallurgy
 and the arts 13–40
 and transition to materials science
  73–79
 as a constituent of materials
  science 78
 as progenitor of materials
  science 75
metals
 chemistry 48–50
 for hard tissue replacement 136
 industries and brand awareness, 93,
  94, 95
 industries, lack of profitability in 82–84
 industries, restructuring of 83–84
 oversupply of 84
MeV electron microscopes 55–69

migration simulations 343
mineralogy 1
mini-mills 91, 92
Mining Record Office 5
Ministry of Aircraft Production 133
Ministry of Technology and AEI HVEMs 58, 68–69
Mint *see* Royal Mint
MIT 42, 44, 48, 76
Modena, University of *xxi*
mokume 17, 20, 21, 22, 25, 26, 27, 28, 29, 30, 31,33, 34
molecular weight, variable 76
molten salt reprocessing 118
monocoque 155
Moon rock 56
Moore's Law 340
Mott, Neville 75
Museum of Economic Geology *ix*, 5

nanosized particles 77
National Academy of Engineering (US) *xix*
National Metallurgist Award (India) *xviii*
National Physical Laboratory *xii*, 14
Naval Research, US office of *xii*, *xviii*
NBS *xix*
nearly-free-electron 311, 313
neolthic man 1
Neumann, John von 74
Newcomen engine *ix*
Newcomen, Thomas *ix*
Newton, Isaac, and transmutation of base metal into gold 4
nickel alloys in medical implants 137
Nokia 90, 95, 99
Norsk-Hydro 95, 97
Northwestern University 73
Notis, M. R. *xx*, 13–40
Notre Dame University *xvii*
Nottingham, University of *xix*
nuclear energy as proportion of power generation 115, 116
nuclear fission, discovery of 133
nuclear fuel 113
 burn-up of 49
 coated 120
 reprocessing 128–129
nuclear industry
 advances in 113–134
 in UK 113
  chronology of 133–134
 materials science and 113–134
 structural integrity in 126–128
nuclear materials *xx*
nuclear power 76
 growth in demand for 116
 in UK 41
nuclear reactor
 coolant 113
 experimental fast 114
 moderator 113
nuclear submarine 114
nuclear waste storage/disposal 118, 119, 129–131
 at Hanford 130
Nucor Steel 90, 91, 202
Nuffield Foundation 42, 45, 46
Nuffield Research Group *xi*, *xvii*, 213, 243
Nutting, Jack 58

O'Kane, W. *xxi*, 105–111
Ohio State University *xix*
open hearth process 91
oral surgery 145
organic liquid moderated reactor 114
Osaka Mint 35
osteoblasts 288–290
 cell cycle 290–292
 grown on Bioglass 145–147
osteoporosis 139
Oxford, University of *xi*, 58
oxygen
 fluxes 253
 interstitials 251
 migration 341–346
 potential 49
 tracer diffusion 246–247
 transport 245–247

Pairpoint Manufacturing Co. 25
Pakistan, CANDU reactor in 115
panel properties 272–274
Paris Exposition
  of 1867 7, 18, 19
  of 1878 19
Paris-Sud, Université de *xvii*
Pashley, D. W. *x, xiv*
Pauli, Wolfgang 75
Pauling's resonant bonds theory 300, 319–328
PBMR 117, 118, 119, 125
PCI 194197
Pebble Bed Modular Reactor *see* PBMR
Peierls, Rudolf Ernst 133
Pepper, J. H. 6
Percy, John *ix, x, xiii*, 68, 9, 14, 35, 36
perovskite 238, 248–250, 252
  related materials 250–252
Perry, Commodore 18
Perspex *see* PMMA
Peters, Tom 83
Pettifor, D. G. *xxi*, 297–335
Pfeil, Leonard 11
phase equilibria 16
PHEMA for artificial cornea 144
Philips resists building HVEM 68
physical metallurgy 14
physical metallurgy 214–215
Pirzada, M. 339–353
Pittsburg 69
planar crystal defects, HVEM image of 55
platinum chloride 23
PMMA used for contact lenses 143
PMMA used to cement implants 139–140
Polak, J. M. *xxi*, 135–151, 283–296
political decision-making 68–69
poly(2-hydroxyl-ethyl-methacrylate) *see* PHEMA
poly(ethylene terephthalate) for vascular grafts 141
poly(ethylene), ultra high weight 140
poly(ethylene), ultra high weight, failure of 140
poly(methyl-methacrylate) *see* PMMA

poly(tetrafluoroethylene) for vascular grafts 142
polyethylene in ACL replacement 139
polyglycolic acid for support of damaged ligaments 139
polymer diffusion *xxii*
polymers 76, 77
  for soft tissue replacement 136
  non-crystalline *xxii*
  synthetic 137
polytetrafluoroethylene *see* PTFE
polyurethane wound dressings 138
polyurethane rubber 180
Portal, Charles Frederick Algernon 113
Portland cement 230
post-irradiation fuel cycle 118
Pratt, Peter *xi*, 56
pre-irradiation 118
prepreg 161
Presley, Elvis 198
pressure driven devices 244–245
pressurised water reactor *see* PWR
Preston, *mokume* used in mayoral chain for 26, 28
Preussag 95, 96, 99
Price, P. *xxi*, 103
Priestley, Joseph 5
Princeton University *xix*
printing and spread of learning 4
process zone 170
Prost Grand Prix *xxi*
prostheses, materials for 135
PTFE for stent 141
PTFE in ACL replacement 139
public policy and decision-making 68–69
Puffing Billy 44
pulmonary artery 142
pulverised coke injection *see* PCI
PUREX reprocessing of nuclear fuel 118
PWR 114, 122
  development in UK abandoned 114
  for submarine propulsion 114
pyramidal core 271
Pyrex glass gas trains 45
pyrocarbon layer in PBMR 121

pyrochlore oxides 341–346
pyrrolidone for contact lenses 144

quantum mechanical simulations 339
Queen Mary College, University of London *xvii*
racing car engineering 155–183
radiation-induced artefacts 63–64
rail, forging iron 27
rake angle of car 178
Rautaruukki 91
RCA electron microscope 56
RDI 190, 195, 197
reactor schools at Harwell and Calder Hall 114
reduction degradation index *see* RDI
reduction tunnelling in haematite observed 61–62
Reed & Barton 25
rejection of implant 137
renewable energy 116–117
reprocessing nuclear fuels 128–129
Richardson, F. E. 41, 42, 43, 45, 46, 51, 56
Rietveld type methods 237
Roberts-Austen, William Chandler *x, xiii*, 9, 11, 13–40
rocket fuels and similarity to electropolishing solution 57
Rockwell *xix*
Rolls-Royce *xxi*
Roman metallurgy 4
Rose, Thomas Kirke 11
Rosenhain Medal *xxii*
Rosenhain, Walter 75
Rosenthal, Daniel 73
Royal Academy of Engineering *xvii, xviii, xx*
Royal Aeronautical Society *xxi*
Royal College of Chemistry *ix*, 9
Royal Commission for the 1851 Exhibition *xxii*
Royal Commission on Scientific Instruction 8
Royal Institution *xi, xix*

Royal Mint *x*, 9, 14, 25, 27, 35
Royal Navy Weapons Development Group 41
Royal Society *xvii, xxi, xxii*, 1
 and nuclear power 116
Royal Society of Arts xvii
 Silver Medal of *xxii*
Royal Society of Canada *xvii*
Royal Society of Chemistry *xvii*
Russian Federation, RBMK reactor in 115

safety in Formula 1 167–169
Samuelson, Bernard 7
Santa Barbara, University of California at *xix*
Savage, G. *xxi*, 155183
Schliemann, H. and excavation of Troy 14
Schrödinger equation 298, 300
Science Museum, London 6
Science Research Council 45
 and AEI HVEMs 58, 60, 68–69
Scipher 90
Scrivener, K. L. *xxii*, 229241
Seagate Technologies xxi
seeding artificial grafts 142
Seitz, Frederick 75
semicrystallinity 76
SGHWR 114
shakudo (alloy) 17, 18, 20, 21, 23, 25, 27, 28, 29, 30, 31, 32, 33
Sheffield, University of 58
shibuich see *gin-shibuichi*
Shimane-3 reactor 116
Shubnikov chair *xx*
Siemens Elmiskop 1 microscope 53
silicon carbide layer in PBMR fuel 121
silicon rubber in artificial heart valve 142
silicones in contact lenses 143
silver, solubility of in liquid silicates 44
silver-based alloys 16
simulation speed 340
sinchu 18
single-crystal superalloys *xix*, 225
skin grafts 138–139

Smallman, Ray 114
Smith, C. S. 13
Smith, Cyril 76
Smyth, W. W. 9
Snelus, George James 11
Societé Française de Métallurgie et Matériaux *xvii*
soda glass 43
sodium-cooled reactor 114
solar energy 97, 116
sold-state chemistry 77
solid state physics as progenitor of materials science 75
South Kensington x
Southampton, University of *xxi*
Southern California, University of *xvii*
specimen
   holders 60
   preparation 57–58, 69–70
Springfields nuclear fuel site 133, 134
Sputnik 74
sputtering deposition 109
SRC *see* Science Research Council
SSAB 91
stainless steel, demand for 82
stamp charging 192–194
stannite 3
Steam Generating Heavy Water Reactor *see* SGHWR
steel
   in medical implants 137
   industry, innovation in 91
   production quantities 185, 186
   demand for 82
steel companies, size a disadvantage in 91
Steele, Brian *xi,* 67
steelmaking 185–212
stent, materials for 141
Stewart Formula 1 team 163
stiffness 262
Stoneham, Marshall 113
Strath Report 114
stress corrosion 64–67
strong steam *ix*

structural integrity in nuclear industry 126–128
sulph-arsenide of tin 3
superalloys, single-crystal *xix*
superconductivity *xx*
superoxide dismutase mimics 138
superoxide radicals 137–138
supersaturation 234
supply chain and the Internet 94
surface exchange coefficients 246–247
surface roughness 348–349, 350
Sussex, University of *xvii*
suture materials 136
Swann, P. R. *xxii*, 53–71
Swinburne Medal *xxii*
synzymes 138
Szirmae, Al 56

Tata Steel *xviii*, 185–212
Technology Partnerships 90
Teflon *see* PTFE
Teledyne 91
TEM 53–71
tempering 14
Tesco 94
testing cores 274–278
tetragonal core 270–271
textile cores 271–272
Th^{230} 49
thermal reactor fuel 118
Thermanox plastic 286–287
thin film direct casting 97
thin slab casting *see* TSC
Thomas Medal *xviii*
Thomas, Sidney Gilchrist 11
Thomas-Fermi response function 306
thoria electrolytes 48
Thyssen 198
Tiffany & Co. 19, 20, 21, 22, 23, 24, 25
Tiffany & Co. 28
Tighe, Nancy 70
tin 3
tin, diffusion in 15
Tipper, Constance (*née* Elam) *xi*, 11

tissue engineering *xxi*, 135–151, 281, 283
Tissue Engineering xxi
Tissue Engineering Centre, Imperial College 283
Tissue Engineering Society of Great Britain *xxi*
tissue regeneration 283–284
titanium
   alloys 103
   demand for 82
   in medical implants 137
Titchfield Thunderbolt 44
TMS Application to Practice Award *xviii*
Tomari-3 reactor 116
Tomlinson, George 133
Tomlinson, J. W. 43, 45
Toronto, University of *xvii*
torsion beam 155
total joint replacement 139–140
Toulouse, University of 55, 56
Toyota Manufacturing System 81, 93
transmission electron microscopy see TEM
transmutation of base metal into gold 4
transport of oxygen 339
trauma 138
trimethylene for support of damaged ligaments 139
Trinity College, Cambridge *xxii*
Troy, excavation of 14
TSC 202–205
Tube Alloys 133
Tufts, James W., Co. 25
tuning Formula 1 cars 178–180
Turnbull, David 74
U^{233} 49
UHMWPE *see* poly(ethylene), ultra high weight
UK Quality Assurance Agency *xi*
UKAEA 115
ultra high performance concrete 239–240
UMIST *xxi*
United Kingdom Atomic Energy Authority *see* UKAEA

universities, role of in business start-ups 101
University College London 41
university departments 213
uranium
   enrichment 118
   oxide 49
   U^{233} 49
uranyl sulphate in HWRs
US Air Force 74
US Biomaterials 147
US Navy xviii
US Office of Naval Research *xii, xviii*
US Steel 56, 57

vanadium alloys in medical implants 137
vascular grafts, materials for 141–142
Victoria & Albert Museum 25
Victoria, Queen 4, 5
virtual crystal approximation 324
virtual materials 337
viscoelastic substances 144
vitreous body of eye, synthetic substitutes for 144
Vodafone 95
Von Hippel Award *xix*

Wagner, Carl 44, 77
Walmart 93
Walsh, Hall & Co. 18
Watt, James *ix*
weldments 64
Wellcome Trust 147
West, D. R. F. 13
Westinghouse 114
Whelan, M. J. 53
Whiting Manufacturing Co. 24–25
Wigner–Seitz cell 319
Wigner–Seitz radius 310, 311, 320, 321
Williams, Evan James 75
Wilson, Alan 75
wind energy 117
Windle, A. *xxii*, 337
Windscale nuclear plant 134

Wisconsin Madison, University of *xx*
Wolfson chair of Metallurgy *xxi*
Wolfson Unit for Solid State Ionics 243
Woodhall, John 70
wound dressings, polyurethane 138

XD composites *xviii*
xenografts 138–139
Xynos, I. D. 283–296

Zacate, M. O. 339353
ZEPHYR nuclear reactor 114
ZETA nuclear reactor 114
ZEUS nuclear reactor 114
Ziman, John 79
zirconia electrolytes 48
zirconium cladding for nuclear reactors 114, 122–125
zone refining 76